T0075312

The aim of this series is to publish a Reference Library, including novel advances and developments in all aspects of Intelligent Systems in an easily accessible and well structured form. The series includes reference works, handbooks, compendia, textbooks, well-structured monographs, dictionaries, and encyclopedias. It contains well integrated knowledge and current information in the field of Intelligent Systems. The series covers the theory, applications, and design methods of Intelligent Systems. Virtually all disciplines such as engineering, computer science, avionics, business, e-commerce, environment, healthcare, physics and life science are included. The list of topics spans all the areas of modern intelligent systems such as: Ambient intelligence, Computational intelligence, Social intelligence, Computational neuroscience, Artificial life, Virtual society, Cognitive systems, DNA and immunity-based systems, e-Learning and teaching, Human-centred computing and Machine ethics, Intelligent control, Intelligent data analysis, Knowledge-based paradigms, Knowledge management, Intelligent agents, Intelligent decision making, Intelligent network security, Interactive entertainment, Learning paradigms, Recommender systems, Robotics and Mechatronics including human-machine teaming, Self-organizing and adaptive systems, Soft computing including Neural systems, Fuzzy systems, Evolutionary computing and the Fusion of these paradigms, Perception and Vision, Web intelligence and Multimedia.

Indexed by SCOPUS, DBLP, zbMATH, SCImago.

All books published in the series are submitted for consideration in Web of Science.

Intelligent Systems Reference Library

Volume 231

Anupam Biswas · Vijay Bhaskar Semwal ·
Durgesh Singh
Editors

Artificial Intelligence for Societal Issues

Editors
Anupam Biswas
Department of Computer Science
and Engineering
National Institute of Technology Silchar
Silchar, Assam, India

Vijay Bhaskar Semwal
Department of Computer Science
and Engineering
Maulana Azad National Institute
of Technology Bhopal
Bhopal, Madhya Pradesh, India

Durgesh Singh
Department of Computer Science
and Engineering
PDPM, Indian Institute of Information
Technology, Design and Manufacturing
Jabalpur
Jabalpur, Madhya Pradesh, India

ISSN 1868-4394 ISSN 1868-4408 (electronic)
Intelligent Systems Reference Library
ISBN 978-3-031-12418-1 ISBN 978-3-031-12419-8 (eBook)
https://doi.org/10.1007/978-3-031-12419-8

This Springer imprint is published by the registered company Springer Nature Switzerland AG
The registered company address is: Gewerbestrasse 11, 6330 Cham, Switzerland

Preface

The recent growth of Artificial Intelligence (AI) with Deep Learning (DL) has provided innovative intelligence solutions for various societal, personal care, health care, assistive technology, and real-world problems. Applications of AI are used to combat some of the seemingly unsolvable social crises facing the world today. In response to epidemics and disease outbreaks like COVID-19 and natural disasters, AI plays a pivotal role nowadays. Disaster awareness and management, demand forecasting, or healthcare informatics AI is used everywhere. The world faces global issues like hunger crisis, clean energy, environmental challenges, and economic empowerment, where AI can holistically address these issues. AI is also used in other societal issues that indirectly pose challenges like cybercrime, agriculture, education, economy, and health. The application of AI also involves the development of prosthesis devices, various robotic agents, brain control devices, etc. This has tremendously impacted the lives of differently-abled people in performing their daily life activities.

The book includes chapters on applications of AI in societal awareness for economic empowerment. Chapters on incorporating AI for smart society and smart education systems and crowd violence detection are included. The book also includes a few chapters on applications of AI responses to COVID-19 management in rural areas and COVID-19 vaccination decisions. The book consists of chapters on emotion detection using AI from text, speech, and image data. The book also comprises chapters on applications of AI in cybercrimes such as phishing URL detection and fraudulent transactions. The chapters on AI in agriculture like green purchasing, paddy grain classification, and plant disease identification are covered. Lastly, the book includes several chapters on applications of AI in health informatics, including global mental health, brain disease, and other disease identification.

This book is organized into four parts, including 18 chapters that report the research in the four main application domains: *Crime and Security, Agriculture and Education, Emotion and Mental Health, and Healthcare Informatics and Management* using AI. Part I of this book includes four chapters. Chapter 1 presents the AI for cyber security, which includes various threats, attacks, and their mitigation. Chapter 2 presents a comprehensive survey on the deep learning model to detect hate speech and bullying in social media. In contrast, Chap. 3 illustrates a deep learning-based system for estimating crowds and detecting violence in videos. Next, Chap. 4 discusses the Role of Machine Learning and Deep Learning in Detecting Fraudulent Transactions to mitigate cyber fraud.

Part II covers four chapters from the domain of *Agriculture and Education*. Chapter 5 presents work for gradation and classification of paddy grains using image processing and deep learning. Chapters 6 and 7 cover users' intentions about online purchasing. Chapter 6 is centered around the role of brand, while Chap. 7 is on the Effect of Online Review ratings on user Purchase Intention. Chapter 8 focuses on a smart education system using Artificial Intelligence.

Part III is devoted to the chapter on the *Emotion and Mental Health* domain. Chapter 9 presents the work on emotion recognition through speech analysis using deep learning. Chapter 10 presents the work on Face Emotion Detection for Autism children using Convolutional Neural Networks. Chapters 11 and 12 present surveys on a prevention of global mental health crisis with Transformer neural networks and survey on the Diagnosis of Mental Illness using Deep Learning.

Part IV of the book covers the six chapters from the *Healthcare Informatics and Management* domain. Chapter 13 presents the Skin Disease Detection and Classification Using Deep Learning. This is an attempt to automate the System of Dermographism for Society. The next Chap. 14 presents work on brain tumor detection using MRI images using deep learning. Chapter 15 is related to COVID-19 Detection in X-rays using Image Processing CNN Algorithm. Chapter 16 presents the Black Fungus Prediction in COVID Contrived Patients Using Deep Learning. Lastly, Chaps. 17 and 18 have provided the fuzzy-based decision framework for COVID-19 vaccination and diagnosis using AI.

The editors would like to express their deep appreciation and gratitude to the contributors who have made their research available for this volume and the anonymous reviewers who have provided an invaluable service in referring to the chapters. We sincerely hope that this work serves as a reference for researchers, as well as a compilation of innovative ideas and solutions for practitioners interested in applications of AI in various societal challenges.

Silchar, India — Dr. Anupam Biswas
Bhopal, India — Dr. Vijay Bhaskar Semwal
Jabalpur, India — Dr. Durgesh Singh
September 2022

Contents

Part I
Crime and Security

Chapter 1
Artificial Intelligence for Cybersecurity: Threats, Attacks and Mitigation

Abhilash Chakraborty, Anupam Biswas, and Ajoy Kumar Khan

Abstract With the advent of the digital era, every day-to-day task is automated due to technological advances. However, technology has yet to provide people with enough tools and safeguards. As the internet connects more-and-more devices around the globe, the question of securing the connected devices grows at an even spiral rate. Data thefts, identity thefts, fraudulent transactions, password compromises, and system breaches are becoming regular everyday news. The surging menace of cyber-attacks got a jolt from the recent advancements in Artificial Intelligence. AI is being applied in almost every field of different sciences and engineering. The intervention of AI not only automates a particular task but also improves efficiency by many folds. So it is evident that such a scrumptious spread would be very appetizing to cybercriminals. Thus the conventional cyber threats and attacks are now "intelligent" threats. This article discusses cybersecurity and cyber threats along with both conventional and intelligent ways of defense against cyber-attacks. Furthermore finally, end the discussion with the potential prospects of the future of AI in cybersecurity.

Keywords Cybersecurity · Cyber-attacks · DDoS · Man-in-the middle · Intrusion detection · Artificial intelligence

1.1 Introduction

Nowadays, it is hard to find a company, institution, or family that is not using the internet and technology. We, as humans, find ourselves overwhelmed with the number of digital devices and apps that we use on a day-to-day basis. Many of us cannot even

A. Chakraborty (✉) · A. Biswas
Department of Computer Science and Engineering, National Institute of Technology Silchar, Silchar, India
e-mail: abhilash21_rs@cse.nits.ac.in

A. Biswas
e-mail: anupam@cse.nits.ac.in

A. K. Khan
Department of Computer Engineering, Mizoram University, Aizawl, India

A. Biswas et al. (eds.), *Artificial Intelligence for Societal Issues*, Intelligent Systems Reference Library 231, https://doi.org/10.1007/978-3-031-12419-8_1

control our technology usage the way we want to. Some people may be addicted to the internet and cannot stop using it, while others may not use it enough to keep up with the rapid changes in technology. Some might spend hours on their phones or computers rather than interacting with others. The internet is a powerful tool, and we cannot resist the urge to use it for everything. There is no doubt that technology has increased our productivity and efficiency in several ways. However, we have to consider the effects it has on our personal and social lives and our mental and physical well-being. With the bright side of the proliferation of Internet-connected devices, a darker side of cyber-crimes has loomed shadow over our lives. The ever-lurking threat of losing our privacy in this open and connected world has raised many questions. This sudden increase in interconnectedness has made us more efficient and made us vulnerable to the dangers of cyber-crime simultaneously. We must be conscious of our connection to cyberspace. The internet has made us conquerors and prisoners simultaneously. The internet has given people the ability to connect in unprecedented ways. However, this has also created vulnerabilities that can be exploited by cybercriminals.

Cyber-crime is a type of crime that uses digital media to commit fraud, steal data, or cause damage. Cyber-crime is an umbrella term that encompasses all forms of cyber-related crimes; in essence, illegal activities that are initiated using computers, such as hacking, phishing, malware distribution, online stalking, and identity theft, among many others. Cyber-crime is one of the most lucrative crimes of the modern age. Every year, cybercriminals make billions of dollars in profits by stealing data, corrupting data, and compromising critical infrastructure. Much like any other crime, cybercrime has evolved dramatically over time and will continue to do so.

The report from "The World Economic Forum" states that if we take a holistic approach to cybersecurity, there will be better protection for business, society, governments, and individuals alike [2]. It also states that the gap between our current state of readiness and what is required to protect cyberspace is vast. There is an urgent need to close this gap before we reach a point of no return. Cyber Security has become a critical aspect for every infrastructure not to get compromised by any kind of attack originating from outside sources such as viruses, malware, or hacking attempts. However, most security breaches or cybercrimes are due to human error, not through an external threat. The responsibility of safeguarding the critical infrastructure and protecting people from cybercrime is a shared responsibility for all.

Although viruses and malware were a concern almost from the beginning of computing, awareness about how important data security has only become apparent as the internet grew in popularity. There has been a recent surge in hacking, and it's exposing these "cybercriminals" to more possibilities on the internet. That being said, there are many threats that arise from this exposure. Hackers can cause things like downtime (e.g., by crashing your website), steal data from our computers/servers, or even commit fraudulent transactions. It has now become a separate branch of criminal investigations called cybercrime.

This is because of the extensive global internet penetration of about 5 billion users, which is approximately 63% of the world's population, and cybercrime has increased at an exponential rate. Internationally, the calculated damages of cybercrime are roughly about 6 trillion USD by 2021 [1], and has become the third-largest economy in the world if it were measured separately. Cybercrime is expected to cost a total of $10.5 trillion by 2025 [31], up from just $3 trillion in 2015 according to Cybersecurity Ventures. This is arguably the most significant transfer of wealth in human history. It throws off incentives for innovations and will be more profitable than anything we've ever seen. Costs of cybercrime can include damage or destruction of data, hacked data being deleted or restored, money stolen, reduced productivity, intellectual property theft, theft of personal and financial data and embezzlement (taking assets for the use for the credited person who is responsible for the crime) and reputational harm. Creating a more secure system from the start, preventing cybercrime from happening, and reducing its impact when it does happen has become a multi-disciplinary affair.

Cybercrime is a growing problem, and it is essential to protect ourselves against it. There are many ways to do so, but the most important thing is to be aware of our surroundings and what we do online with our personal information. Just by being aware of our activities and the risk they might incur, we might be able to avoid most of the threats creeping online. There are many ways that one can avail to protect themselves against cybercrime. One way is to ensure the device's safety and security by using antivirus software, internet security software, and firewall software. Another way is by using strong passwords and changing them often. Lastly, keeping the operating system updated with the latest patches and updates. One can also monitor network traffic for vulnerabilities and set up auto-responders to avoid phishing attacks. Furthermore, we should also manage our social media settings and avoid using unsecured Wi-Fi networks in public places. Even just by minimizing how much personal information we share online, we can avoid the risk of being a target of identity theft, cyberstalking, and many more such threats.

In addition to these conventional methods, which are just a stopgap measure at best, the use of AI is an emerging field in the world of cybersecurity. Nowadays, AI is prevalent in almost any and all fields of science, whether from medicine to business or from the military to law enforcement. The use of AI in science is almost ubiquitous. The use of AI in cybercrime is growing at such a rapid rate that it has become one of the significant areas of concern worldwide. AI is a potent tool that is being used to combat many different types of crime. It will be vital for law enforcement agencies worldwide to find new ways to utilize this technology to keep up with the ever-increasing rate of cybercrime. AI is being applied to crime-fighting in a number of different ways. In the case of cybercrime, AI is being used to help identify potential threats, detect patterns that can lead to previous criminal activity, and detect new forms of existing criminal activity. However, AI is also being used as part of a broader research initiative on cybercrime and its perpetrators. Cybercrime data is collected, analyzed, and used to build sophisticated virtual crime scenes that can predict crimes before they happen.

AI can be used to mine data, identify patterns, and predict future events. It can also be used to detect cyber-attacks and prevent them from happening. In the future, AI systems will be able to detect patterns that are not readily apparent to humans, like a possible cyber-attack, by analyzing network traffic and determining if different strings of data are accessed in the same unusual pattern. AI can do many things, and it will continue to evolve and grow to be used in more everyday aspects of our lives.

The entirety of this chapter is divided into a total of four sections. The first section outlines the concept of cybersecurity along with the threats and attack models that hackers commonly use to compromise a computer system. The second section entails the conventional approaches and methods of mitigating the risks of cyber-attacks. The third section then discusses the AI-based approaches to counter cyber-threats or at least mitigate the risks associated with cyber-attacks. Finally, the fourth and the last section talks about the future scope of AI in cybersecurity.

1.2 Cybersecurity

Cybersecurity is the practice of protecting critical systems and sensitive information from digital attacks. There are many ways to safeguard data and organizational infrastructure, including intrusion detection, malware protection, strict adherence to sound security practices, and many more. A cyber security threat can be a cyber-attack using malware or ransomware to gain access to data, disrupt digital operations, or damage information. There are all kinds of cyber threats, including corporate spies, hackers, and terrorists [28]. In Fig. 1.1, the taxonomy of cybersecurity is presented. Although all have different reasons for attacking, all should be treated with extreme caution as they pose a risk to an organization's and personal data. The rise of the Internet has brought a new era of cyber security concerns. In addition to the threat of criminal hackers and foreign governments, new challenges are being associated with protecting information from internal threats, such as data breaches and insider theft. Cyber security is also an essential cross-cutting concern for sensitive infrastructures, critical assets, and sensitive information. This is why there has been a remarkable rise in cyber security professionals and the industry as a whole and why it is becoming increasingly important to ensure that the defense mechanisms against cyber attacks are comprehensive and robust.

Cybersecurity is a broad term encompassing all measures taken in an effort to safeguard an entity from cyber threats, including securing data and mitigating damage from a cyber security incident. The field of Cybersecurity can be broadly classified into five distinct security areas:

- Critical infrastructure security
- Application security
- Network security
- Cloud security and
- Internet of Things (IoT) security.

Cybersecurity is a complex and ever-changing field. It is essential to understand the different types of cyber threats and how they can be mitigated. Cyberattacks are

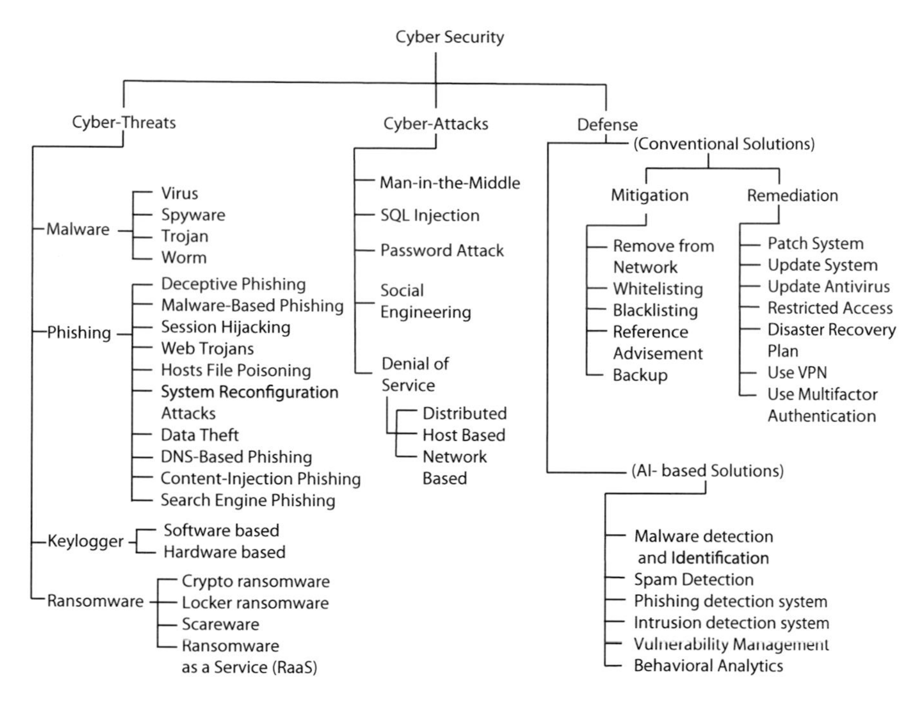

Fig. 1.1 Taxonomy of cybersecurity

becoming a common occurrence in today's society. However, these attacks can be prevented with the proper security measures.

In this article, we will discuss the different types of cyber attacks and threats and deep dive into different defense mechanisms, both conventional and AI-based, and learn about the currently available threat mitigating solutions.

1.2.1 Attacks

Distributed Denial of Service or DDoS: It is a form of cyber-attack where the perpetrator uses multiple systems to flood the target with traffic. The goal is to make it difficult for the target to provide service or access their website. The most common type of DDoS attack is a volumetric attack, which floods the target with an overwhelming amount of data. This can be done by using a botnet, a network of computers that have been infected with malware and are controlled by an attacker without their owner's knowledge. Within the volume attacks category, there are flooding and amplification/reflection attacks. In a flooding attack, traffic is sent in the hopes of exhausting bandwidth, processing capacity, or other network resources. Amplification/reflection attacks seek to force victims to spend money by "overloading" their networks with spam traffic or denying access to certain resources using spam-like messages [27, 36].

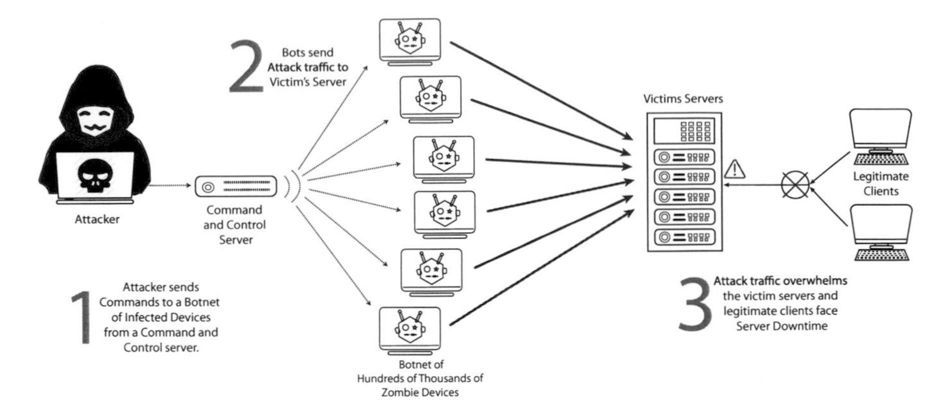

Fig. 1.2 Distributed denial of services attack scenario

There are many tools that can be used to launch a DDoS attack, and the most common is the use of a botnet. The attacker does not need to control the botnet, as they may simply rent it from an online service or purchase it from someone else. Some examples of these services include Blackhole, Stresser, and NitrousDDoS [37] (Fig. 1.2).

Man-in-the-Middle Attack: A form of cyberattack where the attacker secretly relays and possibly alters the communication between two parties who believe they are directly communicating with each other is called a Man in the Middle attack. The attacker can read all messages passing between the two systems and can also inject forged messages. The term "man in the middle" comes from an analogy to espionage: one party thinks they are talking directly to another party, while in fact both their messages are being read by an eavesdropper. The man-in-the-middle attack can be made in a number of ways. One way is for the attacker to physically place themselves between the two parties without either party knowing and then relay messages from one party to the other. This could be done, for example, by having access to a telephone company's network and rerouting calls or by being on a public Wifi hotspot. Another way is for an attacker with system admin privileges on a remote computer to use a man-in-the-middle (MITM) exploit to intercept and control traffic between the client and server (Fig. 1.3).

It is often referred to as "man-in-the-browser" because it uses vulnerabilities in a browser or other software to accomplish the attack. The attack can use social engineering, where the attacker tricks the user into accepting an unsafe connection over HTTPS or other secure protocols, or by exploiting known vulnerabilities in the software, such as Cross-Site Scripting. It is also used to refer to attacks where an unauthorized person gains access to a computer running a browser, uses the browser's interface via webcam and microphone, and recording capabilities, in order to spy on the user. The attacker can then use this information for blackmail or other nefarious purposes.

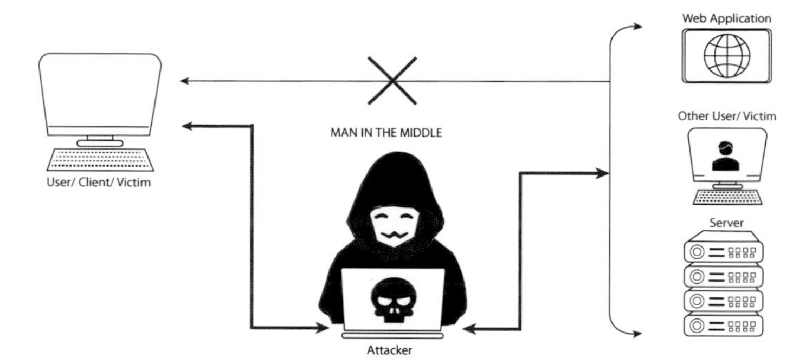

Fig. 1.3 A simple depiction of man in the middle attacks

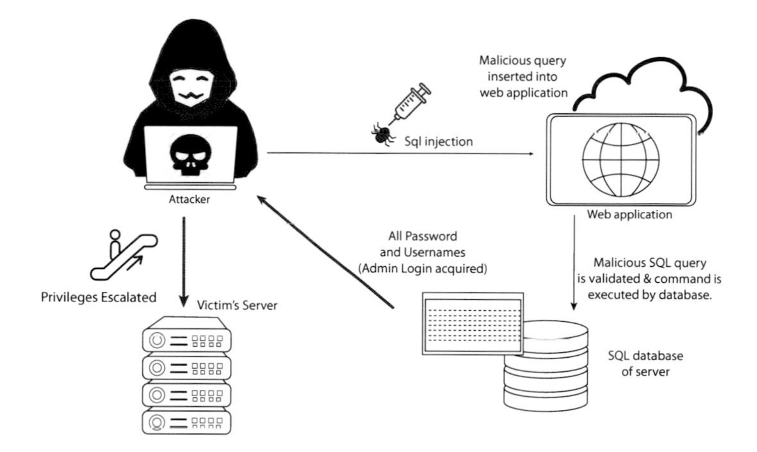

Fig. 1.4 A typical illustration of SQL injection attack

SQL Injection: This type of cyber-attack exploits the security vulnerability in the database. It is a type of code injection technique that can be used to attack data-driven applications. The SQL injection attack is one of the most common and dangerous types of cyber-attacks. It can be used to steal sensitive information from databases, modify or delete data, and disrupt service. SQL injection uses the dynamic nature of SQL (structured query language) to circumvent input validation and access data that is otherwise inaccessible. SQL injection usually involves the use of malformed or clearly erroneous input in a SQL command. For example, capturing the password after inserting into an account table may involve sending "*select password from users*" instead of "*insert into users values (username, password)*". This causes the database to execute the query and return the username, resulting in a list of user names from which the attacker can extract a database user's password. SQL injection may also be used to change data in a table without proper authorization. This may result in loss of confidentiality and availability for that particular table or other tables that reference it (Fig. 1.4).

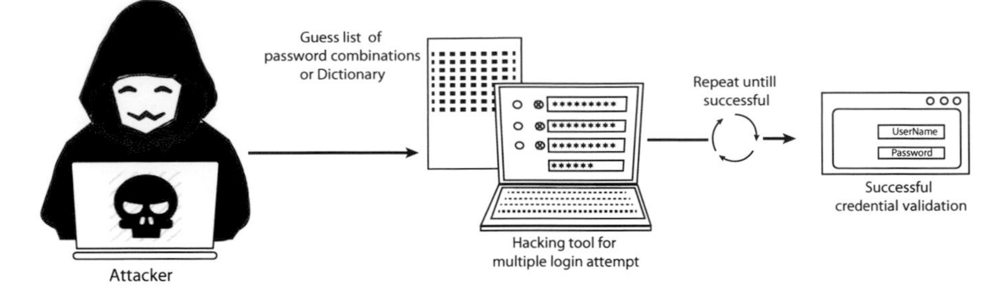

Fig. 1.5 Depiction of generalised principle of a password attack

A Password Attack: A method of cyber-attack where the attacker attempts to guess passwords or steal them outright, usually by hacking into a computer system or network is referred to as a Password attack. The attacker may use a man-in-the-middle attack to intercept the victim's password and then use it to access their account. It is notoriously difficult to crack a password, but with advanced cracking programs and tactics, hackers can eventually breakthrough. There are three types of password attacks: brute force, dictionary, and keylogging. Knowing these can help reduce your risk of attack. A brute force attack is a series of attempts at guessing passwords until the attacker penetrates the system. Dictionary attacks involve trying different combinations of dictionary words before stumbling upon the password. Keylogging is a method that records keystrokes to extract sensitive data like login credentials for use in a password recovery attack. This type of attack can be prevented by using two-factor authentication or by not clicking on links in emails from unknown sources (Fig. 1.5).

Attacks on IoT Devices: The Internet of Things (IoT) is a network of physical objects that are embedded with electronics, software, sensors, and connectivity to enable these objects to collect and exchange data. There are multiple IoT networks in existence, with the most common being the Internet and intranets. The IoT is a new frontier for cybercriminals. They are targeting IoT devices because they are often not appropriately secured. This makes them an easy target for cyber-attacks. As the number of Internet-connected devices continues to grow, it becomes increasingly important to protect these devices from being compromised by malicious hackers and cybercriminals. In general, IoT security requires three main components: IoT Device Software Architecture, Trusted Platform Module (TPM), and Security Standards [14].

According to a study published by Cisco Systems, "Cyber Threats to the Internet of Things: Emerging Risks and Tactical Strategies", it has been found that attackers are targeting the growing number of connected devices on the Internet in order to gain access to private information, cause disruption or steal money from users. The study found that the Internet of Things will significantly impact business and consumer perspectives, and it is important for organizations to consider cybersecurity risks.

Organizations should review their existing network products and services against these new risks.

With the recent pandemic, the work-from-home culture got a colossal thrust, and it seems like the trend of home-based offices is here to stay. With this precedent, residential areas are now becoming a valuable target for various reasons. Most organizations are well prepared to defend against cyber threats, but that is primarily true from inside the organization's infrastructure. The devices that connect to the organization's servers from the employees' homes become a prime vulnerability point for exploitation by the attackers. The volume of these attacks increased by 35% in the first half of 2020 compared to the second half of 2019, as stated by a report from Microsoft. With the popularity of home-based offices, there is a greater risk of these devices being targeted and exploited by hackers. If a hacker focused on compromising one or more homes in an area could wreak havoc within the organization, then it is highly plausible that they would be able to compromise other homes as well. For example, A de-authentication attack on an unsecured wireless network may provide the attacker with a hashed password. Furthermore, this means that the password can ultimately be cracked offline and malicious use of this password is possible.

1.2.2 Threats

Malware: Malwares are a type of malicious software that can be used to steal information, damage or disable computer systems, and gain access to private computer networks. Malware is often disguised as legitimate software or 'adware', or appears as genuine software but performs some hidden function when executed.

Malware can be classified into different categories based on the method of infection:

- Viruses: These are programs that attach themselves to other programs and replicate themselves by infecting other files. They are usually spread through email attachments, downloads from the internet, or by copying infected files onto a CD or DVD.
- Worms: These are self-replicating programs that use a computer's network connections to spread themselves over the internet. They do not need any user intervention in order to spread themselves.
- Trojan horses: These are programs that pretend to be something else, often a valuable item, while performing malicious actions.

Cisco has reported that malware can block key network access, install more harmful software, and transmit data from hard drives. The company is not naming any of the companies involved in the incident, but it does say that Cisco Talos cybersecurity researchers observed a type of malware called Responder being used to launch DDoS attacks. The company notes that Responder was created by a hacker named Peter Severa and can be downloaded from underground forums.

Phishing: It is a cyber-threat that is becoming more and more popular. It is a form of social engineering that tricks people into giving up their personal information. The most common form of phishing is when an attacker sends an email to a victim to trick them into clicking on a link or opening an attachment. The link may lead to a fake website that looks like the real one and asks for login credentials. The attachment may be malware that can steal data from the computer or install ransomware. Phishing emails are often designed to look like they come from legitimate companies and organizations to fool people into thinking they are safe and secure.

Phishing scams are becoming more and more sophisticated, using malware-based phishing, keyloggers, session hijacking, web trojans, DNS-based phishing, and host file poisoning to capture victims [35]. In this type of scam, cybercriminals send emails that appear to be from trusted companies, requesting personal information like account numbers, passwords, and credit card numbers. The goal is to trick the user into thinking they are interacting with a legitimate company in order to gain access to their accounts or steal their identity.

The most common form of phishing is deceptive phishing. This type of phishing uses fake websites to trick people into giving up their personal information, such as passwords, credit cards, and email addresses. There is no solid evidence that most people have any knowledge about and rely on this type of phishing or the effectiveness of these methods. One advantage to using phishing attacks is that geographic boundaries do not restrict them. This is because a person can use a computer anytime and from anywhere. Another benefit to phishing attacks is that they are considered 'low-risk' because the victim does not receive anything in return for giving up their personal information. In contrast, if someone were to use social engineering methods, like taking someone else's identity or creating a fake account with a similar name, to send a message with malicious intent, that would be considered cyberstalking.

Cryptojacking: Another form of cybercrime that involves the use of malware to hijack a computer's processing power to mine cryptocurrency is called Cryptojacking. The malware is injected into browsers, scripts, and ads, which will invisibly use computing power from the device it has infected to mine cryptocurrencies. The term "cryptojacking" was coined in 2017 by cybersecurity company McAfee. It is a portmanteau of "cryptocurrency" and "hijacking." Cryptojacking can be done through various methods, including:

- Malware that infects the victim's device and uses its resources to mine cryptocurrency.
- Websites that use scripts to mine cryptocurrency on their visitors' devices without their knowledge or consent.
- Mining software running on an infected device.

The increase in cryptocurrency value has led to an increase in coin-mining malware. Many of these infect devices and hijack its resources to mine cryptocurrency. There are also websites that use scripts to mine cryptocurrency on their visitors' devices without their knowledge or consent. One of the most common examples is Coinhive, which mines cryptocurrency on a website's visitors' devices. This can result in a decrease in performance and lower battery life for affected devices. Coin-mining malware has also been used to steal cryptocurrency. Attackers have created mining software that steals cryptocurrency, such as Monero, by secretly hijacking devices to use their power. Although the risk of malicious coin-mining malware is high, most of these viruses are easy to prevent by installing the appropriate protection on your devices.

Ransomware: Ransomware attack is a destructive malware that uses encryption to hold data hostage and demands a ransom to decrypt it. It is a type of cyber-attack that has been on the rise in recent years. The term has also been applied to a certain type of malware, dubbed "ransomware-as-a-service," which is a form of ransomware that can be bought and sold on the darknet through so-called "crypto marketplaces".

The ransomware attack can be carried out by an individual or a group of people who are not affiliated with any government or organization. The attackers usually demand payment in Bitcoin, which makes it challenging to trace them. The attack was first observed in May 2017, when an unidentified ransomware program called Petya infected computers worldwide. Ransomware is a type of malicious software that infects vulnerable devices, like a personal computer or server, and restricts or prevents their everyday use until the victim pays ransom to unlock it. Petya uses a stolen NSA-developed malware to spread by exploiting a vulnerability in the Windows operating system and encrypting all of the data on the computer. The software then creates an image with a ransom note that demands $300 worth of bitcoin for the user to regain access to their files. However, unlike other ransomware programs, Petya cannot be detected. The implications of a ransomware attack include:

- Ransomware attacks can disrupt data and operations, slowing business by spreading fear and uncertainty.
- The costs of a ransomware attack are difficult to calculate, but they are likely substantial.
- If a company's information is encrypted, it is more vulnerable to the effects of a future cyberattack.

Keylogger: One of the most commonly used malware to eavesdrop on a user's passwords and other personal information, such as credit card numbers and passwords for email accounts is a Keylogger. If a computer system has a keyboard attached, it is possible for an attacker to use a keylogger to monitor the user's keystrokes through special hardware or software. The goal of this type of attack is usually to steal passwords, personal information, and other sensitive data. Keystroke loggers can be inserted in between the keyboard and the computer's central processing unit, which records keystrokes as decoded by a computer's BIOS or operating system. In some cases, the loggers can be placed on the motherboard itself. Keylogging has been used in espionage, with one notable example being the FBI investigation into John Walker Lindh (the so-called American Taliban) in August 2001. This is done by first installing a keylogger software on the computer and then taking a picture of the victim's keyboard when it is in use. The keylogger software records keys that are pressed, with data about their position on the keyboard and what programs they were used in. The keylogger software runs as a system service and can be configured to start automatically. This allows the software to collect keystrokes in order to create a profile of the user's computer usage patterns and also for the user's personal records. The log files are stored on the hard drive to make subsequent analysis easier and may be saved for future use. The free software is available in a wide variety of languages and can be installed on any computer with a standard browser. It collects keystrokes in two different forms: "scores" generated by the operating system and "patterns" generated by the user's own typing activity. The log files are stored on the hard drive rather than in RAM. The pattern data is encrypted with a key that is specific to the computer and only accessible by the user's password. This means that there isn't an easy way to extract the patterns from a hard drive without a password or breaking into the machine, but if someone could get into your computer, they could read this data in its precise form.

1.2.3 AI as a Tool for Cyber-Attacks

One of the key factors that have allowed the internet to exist is its decentralized nature. The internet is not owned by any one entity, making it difficult for any one entity to shut down or control it. This unique aspect of the internet partly led to its success, and it allowed for new technologies like AI to become possible. However, with AI becoming more prevalent, the internet can quickly become a very different place. For instance, if AI can control the flow of information on the internet, it could be used to manipulate public opinion (e.g., give people false information that leads to herd mentality) or even cause war. Probably one of the more famous factors that led to AI becoming possible was the Singularity. Singularity is a speculative concept in which technological growth becomes so rapid and complete that it crosses a point of no return, triggering runaway technological change. The result is a "post-human" era in which intelligent machines surpass human intelligence. At the time of this writing, the idea of AI being able to create computer viruses has become quite popular-the problem of how to stop such an AI has yet to be faced.

AI can be used to create malware that can evade detection by antivirus software. It can also be used to create fake social media profiles and spread misinformation on social media platforms. AI is used by the military and intelligence communities to identify specific objects in a photo or video. The potential to abuse AI goes hand in hand with its potential to make autonomous decisions such as how many people should die based on a predicted crime rate. AI is being used to predict stock market crashes, a 2019 study showed that over 92% of Forex trading was done by AI and not humans [21]. More than 60% of trades over $10M are currently executed using algorithms, and that number is expected to grow significantly over the next four years.

1.3 Conventional Solutions

The most common defense against cyber threats is network security, including firewalls, intrusion detection systems, antivirus software, and encryption technologies. While network security helps, it is not a solution in and of itself. Experts agree that no system is 100% secure because there will always be vulnerabilities that attackers can exploit. Network security is one component of an overall cybersecurity strategy. Consecutively Cloud security refers to protecting data stored in cloud computing environments such as Amazon Web Services (AWS) or Microsoft Azure from cybercriminals. For example, data security is often provided by encryption when storing large amounts of customer data in a cloud environment. This can be done using the public key infrastructure or equivalent technology.

A cybersecurity strategy is a collection of techniques, policies, and procedures used to reduce the impact of any security breaches. It includes steps to mitigate risk from threats such as cyber-attacks, data breaches, and malicious software. A cybersecurity strategy includes several components: One component of a cybersecurity strategy is risk assessment which determines the likelihood that an event will occur and the potential consequences of the event. A risk assessment will often consider different types of threats and vulnerabilities in a business. For example, an assessment may consider whether an organization has a website that can be hacked or if there are weaknesses in password protection. Once risks are assessed, then mitigation measures can be developed to reduce the likelihood or impact of cybersecurity events. Mitigation measures are defined as means of reducing the risks of cybersecurity events. The most commonly used mitigation measures are security controls, encryption, and patching.

Among numerous policies is the Zero trust policy, which is beneficial for organizations that want to establish more robust control over different aspects of the company's digital security. It ensures that companies can manage access to sensitive information by looking at resources and prior user history. The Zero trust policy makes it possible to reduce the risk of data breaches and maintain privacy for employees. The zero-trust policy applies to the following:- Applications and data management- Email communication- Mobile applications and apps that the company owns or provides- Cloud computing, infrastructure, or storage services providers or terminals used by the company. The zero-trust policy is beneficial to organizations and employees and to individuals' privacy. It helps individuals build a well-protected digital identity and create opportunities for access when needed. A zero-trust policy is an approach to information security in which an end-user can access any other user's computer or application without any trust assumption. The Zero trust policy also referred to as the No Trust Policy, is an approach to information security in which an end-user can access any other user's computer or application without creating a trust relationship. This is done by establishing a "penetration of trusted computing" model that requires all users and devices to meet certain requirements before connecting to network resources. The term is also used in understanding cybersecurity:

a zero-trust approach has been used to create the concept of an Internet of Things (IoT) that relies on trust-less computing protocols.

Here is a list of strategies undertaken to defend against the cyber threats as far as conventional measures are considered:

- Using Firewall and antivirus programs:
  A firewall is software or hardware that separates the computer system from the internet to prevent it from being infected with viruses and other malicious software. Firewalls act as a buffer between the outside world and the network and give organizations greater control over incoming and outgoing traffic. Similarly, antivirus software, when run, detects and blocks any troubling threats. This process often includes scanning the device and/or network for any possible malware that might be present and then removing it. As one might expect, modern antivirus software can assess computers from two different directions. On the simple side is a system scan, which looks through files and prevents any potential harmful threats from causing further damage to the system. It does not get rid of anything, but it does make sure the system does not have malware. The other method is a comprehensive scan. This looks through every file and erases any possible threats, including viruses and malware. On the downside, it requires quite a bit of time to complete because it needs to look at all of the files and erase them if found malicious, but it prevents any potential for future harm.
- Using Secure browser extensions
  A secure browser extension helps stay safe on the internet by blocking phishing and malicious websites that try to steal personal information. It protects against the risk of malware as well as adware too. For instance, to improve online privacy, the "HTTPS Everywhere" Firefox extension forces your browser to use an SSL-encrypted version of a website when it is available. "Privacy Badger" automatically stops third-party websites from following your activity on the web. "AdBlock Plus" helps remove advertisements while browsing.
- Using VPNs
  A VPN is the most effective way to protect data from cyberattacks. It encrypts data and routes it through an encrypted tunnel so that hackers cannot access it. Some VPNs offer a kill switch that stops all internet traffic when the VPN connection is lost. This will prevent any data from leaking out through peer-to-peer connections, which could be helpful in certain situations. The best part of using a VPN is that the data is locked away and secured while traveling between servers. This makes accessing browsing history and data harder for anyone, including the ISPs and hackers. The final destination of the outgoing traffic stays a secret as well. Moreover, by connecting to the VPN, a computer or a device's IP address gets "hidden" behind the one that is given by the VPN server. This is helpful for anonymously browsing online.
- Using Strong and unique passwords
  The use of strong and unique passwords are essential because they make it harder for hackers to guess what password used on different websites or apps are. Using a strong password includes uppercase and lowercase letters, numbers, and symbols

in a unique but easy-to-remember way. Furthermore, a password manager can organize passwords into categories, so we will not forget which websites or apps require what kind of password.

- Using Security paths and updates
  Security updates are essential because they keep computers up-to-date. It is vital to keep software updated for the sake of the device and everyone else on the network, even if it isn't fun. Once a security update has been released, attackers will try and exploit that software and those who do not use it.

1.4 Intervention of AI

The use of AI in cyberattacks is a new and emerging trend. It is not yet clear how this will affect the future of cybercrime. There are several different AI and machine learning techniques used in cybersecurity. The most common ones include strategies that use AI to identify and monitor malicious activities, detect cyberthreats, and protect an organization's networks. For instance, a malware analyst can use machine learning algorithms to train an AI system on how to detect malicious files or identify compromised PCs. An AI system can also monitor the behavior of an individual or group, such as detecting changes in activity on social media or analyzing the traffic patterns of employees to identify those who might be up to something unusual. When integrating AI into cybersecurity, the key challenges for organizations are how to design and manage data that is available across multiple systems and how to structure data to make it accessible for cognitive applications that can incorporate human supervision. Artificial intelligence has permeated many aspects of our professional and personal lives.

1.4.1 Recent Trends

Along with this trend, cybersecurity is also increasingly adopting cognitive technologies. AI-powered cognitive technologies are an essential part of a holistic approach to cybersecurity in which the human element guides the process and plays a pivotal role. In general, cyber defense is a constantly shifting space where the nature of security threats changes with each new development. Cybersecurity professionals who can adopt successful cognitive technologies and guide their human element on a holistic approach will be more successful in defending against cyber-attacks. The industry has also embraced certain trends that have been years in the making, such as blockchain technology's role as an enabler for cyber defense and the increased need for artificial intelligence in cybersecurity. The report predicts that the IT security workforce size will grow as a result of these shifts. Cybersecurity is a vital component of any business and can be challenging to quantify. In its report, Cybersecurity

Trends to Look Out For in 2019, CyberVance discusses the importance of examining cybersecurity strategy trends, particularly how organizations can adopt new technology and protect against cyber-attacks.

In the past, cybersecurity professionals focused primarily on monitoring threats and defending against them. Now they are more concerned with risk assessment and mitigation, which allows them to avoid exploits that could cause harm. As a result, the most critical question to ask is, "What is the risk of this type of exploit?" As we can see, there are many other changes to cybersecurity professionals. They now focus on mitigating risks and assessing probability rather than monitoring threats. These changes create an entirely new world for those looking to enter the field.

A broader classification of the AI techniques used for detecting and mitigating cyberthreats include: Expert Systems and Intelligent agents.

- **Expert Systems**
  Expert Systems are a type of computer device that provides the decision-making power of a person. Knowledge-based systems are made up of two sub-systems, namely the Knowledge Base and the Inference Engine. The Knowledge Base stores the information and is linked to the Inference Engine, which interprets it or draws an inference from the available information to make decisions. Knowledge-Based Systems can make predictions and judgments based on the information in the knowledge base. They may be used in tasks such as medical diagnosis, stock trading, or even prognostication of the future. The knowledge-based system is a computer system that combines a computational engine (Inference Engine) and data storage in order to make predictions about unknown variables based on given known variables. Some examples of such systems are Weather Channel, Google search, Alexa or Siri. Knowledge-based systems take an existing body of knowledge and use it to create predictive models that are used in particular scenarios.
- **Intelligent agents**
  An intelligent agent is a software that exists in an environment that is not controlled by anyone externally. It can respond to fluctuations in its surroundings and continuously pursue its goals over time. They always have multiple ways of achieving those goals. An intelligent agent can be designed to learn all possible actions and then select the best option for accomplishing its goal. Intelligent agents are those that have the ability to learn and adapt to their environment.

Artificial intelligence can be used to detect and stop cyber-attacks by mimicking human intelligence. It can detect behavior patterns to identify potential threat signals that indicate a potential attack. Machine learning can be used in cyber security to detect and prevent targeted attacks on industrial control systems. Machine learning models can be trained to identify anomalous behavior that matches a targeted attack, thus allowing a cyber security system to block the attack before it is executed automatically. Intrusion detection technologies can be improved by incorporating machine learning techniques into them and using neural networks to detect anomalous behavior in traffic.

Machine learning is used in cyber security to help detect and prevent targeted attacks on industrial control systems. Machine learning models can be trained to identify anomalous behavior that matches a targeted attack, thus allowing a cyber security system to block the attack before it is executed automatically. Anomaly detection technologies can be improved by incorporating machine learning as an additional feature of the anomaly detection system. Machine learning can be applied to detect anomalous behavior based on data or anomalous machine-learning models that learn from data. Anomaly detection systems in networks can use machine learning as a metric to determine if anomalous activity is present in network traffic and then take actions such as filtering out the traffic in question or even taking further action. A method of anomaly detection consists of four components: input, training data, model parameters, and output. The input is the sequence of observations for which an anomalous event is predicted. This can be a number of characteristics such as TCP port numbers, HTTP header fields, and IP addresses at a company's edge routers. The training data is a collection of sequences of observations that the system has been annotated with. These sequences likely contain anomalous events. The model parameters define the training algorithm and include: normalization parameters, anomaly detection sensitivity, and detection threshold. These are used to measure how well an anomaly detector can identify an event and whether it is in a state of overconfidence or under-confidence. The model parameters also define how the anomaly detector reacts when detecting a false positive. Finally, the outputs of an anomaly detector include confidence and hypotheses. Confidence measures the likelihood that an anomalous event is occurring in a given sequence. Hypotheses are possible causes for the event with which an anomaly detector can work to identify a pattern or set of patterns in the data.

1.4.2 AI Based Mitigation of Cyberthreats

Malware Detection and Identification: Artificial intelligence being used for malware detection and identification is still in its infancy, but it has the potential to revolutionize the way we deal with cybercrime. AI can help identify malicious files before they reach the end-user and, by doing so, can provide significant security benefits. Many different AI/ML approaches have been used to detect malware, some more successful than others [19]. One approach uses machine learning and data mining to look for malware source code repositories using a technique called "SourceFinder" and analyses them based on characteristics and properties [30]. Another way uses machine learning to look for a particular strings within files that could indicate the presence of malware or malicious code and also classify them [34]. Another approach is to use AI/ML to detect patterns in binary executable files and determine if they are malicious [32, 32]. Another method for detecting and stopping malware is by utilizing visual binary patterns identified in the code and a type of self-organizing network that adapts over time [5].

Systems typically use heuristics - the process of looking for patterns in data - to find malware by crawling through huge data sets and looking for suspicious files that might be indicators of the presence of malware. In other words, systems might use hash values to detect files that have been recently modified, or they might use file size metrics over a period of time to determine that a new file appears to be too large. An antivirus (AV) program is designed to use heuristics and can employ techniques such as pattern matching, statistical analyses, and emulation of known malware signatures.

Coull et al. [9] presented Byte-activation analysis which is a type of neural network where the response to an input is mapped to the activation of each byte. On the other hand the FireEye [20] used the Convolution neural network and here a total of three networks were trained with different parameters, such as training set sizes and dropout settings. Dropout may have been turned off, or on during the process. Training the networks with a particular dataset of 7 million files was done to find the specific parameters. Adversarial Examples are inputs with small, imperceptible variations which cause neural networks to misreport it. Neural networks designed to detect malware automatically may not be immune to these types of attack and is the focus of the work by Demetrio et al. [10]. The purpose of this network is to analyze the structure to create adversarial examples that are misclassified and attempt to challenge it. A similar approach was tried before, but the results found in this study are different from those presented by [22, 23]. Moreover Bose et al. [7] proves that their technique can provide improved insight into each classification from [9, 10], while also exploring new solution areas to look for a better performance. The architecture includes two filters A and B wherein Filter A is designed to detect goodware content, while Filter B is designed to find specific parts in a file which makes it malicious.

A code obfuscation technique is a challenge for signature based techniques used by advanced malware to evade anti-malware tools. To tackle this, Sharma et al. [33] discusses an approach that they used to improve the accuracy of detection of unknown advanced malware and proposed a new method that uses Fisher Score for the feature selection and five classifiers to uncover the unknown malicious. A few other notable works in the same genre are as follows: Chowdhury et al. [8] have published a paper on classifying and detecting malware using data mining and machine learning. They use these methods to classify malicious websites that could potentially lead to infections. Hashemi et al. [18] used KNN and SVMs strategies to detect unknown malware on the data. Malware detection in android devices is done in [26], which used a deep CNN to identify malware in android devices, and [40], where a novel ML approach called rotation forest was used. Ye et al. [39] presents a new deep learning model, which establishes the SAE model for intelligent malware detection that's built on analyzing Windows API calls. The experimental results show that this method can offer a lot more than traditional shallow learning in malware detection and improve overall performance.

Spam Detection: AI can be used to detect spam by analyzing the content of the message and looking for patterns that are typical of spam. This is done by using machine learning algorithms that have been trained on a large number of examples of spam and non-spam messages. In the process of spam detection, AI will often replace a human role and be able to detect that messages are spam without requiring any human intervention. The automated system can also flag messages as potentially being spam and require human review before being classified as such. As an example, Feng et al. [13] presented a system built to filter out some spam emails. It combined a support vector machine and a naive Bayes algorithm.

On the other hand, AI-based spam detection in online review verification is also popular; for instance, A new unsupervised text mining model by Lau et al. [38] was developed in an effort to explore the possibility of detecting false reviews. This method was trained on a semantic language model to identify duplicates in reviews and then compared to supervised learning methods, which have already been successful in the review industry. The dataset was trained with a high-order concept of association. The results were interpreted to extract context-sensitive concept association knowledge among the reviewers and posts.

The Phishing Detection System: An artificial intelligence based system that can detect phishing emails by analyzing the content of the email and comparing it with a database of known phishing emails [4, 29]. The system can also detect if the sender is spoofing another person's identity. The phishing detection system can also be used with voice, video, and image messages. The system activates when a user receives a suspicious email or when they send an email containing personal information. Some of the features of the current phishing detection system are:

- Automatically detects email phishing scams
- Stores emails with malicious content in a quarantine folder
- Triggers user notifications when the system detects a new virus in an email.
- Detailed logs of all email activity
- Detects emails that contain phishing links
- Automatically generates a report of every detected email.

The goal of the phishing detection system is to automatically detect and report emails that contain phishing links. For instance, Feng et al. [12] utilized a neural network to detect phishing websites by using the Monte Carlo algorithm and risk minimization approach. Another approach by Mahajanet al. [25] proposed a system in Phishing Website Detection using Machine Learning Algorithms, which would keep track of various features of legitimate and phishing Uniform Resource Locators (URLs). They deal with machine learning algorithms to detect phishing URLs and use ML techniques to overcome the disadvantages of blacklist and heuristic-based methods, which cannot detect phishing attacks.

Intrusion Detection System: The use of AI for intrusion detection systems is a new and emerging field. It is a branch of computer science that deals with developing intelligent systems to detect, classify, and respond to cyber-attacks. They are designed to identify malicious behavior and stop it before it causes any damage. It can be implemented as a standalone system or as an add-on module to other security software such as antivirus programs. The intrusion detection system is usually configured with a set of rules that define what constitutes an attack, such as the use of certain words in the subject line of an email message or the sending of too many messages in a given period of time. The IDS then compares each packet of data against these rules and takes action if there is a match. The intrusion detection system is usually configured to generate alerts when it detects an event that might indicate an attack or intrusion attempt. IDS responses can be categorized into two main types: An active defense is one in which an IDS initiates a response. For example, it might issue an alert to on-duty personnel about a potential intrusion attempt. An active defense is the closest thing to "real-time" defense because the system initiates an action at the moment of detection rather than waiting for a report from another system. A passive defense is one in which the IDS only responds after receiving and processing information that an intrusion attempt has taken place. An example of this type of response would be when a system that is already installed on-site monitors for changes in network traffic around its perimeter and then initiates a report about these changes to a central monitoring hub and also stores the data in an analytics database.

The goal of an AI-based algorithm is to optimize certain features and improve its classifiers so that it may be able to reduce the number of false alarms that come up while trying to identify an intruder. It can also help to pre-empt and address potential security risks in a company's environment from the moment they are identified. A combination of SVM and a modified k-means was used by Al-Yaseen et al. [3] to create an intrusion detection system. On the other hand Hamamoto et al. [16] used fuzzy logic along with genetic algorithm to detect occurrence any anomaly in network. This was to predict a network's traffic for a given time interval.

A detailed survey of intrusion detection efforts in the last few decades is given in [17]; with many works listed, they conclude that Hybrid Machine Learning techniques have been used widely. Barbara et al. [6] proposed the hybrid Audit Data Analysis and Mining architecture, where the anomaly detection is followed by misuse detection. Farid et al. [11] improved anomaly intrusion detection using the Self Adaptive Bayesian Algorithm, which is designed to be used in large amounts of data. Another approach integrated Correlation-Based Feature Selection to select the best feature set. Resulting in improvement of the detection rate of the reduced data-set, as it selected the best feature set and removed unimportant data-sets [15]. Chowdhury et al. [24] proposed a new technique to reduce the dimensionality of data. Instead of using a traditional neural network, they use a triangular approach to calculate and visualize data.

1.5 Conclusion

AI can detect and stop cyber threats in real-time with limited resources. The constantly evolving nature of cyber-attacks means that humans shall struggle to keep up with the intel. However, using machine learning, AI can chomp down data for quick analysis and provide excellent security coverage without taking much time or energy away from the existing tasks. Machine learning allows Human analysts to focus on interpreting the results from deep analysis and devising novel techniques for fighting cyber-crime.

AI is not the elixir for all forms of security. Although AI-based approaches are becoming more common and cost-effective in most aspects of cybersecurity, they do not provide complete prevention or remediation measures. When a human opponent with an unfaltering stance attacks an intelligent system, there are limits to what an AI can do. It is essential to know that AI is not a factotum and will not be able to handle everything on its own, at least not right now. It actually needs expert human training and supervision to improve over time for the best results. Research shows that artificial intelligence has seemingly positively affected cybersecurity and risks. Hence the continuation of AI and machine learning will take the cybersecurity field to a new level of intelligence.

References

1. "cybersecurity ventures official annual cybercrime report" (2022). https://cybersecurityventures.com/annual-cybercrime-report-2017/. Accessed 19 May 2022
2. "global cybersecurity outlook 2022" (2022). https://www3.weforum.org/docs/WEF_Global_Cybersecurity_Outlook_2022.pdf. Accessed 19 May 2022
3. Al-Yaseen, W., Othman, Z., Ahmad Nazri, M.Z.: Multi-level hybrid support vector machine and extreme learning machine based on modified k-means for intrusion detection system. Expert Syst. Appl. **67**(01) (2017). https://doi.org/10.1016/j.eswa.2016.09.041
4. Reshma Banu, M.A., Akshatha Kamath C., Ashika S., Ujwala, H.S., Harshitha, S.N.: Detecting phishing attacks using natural language processing and machine learning. pp. 1210–1214 (2019). https://doi.org/10.1109/ICCS45141.2019.9065490
5. Baptista, I., Shiaeles, S., Kolokotronis, N.: A novel malware detection system based on machine learning and binary visualization. pp. 1–6 (2019). https://doi.org/10.1109/ICCW.2019.8757060
6. Barbara, D., Couto, J., Jajodia, S., Popyack, L., Wu, N.: Adam: Detecting intrusions by data mining. pp. 5–6 (07 2001)
7. Bose, S., Barao, T., Liu, X.: Explaining AI for malware detection: analysis of mechanisms of malconv. In: 2020 International Joint Conference on Neural Networks (IJCNN), pp. 1–8 (2020). https://doi.org/10.1109/IJCNN48605.2020.9207322
8. Chowdhury, M., Rahman, A., Islam, M.R.: Malware analysis and detection using data mining and machine learning classification. pp. 266–274 (2018). https://doi.org/10.1007/978-3-319-67071-3_33
9. Coull, S., Gardner, C.: Activation analysis of a byte-based deep neural network for malware classification. pp. 21–27 (2019). https://doi.org/10.1109/SPW.2019.00017
10. Demetrio, L., Biggio, B., Lagorio, G., Roli, F., Armando, A.: Explaining vulnerabilities of deep learning to adversarial malware binaries (2019)
11. Farid, D., Zahidur Rahman, M.: Anomaly network intrusion detection based on improved self adaptive bayesian algorithm. J. Comput. **5** (2010). https://doi.org/10.4304/jcp.5.1.23-31
12. Feng, F., Zhou, Q., Shen, Z., Xuhui, Y., Lihong, H., Wang, J.: The application of a novel neural network in the detection of phishing websites. J. Ambient. Intell. Humanized Comput. (2018). https://doi.org/10.1007/s12652-018-0786-3
13. Feng, W., Sun, J., Zhang, L., Cao, C., Yang, Q.: A support vector machine based naive Bayes algorithm for spam filtering. pp. 1–8 (2016). https://doi.org/10.1109/PCCC.2016.7820655

14. Guan, Z., Li, J., Wu, L.: Achieving efficient and secure data acquisition for cloud-supported internet of things in smart grid. IEEE Internet Things J. **4**(6), 1934–1944 (2017). https://doi.org/10.1109/JIOT.2017.2690522
15. Hall, M.: Correlation-based feature selection for machine learning. Dep. Comput. Sci. **19** (2000)
16. Hamamoto, A., Carvalho, L.D.H., Sampaio, L., Abrao, T., Proença, M.: Network anomaly detection system using genetic algorithm and fuzzy logic. Expert Syst. Appl. **92** (2017). https://doi.org/10.1016/j.eswa.2017.09.013
17. Hamid, Y., Muthukumarasamy, S., Ranganathan, B.: Ids using machine learning -current state of art and future directions. Br. J. Appl. Sci. Technol. **15**, 1–22 (2016). https://doi.org/10.9734/BJAST/2016/23668
18. Hashemi, H., Azmoodeh, A., Hamzeh, A., Hashemi, S.: Graph embedding as a new approach for unknown malware detection. J. Comput. Virol. Hacking Tech. **13** (2017). https://doi.org/10.1007/s11416-016-0278-y
19. Hossain Faruk, M.J., Shahriar, H., Valero, M., Barsha, F., Sobhan, S., Khan, A., Whitman, M., Cuzzocrea, A., Lo, D., Rahman, A., Wu, F.: Malware detection and prevention using artificial intelligence techniques (2021). https://doi.org/10.1109/BigData52589.2021.9671434
20. Johns, J.: "representation learning for malware classification" (2017). https://www.fireeye.com/content/dam/fireeye-www/blog/pdfs/malware-classification-slides.pdf. Accessed 19 May 2022
21. Kissell, R.L.: Chapter 2 - algorithmic trading. In: Kissell, R.L. (ed.) Algorithmic Trading Methods, 2nd edn., pp. 23–56. Academic Press (2021). https://doi.org/10.1016/B978-0-12-815630-8.00002-8, https://www.sciencedirect.com/science/article/pii/B9780128156308000028
22. Kolosnjaji, B., Demontis, A., Biggio, B., Maiorca, D., Giacinto, G., Eckert, C., Roli, F.: Adversarial malware binaries: evading deep learning for malware detection in executables (2018). https://doi.org/10.48550/ARXIV.1803.04173, https://arxiv.org/abs/1803.04173
23. Kreuk, F., Barak, A., Aviv-Reuven, S., Baruch, M., Pinkas, B., Keshet, J.: Deceiving end-to-end deep learning malware detectors using adversarial examples (2018). https://doi.org/10.48550/ARXIV.1802.04528, https://arxiv.org/abs/1802.04528
24. Luo, B., Xia, J.: A novel intrusion detection system based on feature generation with visualization strategy. Expert Syst. Appl. **41**, 4139–4147 (2014). https://doi.org/10.1016/j.eswa.2013.12.048
25. Mahajan, R., Siddavatam, I.: Phishing website detection using machine learning algorithms. Int. J. Comput. Appl. **181**, 45–47 (10 2018). https://doi.org/10.5120/ijca2018918026
26. McLaughlin, N., Doupé, A., Ahn, G., Martinez-del Rincon, J., Kang, B., Yerima, S., Miller, P., Sezer, S., Safaei, Y., Trickel, E., Zhao, Z.: Deep android malware detection. pp. 301–308 (2017). https://doi.org/10.1145/3029806.3029823
27. Molina Valdiviezo, L., Furfaro, A., Malena, G., Parise, A.: A simulation model for the analysis of DDOS amplification attacks (2015). https://doi.org/10.1109/UKSim.2015.52
28. Obotivere, B., Nwaezeigwe, A.: Cyber security threats on the internet and possible solutions. IJARCCE **9**, 92–97 (2020). https://doi.org/10.17148/IJARCCE.2020.9913
29. Peng, T., Harris, I., Sawa, Y.: Detecting phishing attacks using natural language processing and machine learning. pp. 300–301 (2018). https://doi.org/10.1109/ICSC.2018.00056
30. Rokon, M.O.F., Islam, R., Darki, A., Papalexakis, E., Faloutsos, M.: Sourcefinder: finding malware source-code from publicly available repositories in GitHub (2020)
31. Sausalito, C.: "cyberwarfare in the c-suite." (2022). https://cybersecurityventures.com/hackerpocalypse-cybercrime-report-2016/ (Nov 13, 2020); Accessed 19 May 2022
32. Schultz, M., Eskin, E., Zadok, F., Stolfo, S.: Data mining methods for detection of new malicious executables. pp. 38–49 (2001). https://doi.org/10.1109/SECPRI.2001.924286
33. Sharma, S., Challa, R., Sahay, S.: Detection of advanced malware by machine learning techniques (2019)
34. Shrestha, P., Maharjan, S., Ramirez-de-la Rosa, G., Sprague, A., Solorio, T., Warner, G.: Using string information for malware family identification. pp. 686–697 (2014). https://doi.org/10.1007/978-3-319-12027-0_55

35. Syiemlieh, P., Golden, M., Khongsit, Sharma, U., Sharma, B.: Phishing-an analysis on the types, causes, preventive measures and case studies in the current situation (2015)
36. Taghavi Zargar, S., Joshi, J., Tipper, D.: A survey of defense mechanisms against distributed denial of service (DDOS) flooding attacks. IEEE Commun. Surv. Tutor. **15**, 2046–2069 (2013). https://doi.org/10.1109/SURV.2013.031413.00127
37. Tandon, R.: A survey of distributed denial of service attacks and defenses (2020). https://doi.org/10.48550/ARXIV.2008.01345, arXiv:abs/2008.01345
38. Lau, R.Y., Liao, S.Y., Kwok, R.C.W., Xu, K., Xia, Y., Li, Y.: Text mining and probabilistic language modeling for online review spam detection. **2**, 1–30 (2011). https://doi.org/10.1145/2070710.2070716
39. Ye, Y., Chen, L., Hou, S., Hardy, W., Li, X.: DeepAM: a heterogeneous deep learning framework for intelligent malware detection. Knowl. Inf. Syst. **54**, 1–21 (2018). https://doi.org/10.1007/s10115-017-1058-9
40. Zhu, H.J., You, Z.H., Zhu, Z., Shi, W.L., Cheng, L.: DroiDdet: effective and robust detection of android malware using static analysis along with rotation forest model. Neurocomputing **272**, 638–646 (2018). https://doi.org/10.1016/j.neucom.2017.07.030

Chapter 2
A Survey on Deep Learning Models to Detect Hate Speech and Bullying in Social Media

Carol Eunice Gudumotu, Sathvik Reddy Nukala, Kartheeka Reddy, Ashish Konduri, and C. Gireesh

Abstract Any statement that is vituperative towards an individual or a group based on their traits like race, ethnicity, gender, sexual orientation, color, religion, nationality, or another attribute is described as hate speech. Hate speech and bullying, spreading uncontrolled might undermine society's peace and harmony, becoming a societal issue. Especially when hate speech is used to hurt people or to hurt the respect of individuals, groups, or countries. This complicates the task since social media posts contain paralinguistic tools (e.g., emoticons and hash tags) and a lot of poor quality written text that does not follow grammatical norms. With the recent advancements in NLP, it is possible to analyze unstructured composite natural language content. The chapter first focuses on discussing various deep learning architectures such as DCNNs, Bi- LSTMs, Transformers and models like BERT and how they are applied in identifying hate speech in social media. The chapter examines the capacity of deep learning algorithms to capture hate speech on public media systematically. The chapter also reviews the accuracy of models on publicly available standard datasets. The findings of this study pave the way for more research into the discovery of spontaneous abusive conduct on social media in the future.

Keywords Hate speech detection · Bullying · Deep learning · Natural language processing · Social media networks · Transfer learning · BERT · LSTMs · DCNNs

---

C. E. Gudumotu (✉) · S. R. Nukala · K. Reddy · A. Konduri · C. Gireesh
Vasavi College of Engineering, Ibrahimbagh, Hyderabad, Telangana, India
e-mail: gcaroleunice@gmail.com

C. Gireesh
e-mail: c.gireesh@staff.vce.ac.in
URL: https://www.vce.ac.in

A. Biswas et al. (eds.), *Artificial Intelligence for Societal Issues*, Intelligent Systems Reference Library 231, https://doi.org/10.1007/978-3-031-12419-8_2

2.1 Introduction

Hate speech needs to be monitored so that it does not hurt vulnerable communities. Using derogatory language towards a person on the basis of their race, gender, religion, nationality etc. constitutes hate speech. Bullying is an old phenomenon. Earlier school children have experienced this, but now due to the advent of technology, bullying has taken up a digital stage. Now, hate speech on online social networks is one of the forms of cyberbullying and now it is not limited to any particular age group.

Facebook (now Meta) has also taken responsibility to detect hate speech [1]. They have demonstrated recent AI advancements in two significant areas by:

- Increasing the semantic comprehension of language allows the systems to identify more delicate and complicated meanings.
- Increasing the breadth of the tools' understanding of the content so that the systems look at the image, text, comments, and other aspects as a whole.

They have recently introduced new technologies for semantic language comprehension, such as XLM, Facebook AI's way of self-supervised pre-training across several languages. They aim to develop these systems using new models like XLM-R, which combines RoBERTa, Facebook AI's advanced self-supervised pre-training approach.

They have constructed a pre-trained universal content representation to widen how their technologies perceive content for integrity challenges. This complete entity understanding system is currently being utilized on a large scale to evaluate content to detect whether it includes hate speech. They have recently enhanced the overall entity understanding system by employing post-level, self-supervised learning.

According to the May 2020 Community Standards Compliance Report, AI detects 88.8% of hate content removed from Facebook (now the Meta), up from 80.2% in the previous quarter. Facebook (now Meta) decided to take action on 9.6 million pieces of content for breaching its hate speech policy in the first quarter of 2020, an increase of 3.9 million.

This article explores three deep learning methods for alleviating hate speech and bullying on social media.

2.2 Methodology

This chapter mainly explores three important methods for hate speech and bullying detection. They are Deep Convolutional Neural Networks, Long Short Term Memory and Bidirectional Encoder Representations from Transformers or BERT.

- Under the Deep Convolutional Neural Networks (DCNN) section, this chapter surveys three prominent architectures of DCNN that can be used for identifying hate speech. The work of the authors suggests that DCNN architectures have provided better results when compared to traditional classification algorithms for the task.

- Under the LSTM Section, this chapter has surveyed five prominent architectures of LSTM to discover hate speech and bullying in social media.
- The BERT section provides survey of two different fine tuning strategies. The key novelty is to apply the bidirectional training of Transformer, a popular attention model, to language modelling. Two-way trained models can have deeper intelligence about language context and flow than one-way language models.

2.2.1 Convolution-Based Methods

Badjatiya et al. [2] described the goal of detecting hate speech on Twitter by categorizing tweets as racist, sexist, or neither. This process can be arduous due to the natural language structures' complexity. Their suggested method explores deep learning architectures. They test various classifiers like Random Forest, Logistic Regression, Gradient Boosted Decision Trees (GBDTs), SVMs, and Deep Neural Networks (DNNs). Using three deep learning architectures: FastText, Convolutional Neural Networks (CNNs), and Long Short-Term Memory Networks (LSTMs), these classifiers' feature spaces are defined by task-specific embeddings. They compare baselines to feature spaces composed of Bag of Words vectors (BoWV), char n-grams, and TF-IDF vectors.

This paper's key contributions are as follows:

- Using deep learning algorithms for detecting hate speech.
- Explore char n-grams, TFIDF word meanings, and task-specific embeddings created with BoWV and LSTM with GloVe, FastText and CNN.
- Their approaches outperformed methods by 18 F1 points.

To test their hypothesis, they utilized a dataset [3] that had 16000 annotated tweets. There are 3383 sexist tweets, 1972 racist tweets, and the remainder are neither racist nor sexist. For the embedding-based methods, they used word em beddings pre-trained by GloVe [5]. They trained GloVe embeddings on a massive corpus of tweets with 27B tokens, 2B tweets, and 1.2M vocab. They have tried experimenting with different word-embedding sizes for the job using embedding size=200 and discovered that different sizes provided similar outcomes. Using cross validation of 10-fold, they derived recall, F1 score, and weighted macro accuracy. For LSTM and CNN, they used 'adam.' For 'RMS-Prop,' they used FastText.

They discovered that combining embeddings obtained from deep neural network models with gradient boosted decision trees resulted in the most significant accuracy scores. The suggested approaches, which rely solely on neural networks, outperform baseline methods.

They tested with three broad representations for the baseline techniques.

1. Character n-grams: It detects hate speech using character n-grams.
2. TF-IDF which is a standard text categorization characteristic.

3. BoWV stands for “Bag of Words.” -To represent a sentence, the vector method employs the mean of Glove embeddings.

They tried several different classifiers for BoWV and TF-IDF methods. CNN outperformed LSTM, which surpassed FastText. When used with GBDT, initializing with arbitrary embeddings performs somewhat better than initializing with GloVe embeddings. The neural network method has been replaced by a method where the DNN uses average word embeddings learned as a function of the GBDT. The best approach used was sum of LSTM, Random Embedding and GBDT.” In this methodology, they initialize Twitter embeddings to arbitrary vectors. LSTMs are trained using backpropagation and the learned implants are used to train GBDT classifiers.

Zhang et al. [4] presented a new solution by combining convolutional and long short-term memory networks. The technique was evaluated against several baseline methods and on the compilation of publicly available datasets to date. Deep learning methods for hate expression classification use deep artificial neural networks (DNNs) to learn representations of abstract features from input data through many layers superimposed on each other. Convolutional Neural Network (CNN) and Long Short-Term Memory Network (LSTM) are the most common network designs (LSTM). CNN is well-known for its effectiveness as a network for acting as ‘feature extractors.’ In contrast, LSTM is a recurrent solid network for modeling ordered sequence learning issues. CNN naturally extracts character or word combinations like phrases and n-grams in the context of hate speech categorization. At the same time, the LSTM learns the long-range dependency of characters and words in tweets.

CNN+LSTM networks are effective in performing tasks like gesture and activity recognition, and they are robust enough to capture long-term dependencies between features extracted by CNN. The combination of CNN network and LSTM captures matching word n-grams as feature templates for classification. These networks would be helpful in hate speech classification tasks.

From the sentence-*“These Muslim refugees are not welcome in my Country; they should all be deported ...”*.

Pairs such as (Muslim refugees, deported) and (Muslim refugees, not welcome) can be captured. This was the hypothesis stated by them. They used minor preprocessing in the approach to normalizing the text of the tweet.

The preprocessing contains the following steps:

- Delete the following characters: **—:,;&** .
- Transform ‘#’ into words, an example stated by them is: “‘#refugeesnotwelcome”’ becomes ‘refugees not welcome.’ It is because ‘#’ are frequently used to construct phrases. To separate such hashtags, they utilized a dictionary-based lookup.
- For minimizing word inflections, use of lowercase and stemming is done
- Tokens with a document frequency of less than 5 are eliminated.

Then, they send the preprocessed inputs into the CNN+LSTM Architecture. The word embedding layer is the first layer in this architecture that will convert each text message (or “sequence”) into a real vector domain (Fig. 2.1).

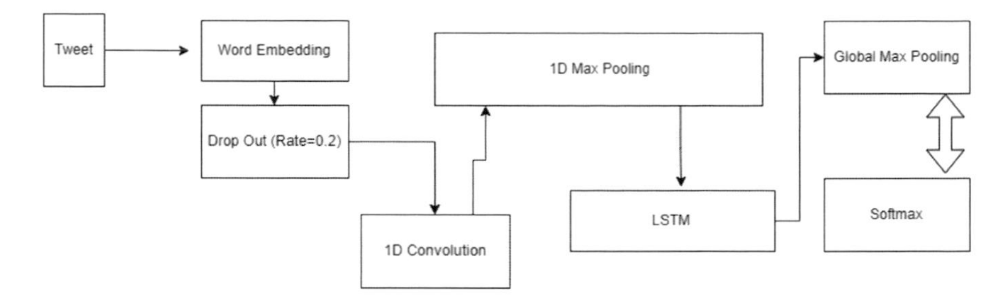

Fig. 2.1 The CNN+LSTM network architecture (Courtesy of Zhang et al. [4])

In order to avoid over fitting, an input feature space with dimensions of 100 × 300 is sent by the embedding layer to a failure layer with a rate of 0.2, to regularize learning. A 1-Dimensional convolution layer with 100 filters is fed with the output. The window size of the 1D convolution layer is 4 and it is used to fill the input so the length of the output is equal to the length of the original input. For activation, the rectified linear unit (ReLU) function is utilized.

As a result, the input feature space is reduced to 100 × 100 interpretation. It is then further down sampled along the word measurement by the 1D max-pooling layer. The pool size is 4, which gives a 25 × 100 shape output. Each dimension of 25 is an "extracted feature". These are sent into the LSTM layer. For each time step, the LSTM Layer produces 100 hidden units.

Following that, a overall max-pooling layer takes the utmost value in each time step dimension, resulting in a 1 × 100 vector. Finally, to estimate the probability distribution across all potential classes (n), a soft-max layer uses this vector as the input, depending on the dataset. They employed a total of seven avail- able datasets for the experiment. They utilize an SVM Model for the baselines, and each tweet is first pre-processed. The Basic feature set includes a variety of various sorts of features such as surface characteristics, language features, and sentiment features. They also introduce new features, which they refer to as the Enhanced feature set.

They created four baseline models to investigate the impact of feature selection:

- SVM: On the basic feature set, this baseline model uses the linear SVM model
- SVMfs: On the basic feature set, this baseline model applies the feature selection process.
- SVM+: On the enhanced features, this baseline model uses the linear SVM model.
- SVMfs+: On the enhanced feature set, this baseline model applies feature selection.

The findings of SVM-based models have shown that feature selection might improve model learning accuracy substantially. They modified their CNN+LSTM network to generate a new baseline. For generating the new baseline model, they have removed the drop-out layer which is among the embedding layer and the convolutional layers and also the global max pooling layer which is between the LSTM and the soft-max layers. They demonstrate that these two layers can help to improve task accuracy. CNN + LSTMbase will be the name given to this baseline model.

It can be concluded that adding the drop-out layer and the global max- pooling layer is essential for learning since the suggested CNN+LSTM model has a higher F1 score. The superiority of the network design is responsible for the higher results achieved by CNN+LSTM models utilizing solely word-based data. So, combining CNN and LSTM along with drop-out layer and pooling has given an improved feature extraction that is useful for learning.

Finally, they presented a technique for categorizing hate speech by using deep neural networks that combines Convolutional Neural Networks, Long Short Term Memory, drop-out, and pooling, which has been shown to experimentally enhance classification accuracy.

Cao et al. [5] presented “DeepHate,” an unique deep learning model that detects hate speech in social media platforms by combining text representations that are multi-faceted like topical information, word embeddings, and emotions. They assessed DeepHate by carrying out thorough tests on three big publicly available real-world datasets. “DeepHate” beats baseline methods on the given task, which is hate speech detection, according to their experiment results. They also conducted case studies which have given insights into the key elements that help in the detection of hate speech on online social networks.

To describe a post p, the suggested model initially employs several forms of feature embeddings. Following that, semantic, topic and sentiment are the three forms of latent textual representations to be learned when the feature embedding was introduced into neural networks. Then use a feed-forward network to integrate the latent representations. Finally, the combined representation is used as input to the softmax layer to predict the probability distribution for all classes. For both pre-trained word embeddings and sentiment embeddings, they utilize the same embedding size d = 300. They added a dropout layer following the embedding layer with a dropout of 150 % for regularization and a dropout of 20 % for the fully linked layer.

Each CNN layer has different filter window sizes, which are three in number, each with filters that are 50 in number. In LSTM, the count of hidden states is made to be 200. To train the model, they utilize the ADAM optimizer with 0.01 as the learning rate. They tested “DeepHate” on three publicly accessible datasets and one rebuilt dataset created by combining tweets of Twitter from the three datasets.

Then three CNN models are trained with various input embeddings for base- lines:

1. Word embedding represented as (CNN-W)
2. Character embedding represented as (CNN-C)
3. Character-bigram embedding represented as (CNN-B).

Another model that has been extensively studied in past hate speech detection research is the LSTM model. Similarly, three LSTM models are trained with distinct input embeddings:

1. Word embedding represented as (LSTM-W)
2. Character embedding represented as (LSTM-C)
3. Character-bigram embedding represented as (LSTM-B).

Table 2.1 Survey summary of convolution-based architectures to solve hate speech and bullying detection

Architecture proposed by	Model details	Objective	Dataset	Result
Badjatiya [2]	LSTM (Long Short Term Memory)+ Random Embedding + GBDT (Gradient Boosted Decision Trees)	To detect hate speech on Twitter	They worked on a dataset [3] that had 16000 annotated tweets. There are 3383 sexist tweets, 1972 racist tweets, and the remainder are neither racist or sexist	Deep learning methods outperform state-of-the-art char/word n-gram methods by ~ 18 F1 points.
Zhang [4]	CNN + LSTM Model	To categorize hate speech by using deep neural networks	A collection of publicly available datasets.	Neural network based methods consistently outperformed SVM and SVM+, by as much as 9% on the WZ-LS dataset. For SVM+, adding enhanced features leads to incremental improvement over SVM
Cao [6]	Model name: *DeepHate* Based on: 3 CNN Models, 3 LSTM Models	To introduce a deep learning model that detects hate speech in social media platforms	Three big publicly available real-world datasets	CNN and LSTM models with word embedding perform better than character-level embedding input

CNN and LSTM models using word embedding input out-perform character level embeddings, suggesting that semantic information in words can give high performance for hate speech identification (Table 2.1). The architecture of "DeepHate" is given in Fig. 2.2.

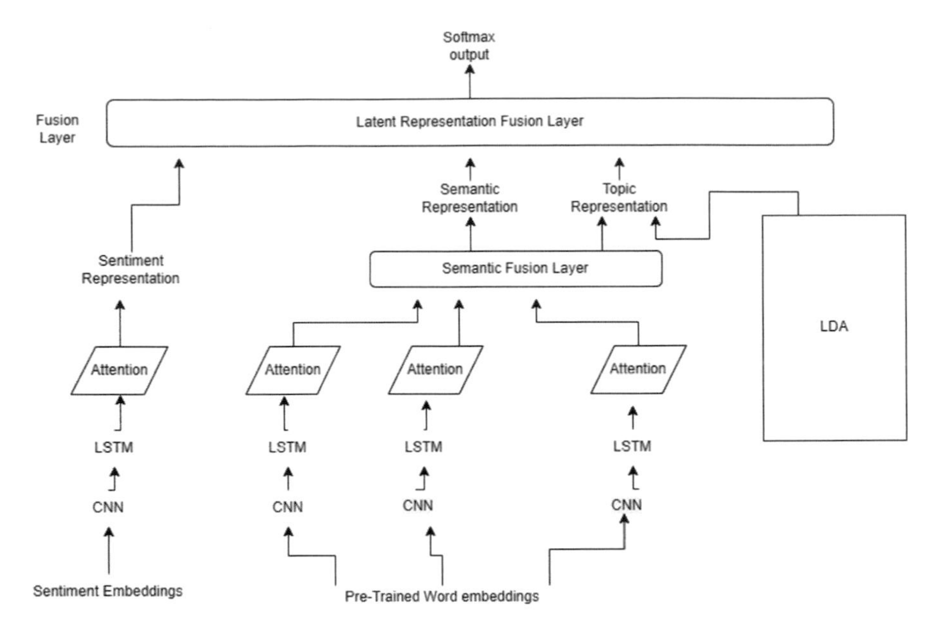

Fig. 2.2 Deephate architecture (Courtesy of Cao et al. [5])

2.2.2 Sequential Deep Learning Based Methods

These are specific forms of neural networks that are designed to perform effectively when there is a long-term reliance and a sequence data collection. When a network is required to recall information over a longer length of time, these networks can be beneficial. This characteristic qualifies LSTM for textual data processing. An LSTM is made up of a group of alike cells, each of which processes input in a unique way. Aside from external input, each cell also gets input from the cell before it in the chain. This cell layout allows the LSTM to recall earlier data for a long period of time.

The normal version of an LSTM can recall or refer to the information it has encountered so far. However, it lacks proof of information available after the point travelled till it. When working with sequence data, particularly text, this becomes a significant disadvantage. Another variant of LSTM is bidirectional LSTM, which may remember data in 2 directions. Back propagation is done in two methods in bidirectional LSTM. From the front and back, respectively. Bi- LSTM is a strong tool for analyzing data because of this procedure.

Chayan et al. [6] proposed that hate speech is defined as the exchange of verbal/nonverbal data between users who have a high level of intolerance and hostility. Hate speech may take many forms, such as user engagement on social media platforms that may contain un parliamentary language. Abusing an individual or group because of their beliefs, sex, race, or political alliance is another example. People's self-esteem is often lowered as a result of these exchanges of harsh words, which can have a detrimental influence on society. Their research aims to create a deep learning model for categorizing social media material as hostile or normal. The medium for detecting hate speech was decided to be Twitter. The models were trained using an open source dataset that was made publicly available. Using the dataset, their study primarily constructs a LSTM and a Bi Directional LSTM.

They used Kaggle, a platform, to gather information. Hate speech and non hate speech were divided into two categories in the labelled data set. Hate speech is labelled with a 1 while non-hate speech is labelled with a 0. The special symbols were deleted from the texts. The texts were then changed to lowercase. They also employed stemming to break the words down into their component parts. They looked at the amount of points for hate speech and non hate speech in dataset and discovered that it was significantly unbalanced.

If we use classification algorithms to this data set in its current condition, we will almost certainly receive biased results. As a result, they decided to up- sample the minority class by picking individuals at random from the class and adding it back to data set. This method supplied them with a well balanced data collection, but total quantity of the tweets skyrocketed. They separated the data into two categories: training and testing. For training and testing, they retained a 67:33 ratio. They trained an LSTM and a Bi-LSTM using the training data set. To prepare the data for the algorithms, they used one hot encoding. After that, they added padding. After that, they used word embedding.

For LSTM, the accuracy is 0.9785, precision is 0.9598, recall is 0.9986 and F1 Score is 0.9785. For Bi-LSTM, the accuracy is 0.9781, precision is 0.9582, recall is 0.9990 and F1 Score is 0.9781. The results indicate that LSTM outperformed Bi-LSTM in expressions of accuracy, precision, and f1 score.

Pedro et al. [7] proposed that it's critical to be able to monitor objection- able content and allow moderators to take whatever action they see fit. It is essential when seeking to safeguard groups such as immigrants, females, LBTQ, or anybody else who is the target of hate.

The hate speech identification challenge has three sub-tasks, according to the HASOC2019 task description. The following are the sub-tasks:

- **Sub-task A**:
  NOT: These are postings that do not contain any sentences that are deemed to be offensive in nature.
  HOF: Hateful, insulting, or vulgar language is included in these entries.
- **Sub-task B**:
  Hate speech (HATE): Hate speech phrases can be found in posts that fall in this category. These include describing negative characteristics or ascribing flaws to people because they belong to a specific group (for example, impoverished people are stupid). Hateful remarks directed towards groups of individuals based on their political stance, race, sex, gender, socioeconomic standing, physical condition are also included.
  Offensive(OFFN): Offensive material is present in posts that fall under this category. This refers to comments that degrade, dehumanize, or abuse a person. Posts

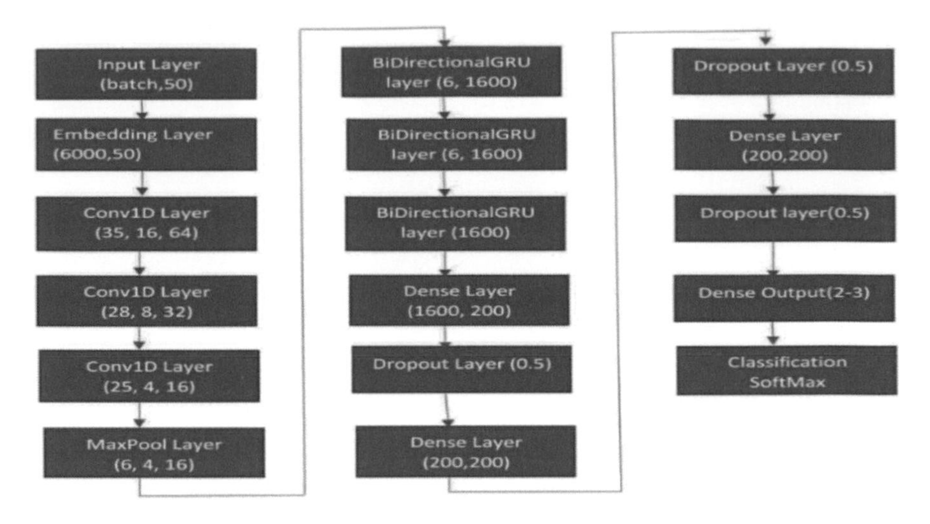

Fig. 2.3 Architecture of the proposed hate speech detection system (Courtesy of Pedro et al. [8])

that threaten people with fierce activities are too included in this category.
Profane (PRFN): Posts in this category include vulgar or inappropriate language but do not contain direct insults or abuse. Swearing (e.g., shit, fuck) and cursing are common topics in this class.

- **Sub-task C**:
  Targeted Insult(TIN) Targeted Insult(TIN): Insulting and threatening posts directed at individual, a group, or others.
  Un-targeted(UNT) Un-targeted(UNT): Posts that aren't aimed at a specific person or group, yet nevertheless include offensive language (Fig. 2.3).

This model works by stacking few convolutions on top and changing the middle layers. In this case using a Bi-LSTM to disparity with LSTM, GRU. They began with input layer with number of groups multiplied by span of text, and then utilized a self-trained embedding layer. They employed a max pooling layer of size four at the end of the following step, which is made up of convolutions of 16, 8, and 4 to minimize the size of input as much as possible without sacrificing too much data.

For the final classification stage, they utilize three bi-directional LSTMs with 1600 neurons. Then they combined dense and dropout layers, with a 0.5 dropout probability. Finally, a soft max layer with 2 neurons analogous to 2 categories was employed (NOT and HOF). These tests were carried out in two stages. On Subtask 'A', it's overall average F1 and weighted average F1 score in run1, run2 were 0.6279, 0.6094 and 0.6963, 0.6779 respectively.

Mihai et al. [8] stated that they used a single-layer, unidirectional LSTM model. As soon as they viewed the early findings, they made changes to our model. They also tried a two layer LSTM model but in the end landed on a 1 layer LSTM model with one basic embedding layer, mostly utilizing the Keras (Chollet et al. 2015) library.

Based on their findings, they determined that minimal pre-processing of the data set may enhance performance; as a result, the following pre-processing procedures were implemented:

- Use username markers instead of usernames.
- delete special letters and punctuation (@ / , . : ? ! $)
- convert to lowercase letters.

They decided not to exclude hashtag because, while the usernames may not always communicate data relevant to the tweet, hashtags are commonly used for reasonable purposes and must be taken into account when attempting to categorize data from Twitter. All data was represented using character-based representations in their model. They engaged one embedding layer with an input dimension of five thousand and the output dimension of twenty eight; the input length was considered by identifying lengthiest item in the data set and filling all representation to that length. They also used a 64-element layer LSTM with a dropout coefficient of 0.1 and a binary cross entropy loss function and a sigmoid activation function in their model. Their model was skilled on the English dataset for fifty epochs and the Spanish dataset for twenty epochs. The F1 score of 1 Layer LSTM is 0.31, 2 Layer LSTM is 0.42 and 1-Layer LSTM w/ Embedding is 0.69.

They started by evaluating a model prototype using a development dataset, focusing on the outcomes of subtask 'A' in English to determine which of the initial methods were most successful. After making this conclusion, they built on our best functioning model (basic LSTM model with embedding layer) and applied it to all jobs in both languages. For each of our models, they computed the F1- score and utilized it in their assessments. On the development set, the 1 Layer LSTM model with Embedding outperformed the other 2 models substantially, with an F1-score of 0.69.

They developed a basic LSTM model and used it to detect hate speech, determine aggressiveness, and determine whether speech is targeted or generic, with F1 scores 0.466, 0.462 for Subtask A, B in English and 0.617, 0.612 for tasks A, B in Spanish, respectively. When compared to Spanish data, they discovered that the model performed better on English data. With further pre-processing of Spanish data, they were unable to decrease the disparity. The disparity in its performance might be explained by its nature of online Spanish dialogue, which may have more accent or dialect differences than English.

Ghosh et al. [9] proposed a manner to tokenize the given text by converting it into tokens. Here the tokens are unique. Here the proposed LSTMs is the combination of layers which are stated as follows: embedding layer (an extra one), Dropout layer of factor 0.2, dense Layer along with the base LSTM Model. 5.16 million Macs are present in this model which makes it more valuable as it decreases the computation time.

The dataset used mainly focuses on the following classes: threat, insult, toxic, severe toxic, identity hate, and obscene in order. So the model when used on this dataset categorizes it into bully or non-bully with one of the classes as discussed

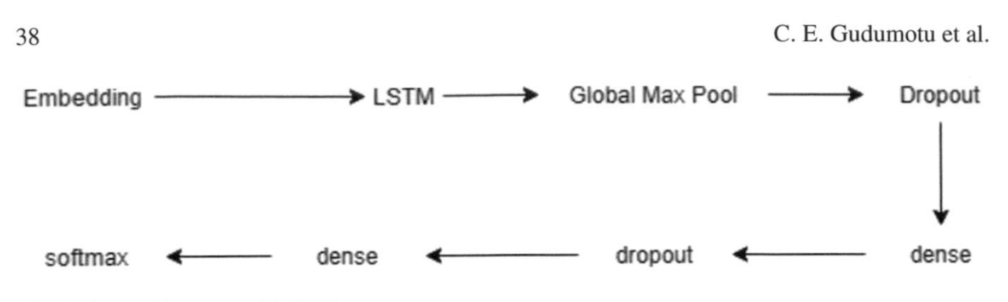

Fig. 2.4 Architecture of LSTM

above. Vanishing Gradient Problem can be solved by using LSTMs and it contains the cells which are classified into input, output and forget gates. As far as Accuracy is concerned the Bi-LSTM overweighs other models like LSTMs and D-CNN. So If our focus is more on accuracy and performance then it is best recommended to use the Bi-LSTMs. However this model lacks to detect sarcasm and performance can go down a little. The most important problem is to handle the long text sequences well and Bi-LSTM will come to best use in such cases (Fig. 2.4).

Roy et al. [10] proposed a model with LSTM and D-CNN to improve the hate speech classification on twitter. This model trains only tweet texts and reduces the process of feature extraction. They initially used the DCNN model which achieved F1-score 0.98 considering the best case scenario, still there are few misclassifications around 2%. To handle this scenario they further extended the model with LSTM. The LSTMs are used when the model needs to store context of long-sequence. As proposed in the model LSTM has input, forget, output and memory gates.

Before training the model the data set is split into training and testing in a 3:1 ratio, where a larger part is used for training and the rest for testing. They had run the CNN and LSTM model for 50 epochs with considering batches of tweets of size 50. They have tested with different sizes of tweets for CNN and fixed text size as 30 words. Here the First layer used is embedding layer. A pre- trained word vector - GloVe is used to initialize the weight of the embedding layer and for CNN they have used most of the parameters as default. After classification, the soft-max activation function is used to predict and calculate accuracy (Table 2.2).

2.2.3 Transformer-Based Methods

Bhatia et al. [17] proposed a review of related work on detecting cyber- bullying using various strategies is offered, which adds to our understanding of the current state of the art in this field while also offering computational reasoning. Because there are so many different types of cyber bullying, researchers have had a hard time detecting it. This challenge of identification becomes significantly more difficult when sarcasm is included. They divided methodology into: Dataset Collection, Pre-processing, Model Architecture and model approach. The hate-speech data set used is collected from various open-sources. Encase [18], FormSpring [19] and Waseem [20] are datasets used.

Table 2.2 Survey summary of Sequential Deep Learning-Based Architectures to solve Hate speech and bullying detection

Architecture proposed by	Model Details	Objective	Dataset	Result
Ghosh [9]	1D-CNN LSTM Bi-LSTM	Detect cyber bullying using DL methods	Has 159k instances which has bullying and non-bullying sentiments	Accuracies of the models: 1D-CNN:0.9633 LSTM:0.9412 Bi LSTM: 0.9745
Roy [10]	LSTM Model	To detect hate speech on Twitter	Has 29,720 Non Hate Speech and 2,242 hate speech tweets	Gives 0.53 recall value
Alonso [8]	Hate speech detection model	Detection of hate speech in social media	HASOC-2019 hate speech corpus	Sub-task A-Weighted average F1-scores: run 1:0.6963 run 2 0.6779
d'Sa [11]	Feature-based & Fine-tuning	Hate speech classification	Twitter corpus (Davidson et al., 2017)	An improvement of 11.6% in terms of macro-average F1-measure (84% vs. 72.4%) and absolute improvement of 4.8% in terms of weighted F1-measure (94.4% vs. 89.6%) is seen
Manolescu [7]	LSTM model	Hate Speech Detection in English and Spanish	Development data set	Achieved F1-scores of 0.466 and 0.462 for Subtask A and B in English, and scores of 0.617 and 0.612 for Tasks A and B in Spanish.

The datasets were inconsistent and had to be carefully considered before it was used for model training. They also included slang words [21] which are to be converted into conversational English before training. In preprocessing initially they cleaned URLs, emojis, user tags, re-tweet tags, hashtags from data. Then handled stop-words and lemmatization and for slang words used slang corpus. The data was experimented with a transformed model BERT which has seen best results in hate speech detection [22]. They used the 'bert-base-uncased' model for pre-training word embedding. The BERTbase comprises of 12 layers, 768 hidden layers, and 12 heads of several attention spans that find broad interactions between words compared to a single attention span.

In the pre-trained BERT model, they converted the final layers using a dense ReLU layer followed by an outlet between 0.3. The dense layer helps in achieving less training loss. Finally, the soft-max extraction layer is used for binary separation and classification. The outcomes obtained by the following species in the use of the slang corpus in screening during processing and outside of the complete database are recorded.

The survey compares the results with GloVe and custom models with and without slang corpus filters.

Thcy observed with slang corpus they have achieved a good increase in all models up to 8%. This is a useful discovery as it has shown reduction of false positives. This is important with regard to the purposes of the application where our aim is to reduce false profits and not to unreasonably accuse anyone of cyber bullying.

Zahiri et al. [23] proposed two steps for hate speech detection tasks as annotated data collection and model development. To achieve this purpose, the authors structured abusive and hate tweet detection as a classification problem and created numerous NLP algorithms. Similarly they have proposed a classification model for hate speech detection.

The model proposed is the CRAB model, which classifies the tweets data into defined classes. This model consists of four layers as:

- Representation layer
- Tokenwise class representation layer
- Sentence-wise class representation layer
- Aggregation layer.

In the representation layer, to vectorize the input data, they used the BERT encoder. By examining the left and right sides of the token context at all levels, BERT can create more complex textual representations. CRAB collects all of the BERT embedding created by the transformer's last block. Then in the token-wise class representation layer they classified the data as 4 classes (Normal, Abusive, Spam, and Hateful). In sentence-wise class representation layer is used to learn sentence level class depiction during the training process. The Aggregation Layer's goal is to combine information from preceding layers. To forecast class labels, token level and sentence level similarity scores are aggregated and given to a soft-max layer (Figs. 2.5 and 2.6).

The model is trained with tweets data collected by Founta et al., it is classified data into 4 classes. The tweets are heavily imbalanced, they applied sampling and create train, validation, and test sets with proportions of 80%, 10%, and 10%. In data emoji's, emoticons, hashtags, and website URLs abound in tweets. They used a pipeline that maps emotionally comparable emoji's and emoticons into the same special tokens to further filter tweets while preserving as much meaningful information as possible. To avoid loss of information they are not removing stop words and lemmatization. Then the BERT is initialized with different batch sizes like 32, 64, and 128 to experiment to get better results.

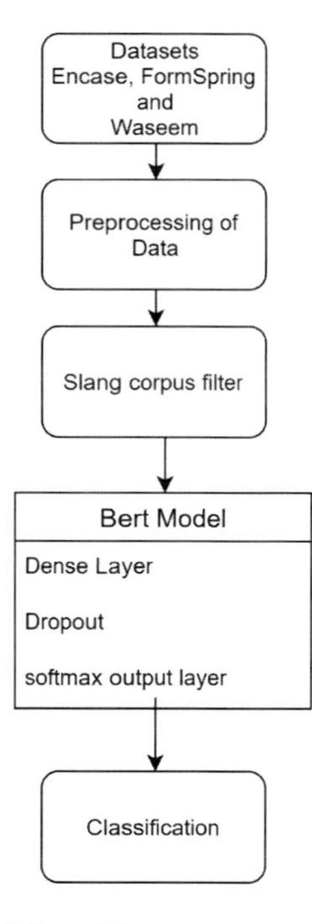

Fig. 2.5 Flowchart of proposed BERT model

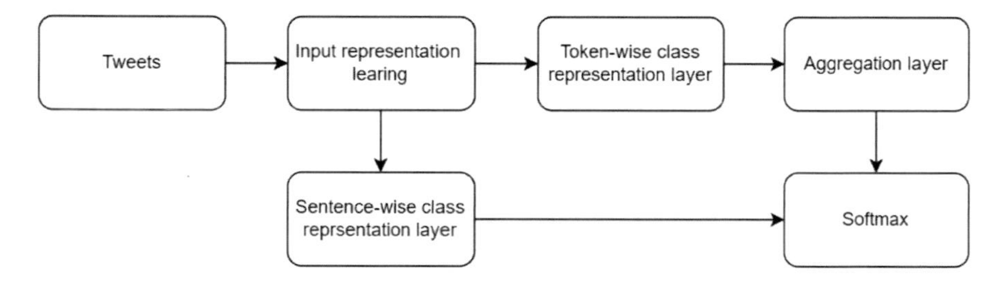

Fig. 2.6 Architecture of Class Representation Attentive BERT model (Courtesy of Zahiri and Ahmadvand [23])

In the hate speech detection procedure, this model uses both word and class info from tweets. On relative F1, the proposed model exceeded the state- of-the-art BERT-based baseline by 1.89% (Table 2.3).

Table 2.3 Survey summary of BERT Architectures to solve Hate speech and bullying detection

Architecture proposed by	Model Details	Objective	Dataset	Result
Bhatia [17]	BERT, GloVe (Global Vectors for World Representation)	To detect Cyberbullying using Natural language processing on Twitter	WASEEM dataset (2016) ENCASE Horizon 2020	GolVe twitter has accuracy 0.91 BERT has an accuracy of 0.90
Zahiri [23]	CRAB model	To detect hate speech in tweets using CRAB model	Tweets collected by Founta et al. The corpus in comprised of 4 classes: Normal, Abusive, hateful and Spam	On relative F1, the proposed model exceeded the state-of-the-art BERT-based baseline by 1.89%

2.3 Conclusion

In this chapter, we provide a critical overview of how deep learning models particularly convolution-based, sequential deep learning- based and transformers in particular are used to automatically detect hate speech in texts. In this paper, state-of-the-art architectures based on the above three methods are discussed and a survey of the same is done at the end of each section in Sects. 2.2.1, 2.2.2 and 2.2.3.

In the attribute-based approach, the word embedding sequence is used as input to the classifier. Classifiers DCNN, Bi-LSTM were checked. The Bi-LSTM categorizer achieved slightly better than the DCNN. The fine-tuning method is a one-step method that involves tailoring a previously trained BERT model to a disliked speech classification task. Matched to the feature-based method, the BERT fine-tuning approach improved the accuracy, perception and scale of F1.

Each Sects. 2.2.1, 2.2.2, 2.2.3 describes in depth how recent Deep Learning based methods can tackle the problem of hate speech and bullying on various social media platforms. Some of the architectures mentioned in Sects. 2.2.1, 2.2.2 and 2.2.3 are designed and developed based on publicly available corpus. We observe that deep learning based methods provide an advantage over traditional machine learning models. The chapter's purpose is to highlight the same.

References

1. AI advances to better detect hate speech, ML Applications, Integrity, Meta AI: https://ai.facebook.com/blog/ai-advances-to-better-detect-hate-speech. Accessed 16 Jan 2022
2. Badjatiya, P., Gupta, S., Gupta, M., Varma, V.: Deep learning for hate speech detection in tweets. In: Proceedings of the 26th International Conference on World Wide Web Companion, pp. 759–760 (2017)
3. Waseem, Z., Hovy, D.: Hateful symbols or hateful people? Predictive features for hate speech detection on twitter. In: NAACL-HLT, pp. 88–93 (2016)
4. Zhang, Z., Robinson, D., Tepper, J.: Detecting hate speech on twitter using a convolution-gru based deep neural network. In: European Semantic Web Conference, pp. 745–760. Springer, Cham (2018)
5. Cao, R., Lee, R.K.-W., Hoang, T.-A.: DeepHate: Hate Speech Detection via multi-faceted text representations. In: 12th ACM Conference on Web Science (2020). https://doi.org/10.1145/3394231.3397890
6. Ribeiro, A., Silva, N.: INF-HatEval at SemEval-2019 Task 5: convolutional neural networks for hate speech detection against women and immigrants on twitter. In: Proceedings of the 13th International Workshop on Semantic Evaluation, pp. 420–425 (2019)
7. Manolescu, M., Löfflad, D., Saber, A.N.M., Tari, M.M.: TuEval at SemEval-2019 Task 5: LSTM approach to hate speech detection in English and Spanish. In: Proceedings of the 13th International Workshop on Semantic Evaluation, pp. 498–502 (2019)
8. Alonso, P., Saini, R., Kovács, G.: The North at HASOC 2019: hate speech detection in social media data. In: FIRE (Working Notes), pp. 293–299 (2019)
9. Ghosh, S., Chaki, A., Kudeshia, A.: Cyberbully detection using 1D-CNN and LSTM. In: Proceedings of International Conference on Communication, Circuits, and Systems, pp. 295–301. Springer, Singapore (2021)
10. Roy, P.K., Tripathy, A.K., Das, T.K., Gao, X.Z.: A framework for hate speech detection using deep convolutional neural network. IEEE Access **8**, 204951–204962 (2020)
11. d'Sa, A.G., Illina, I., Fohr, D.: Classification of hate speech using deep neural networks. Rev. d'Information Sci. Tech. **25**(01) 2020
12. Devlin, J., Chang, M., Lee, K., et al.: BERT: pre-training of deep bidirectional transformers for language understanding (2018). CoRR arXiv:abs/1810.04805
13. Mozafari, M., Farahbakhsh, R., Crespi, N.: A BERT-based transfer learning approach for hate speech detection in online social media, arXiv:1910.12574v1 [cs.SI] , 28 Oct 2019 hate speech detection in online social media, arXiv:1910.12574v1 [cs.SI] . Accessed 28 Oct 2019
14. Waseem, Z., Hovy, D.: Hateful symbols or hateful people? Predictive features for hate speech detection on twitter. In: Proceedings of the NAACL Student Research Workshop, pp. 88–93. Association for Computational Linguistics, San Diego, California (2016). https://doi.org/10.18653/v1/N16-2013
15. Davidson, T., Warmsley, D., Macy, M.W., et al.: Automated hate speech detection and the problem of offensive language (2017). CoRR arXiv:abs/1703.04009
16. Waseem, Z., Thorne, J., Bingel, J.: Bridging the Gaps: Multi Task Learning for Domain Transfer of Hate Speech Detection, pp. 29–55. Springer International Publishing, Cham (2018). https://doi.org/10.1007/978-3-319-78583-7_3
17. Bhatia, B., Verma, A., Katarya, R.: Analysing cyberbullying using natural language processing by understanding jargon in social media. arXiv:2107.08902 (2021)
18. Founta, A.M. et al.: Large scale crowdsourcing and characterization of twitter abusive behavior. In: Twelfth International AAAI Conference on Web and Social Media (2018)
19. Saravanaraj, A., Sheeba, J.I., Pradeep Devaneyan, S.: Automatic detection of cyberbullying from twitter. Int. J. Comput. Sci. Inf. Technol. Secur. (IJCSITS) (2016)
20. Waseem, Z., Dirk, H.: Hateful symbols or hateful people? Predictive features for hate speech detection on twitter. In: Proceedings of the NAACL Student Research Workshop (2016)
21. Raisi, E., Huang, B.: Cyberbullying identification using participant-vocabulary consistency (2016). arXiv:1606.08084

22. Mozafari, M., Farahbakhsh, R., Crespi, N.: A BERT-based transfer learning approach for hate speech detection in online social media. In: International Conference on Complex Networks and Their Applications. Springer, Cham (2019)
23. Zahiri, S.M., Ahmadvand, A.: Crab: Class representation attentive BERT for hate speech identification in social media (2020). arXiv:2010.13028

Chapter 3
A Deep Learning Based System to Estimate Crowd and Detect Violence in Videos

Y. H. Sharath Kumar and C. Naveena

Abstract One of the major concerns throughout the world in all the places of large gatherings during an event is crowd control. The event can be of any form ranging from gatherings of few hundreds to millions. When large number of people assemble or move in an area which might not be suitable to handle the number it can create congestion which would lead to unexpected events such as stampede, riots or emergencies in an unfavourable situation. It would be extremely time consuming and hectic for manual monitoring. Henceforth an attempt is made to build an automated system using deep learning that would count the number of people assembled and compare it with the thresholds provided by the security management and also would try to identify physical violence in the region of interest. This deep learning system would also help in maintaining law and order during protests, curfew and imposed social distancing. These would alert the security personnel to take necessary actions. The system is built to estimate crowd and detect violence based on the street camera's feed which are usually positioned at the strategic places of crowd gatherings. The crowd estimation model is built with Convolutional Neural Network using different custom layers. All the results which are provided by this system is based on various methods crowd estimation and violence detection. A real time alert system in practicality has it's own challenges and limitations. Therefore, this paper focuses on the model which could help the system perform better.

Keywords Detection · Violation · CNN

Y. H. Sharath Kumar (✉)
Department of ISE, Maharaja Institute of Technology Mysore, Mandya, India
e-mail: sharathyhk@gmail.com

Y. H. Sharath Kumar · C. Naveena
Department of CSE, SJBIT, Bangalore, India
e-mail: naveena.cse@gmail.com

C. Naveena
Vivesvaraya Technological University, Belagavi, India

A. Biswas et al. (eds.), *Artificial Intelligence for Societal Issues*, Intelligent Systems Reference Library 231, https://doi.org/10.1007/978-3-031-12419-8_3

3.1 Introduction

With the increased number of population mainly in the cities it has become quintessential to monitor the crowd for numerous reasons such as thefts, fights, stampede, harassment etc. One of the major reason for all these unfortunate events to occur is large crowd gatherings in a small or unsuitable area. It would be very difficult for security department to monitor every surveillance camera in real time; also it would require a lot of manual work and efficient monitoring but if the department is under staffed it would cause problems. Stampede has become a major reason for deaths at religious events, protests and rallies; whereas in political campaigning and other such events riots are caused which should be detected at the earliest and actions must be taken. This paper is an attempt to build a system to cater than need. This would be of great help to analyses the crowd shape and size and allocate the required staff for the task.

The existing systems are not conducive to be used in every places as few of the systems are expensive and needs high end monitoring systems to adapt to their program. These surveillance systems must handle the computation which can be complex and heavy on the system. Many real time system are built to learn continuously with the new data but these are on the highest end of technology and infrastructure. The proposed system is a simple model which can adapt to most of the locally available systems and provide reasonable accuracy which is sufficient to achieve the desired goal. The trade-off between the computation power, time and accuracy has been one of the major focus in our paper.

3.2 Related Work

The related works of others with regards to crowd estimation were of different varieties each varied based upon various scenarios, environment and parameters. The paper [1] keeps track of both the crowd density and their direction. The UCF crowd dataset was used in this paper. The direction was determined with the help of feature extraction methods like simple blob detector, SURF, MSER, SIFT. The density was set to a threshold of a factor of 0.6 above which it classified as densely populated area but fails to give the exact count. The paper [2] partitions the frame to grids and then applies gabor kernel to determine the density and maps into 3 categories such as empty region, low and high density. The papers [3, 4] is based on detection and estimation of crowd density by the count of the heads in crowd with the input source as CCTV surveillance with respect to their region of interest, though the approach works well it has many real world limitations and challenges. Few papers proposes a different novel method built with the help of LSTM (Long short-term memory). The paper [5] uses of LSTM along with RNN (Recurring neural network). The paper [6] uses of LSTM along with CNN-RNN Crowd Counting Neural Network (CRCCNN). This method showed improvements than the previous other methods and not only produced density but also produced the count of people with better accuracy. This paper used the ShanghaiTech dataset which was promising for our study. The paper [7] used Multi-column convolution neural network (MCNN) to produce density maps. The extension of this method is in the paper [8] where Dilated Convolutional Neural Network was used. The paper [9] concentrated on the indoor crowd estimation whereas most of the papers focus on outdoor environment. Overall this paper attempts to obtain accurate estimation of crowd indoor and outdoor and in the methods of the papers [7, 8] and focus is not only on classification based on density but also to provide the accurate count of the crowd. This paper further explores these methods in this paper.

The related works of others with regards to violence detection were of different varieties each varied based upon various scenarios, environment and parameters. The papers [10, 11] focuses on the macro level anomaly detection in the crowd movement, it is achieved by considering the speed and direction of their movement and the region or space designated for their movement. This is a wide-scale approach whereas our focus is on static micro camera feed mostly for a small region of interest. The paper [12] proposes a holistic approach for real time violence detection by partitioning captured frames into grids and further mapping them as a scene of friction and congestion among people in the frame and by training the model to detect violence in such cases. The paper [13] proposes a novel method using LSTM and CNN to detect violence from the hockey fights dataset [14] and movie fight dataset [15]. This method performed better than the previous comparable methods in it's approach. This paper attempts a different method to build a model for a more generalized crowd violence dataset [16].

3.3 Methodology

3.3.1 Crowd Estimation

There are many datasets that are open sourced online for the purpose of developing deep learning models for crowd count estimation [17]; but all of these data sets have the annotated bodies of people and the drawbacks of these are in detecting the relative distance between any two people from a live visual. Therefore, a better option is to find out the density. One of the dataset that has this option is the ShanghaiTech dataset and it also provides the true count in it's dataset which would help in determining accuracy. The ShanghaiTech dataset is made up of two parts. Part A consists of 500 images and Part B is made up of 700 images. Both these parts have unique images. An image from both these parts and their heatmap can be seen in Fig. 3.1. Each part has 3 folders, images, ground truth and groundtruth-h5. Images folder has the jpeg files, the ground truth folder has matlab files contain annotated head (coordinate x, y) for that image. And the groundtruth-h5 folder is having the density map of that

Fig. 3.1 A sample of dataset [17]

image. The density map of the image is calculated by using both matlab file and the jpeg file by using gaussian filter on the places where heads are annotated, pixels.

The crowd estimation model follows [8] with few changes in it's model. The original model used VGG-16 for first 10 layers of it's 16 layered model. This paper attempts in building 3 different related models and experiment. The first model for crowd estimation is built up of six convolution layers and two max pooling layers. All convolution layers have a kernel size of 3×3 and for maxpooling layer 2×2 filter size is used. The first two layers have 64 kernels followed by another two convolution layers having 128 kernels, after every 2 of such layers a maxpooling layer is present. The model consists of two dilated convolution layers which have 128 and 64 kernels respectively with a dilated rate of two. The output image is reduced to 1/4th the size of the input image using two maxpooling layers. The Rectified Linear Unit (ReLU) activation function used is used. The model can be visualised as in Fig. 3.2. After the 6 layered model, 8 is attempted by adding a 256 kernel layer and a dilated convolution layer and 12 layered model by adding 256,256,512 kernel layers and 2 dilated convolution layers.

The Input video is converted into frames using various functions of opencv(). The RGB image is converted into a tensor which is a form of n-dimensional array. The tensor is normalized and then fed into the CNN model. The output is also in the form of a tensor which is converted into a numpy array. Density map is obtained by converting the numpy array into an image file. The Mean Absolute Error (MAE) and the computation time taken for each model is discussed in the results and analysis section of this paper.

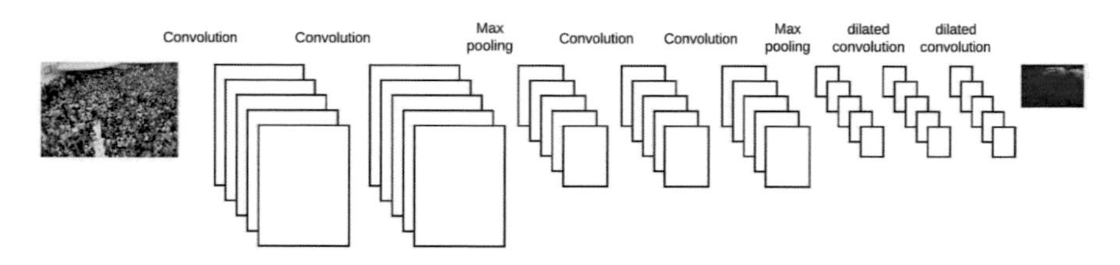

Fig. 3.2 A 6 layered crowd estimation model

3.3.2 *Violence Detection*

A model is built to detect violence in a video by using image processing technique with the help of Keras from Tensorflow library. The code for the model is of python language. The model is trained from a video dataset called "real-life-violence-situations-dataset" [16] obtained from the kaggle datasets library which consists of 2000 videos in total; where it is classified into 2 classes namely violence and non violence with 1000 videos in each class. This dataset consists of all the features of both the hockey fights dataset [14] and movie fight dataset [15] and much more varieties. The videos are ranging from the short duration of few seconds to few minutes and majority of the videos are between 25 and 30 frames per second. The model is trained using images or frames obtained from the video at the rate of 10 frames per second which had better data when compared to our trial using 2, 3, 5, 7, 12, 15 frames per second. The train-test split was in the 80–20 ratio which is one of the standard general split. Totally for training 23,000+ images are used and for testing 5700+ images are used with the goal of binary classification. Few sample images can be seen in the Fig. 3.3.

Each frame undergoes data augmentation and resizing before subjected to Convolutional neural network (CNN) model. Frames are rescaled at 1./255 with shear range of 0.1 and zoom range of 0.1 along with horizontal flip. The Hyper parameters of image width $\times$ image height are taken as 200×200 and the batch size as 32. The batch sizes of 8, 16, 32, 64, 128 and 256 was tried and batch size of 32 was efficient.

The model consists of multiple CNN layers, it can be seen in the Table 3.1. All the convolutional layers has it's kernel size as 3×3 and 2×2 filter size for MaxPooling layer. First 2 convolutional layers consists of 32 kernels, followed by a MaxPooling layer. Input shape is fed to the first convolutional layer. Next 2 convolutional layers consists of 64 kernels, followed by a MaxPooling layer. Next 2 convolutional layers consists of 128 kernels, followed by a MaxPooling layer. Next 2 convolutional layers consists of 256 kernels, followed by a MaxPooling layer. All these layers have a Rectified Linear Unit (ReLU) function which is an activation function. Input is flattened and fed into 2 fully connected dense layers. Each fully connected dense layer consists of 256 neurons followed by a dropout layer with value of 0.5. The output layer has a single neuron with sigmoid activation function where if the output is greater 0.5 it's classified to the non violence class if not then violence class. The model is trained using Stochastic Gradient Descent (SGD) optimizer with momentum. RMSProp and

Fig. 3.3 Few samples of violence detection dataset [17]

Table 3.1 The violence detection model summary

Layer (type)	Shape
conv2d (Conv20)	198, 198, 32)
conv2d_1 (Conv2D)	196, 196, 32)
nax_pooling2d (MaxPooling2D)	98, 98, 32)
conv2d_2 (Conv2D)	96, 96, 64)
conv2d_3 (Conv2D)	94, 94, 64)
max_pooling2d_1 (MaxPooling2)	47, 47, 64)
conv2d_4 (Conv2D)	45, 45, 128)
conv2d_5 (Conv2D)	43, 43, 128)
max_pooling2d_2 (MaxPooling2)	21, 21, 128)
conv2d_6 (Conv2D)	19, 19, 256)
conv2d_7 (Conv2D)	17, 17, 256)
max_pooling2d_3 (MaxPooling2)	8, 8, 256)
flatten (Flatten)	16,384)
dense (Dense)	256)
dropout (Dropout)	256)
dense_1 (Dense)	256)
dropout_1 (Dropout)	256)
dense_2 (Dense)	1)
activation (Activation)	1)

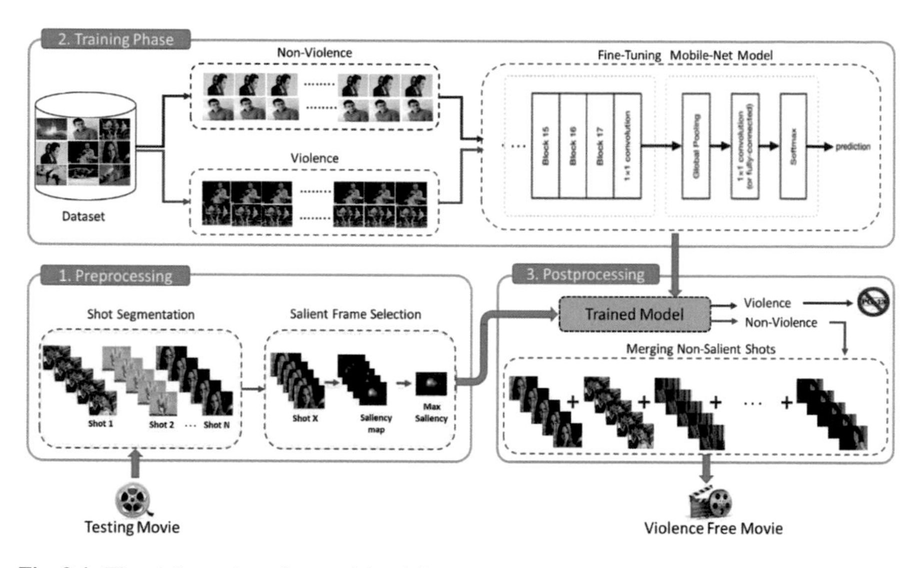

Fig. 3.4 The violence detection model training

Adam optimizer with different learning rates (lr) was tried but SGD optimizer with the learning rate of 0.001 and momentum of 0.9 performed better.

In the Fig. 3.4, the code to obtain the best model based upon the early stopping criterion is given. Early stopping function is used to stop the epochs if there is no further minimization in validation loss with a patience of 10 epochs. The best model is saved with the help of model checkpoint function where it saves the model with the highest validation accuracy as an h5 file. The number of epochs was set to 100 with a patience of 10 epochs. The validation steps and steps per epoch is dependent on the batch size.

3.4 Implementation

The System to estimate crowd and detect violence has a Graphical User Interface (GUI) built using web technologies such as HTML and Bootstrap library for the front end and Flask framework which is a micro framework of web written in python for the back end, the data is stored using file structures. The web page is divided into 2 major blocks one for crowd estimation and another for violence detection. Both of these accept images and videos of given formats and produces the output. If in case of any error it shall display it to the user. This is a prototype but can be extended to implement for the real time camera feed.

The crowd estimation system in order to provide the density of crowd accepts threshold or limit number from the user, which determines the sparse or dense crowd gathering and can produce an output. Along with the overall threshold number, horizontal sectional threshold number is also taken for 4 divisions which is an idea based on camera sight, where farther the distance from the camera more the number of people can gather within safety limits but not near the camera because the area covered by the camera coverage narrows down limiting less number of people. The Fig. 3.5 explains the flow of events of the system where the user inputs a file or data for either crowd estimation or violence detection. The system performs file format validation. The process continues only if the format is suitable; if not, an error is displayed and user is redirected to the home page. In case of crowd estimation the validation of thresholds is also done as the fields must not be left empty or given an alphanumeric value or a negative integer or float value. The system also detects corrupt file and if the file is corrupt it redirects it to home page by revoking output request. Once the validation checks are cleared the input videos are converted into frames and pre processed and fed into the model. If the input is an image it directly be subjected to pre process stage. The crowd estimation output consists of original image, its horizontal quadrant images and their heatmap in a table followed by the count, with appropriate font color. The results can be downloaded in the portable device format (PDF) file. For the violence detection system the output is of the same file format as it's input. This is further discussed in results and analysis section of this paper.

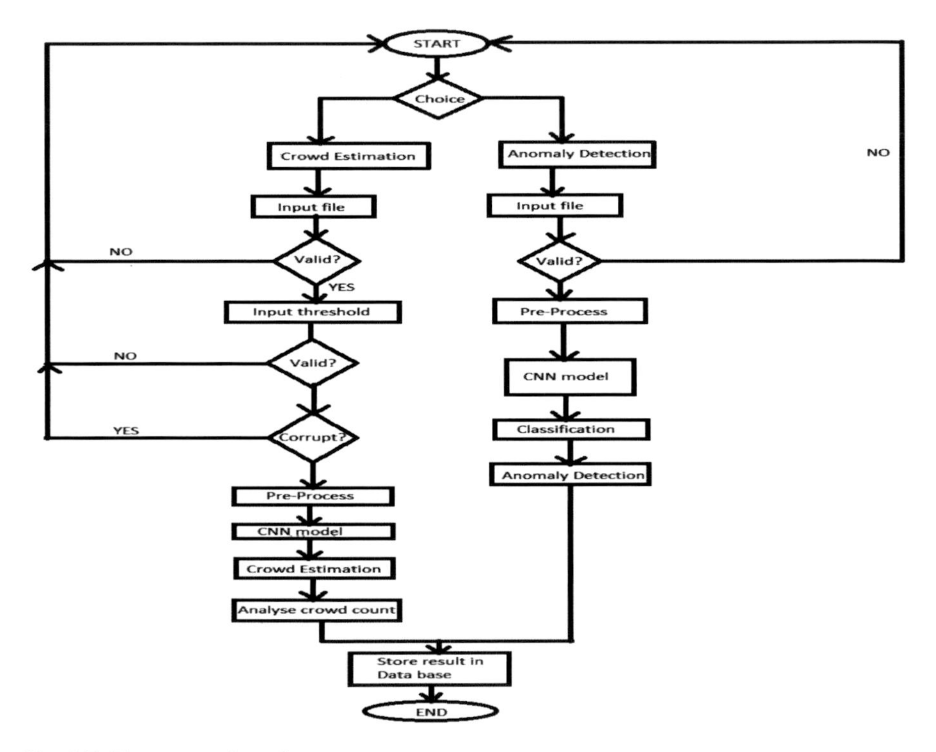

Fig. 3.5 The process flow chart

3.5 Results and Analysis

In the Fig. 3.6, we have compared the trade-off between 3 parameters namely computation time (CPU), number of layers in the model and Mean absolute error (MAE). From this experimental result it can be notice that if the infrastructure is basic and cannot wait for the result for a long time, then a lesser a layered custom model can be deployed with the increase in MAE. These are the systems which do not require highly accurate count. If the infrastructure is advanced with more graphics processing unit (GPU) or Tensor Processing Unit (TPU) cores one can use the VGG based model for higher accuracy. Henceforth this study provides a view over the trade-offs for the system.

The output produced by the crowd estimation model it can be that there is a 5 × 3 table. In the table first column consists of the original input given by the user split into horizontal quadrants and also the original image in the last row. The second column consists of heat map generated with respect to the image. The last column provides us the actual count which can be of 3 colors given by the color coding. The green color represents safe level, orange represents to be alert and red means danger. The coloring is based on the thresholds provided by the user if the count is below the threshold number then green, if above by 10% then orange and beyond that is red; these can be seen in the Fig. 3.7. This system provides both the count of the crowd

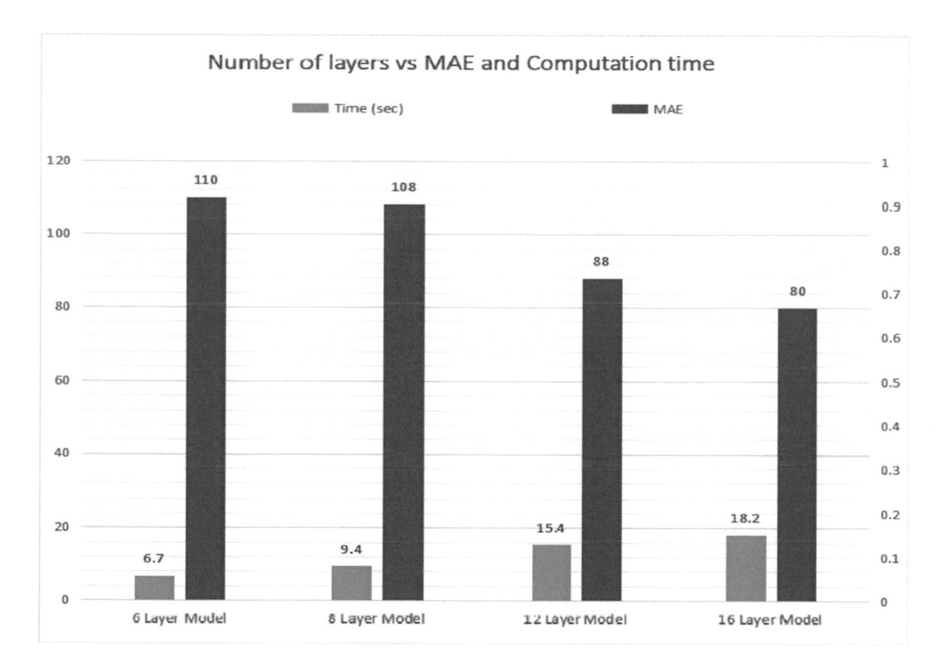

Fig. 3.6 Comparison between number of layers, mean absolute error (MAE) and computation time in the crowd estimation model

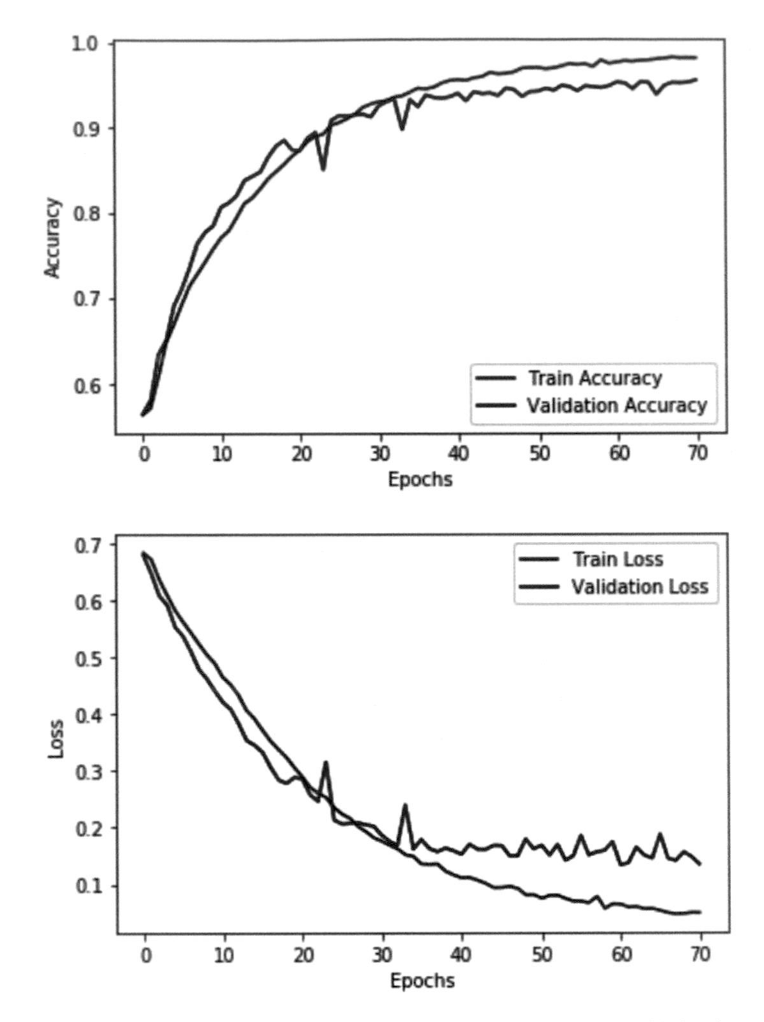

Fig. 3.7 The accuracy versus validation accuracy (top) and the loss versus validation loss (bottom) graph for violence detection model

and also the alert for the user. Hence can be used for strategic and statistical purposes and for the crowd control by security.

The violence detection model had an early stopping criterion and a model checkpoint function was used to save the best possible model weights before the divergence of the curve. The model achieved training accuracy of 98% and 95.55% validation accuracy with training loss of 04% and validation loss of 11%. The model performs very well given that it is a binary classification model. The model classified different videos from different sources with an average accuracy of 95% and it also classified graphical videos and low resolution images with the same accuracy. The testing was done with different set of frames ranging from cartoons, video games with lower graphics to higher graphics, pictures from different textbooks and novels, street fights, protests, riots, fights between 2 groups of sport team supporters and political campaign fights, scenes from the movies etc. and can also differentiate between close movement of people, hug against physical violence; few of them can be seen in Fig. 3.8. The location of these videos were also of different varieties indoor, streets, parks, fields, stadiums, etc. The violence detected in videos or images would result in message stating violence detected in the frame as shown in the Fig. 3.9.

Fig. 3.8 Few results of the violence detection model [17]

Fig. 3.9 An output frame of the violence detection model [17]

3.6 Future Enhancement

For the crowd estimation in future, attempts shall be made to minimize the MAE and provide the count with much higher accuracy using different image processing techniques and advancements in the neural networks. MCNN and LSTM are other promising methods to estimate crowd and shall be further studied. For the violence detection various other image processing models can be used and experimented along with LSTM or few other methods. The proposed model can also be subjected for different hyper parameters and a locally built dataset for each surveillance system would helpful in building a system with higher classification accuracy. A self-learning system can be built to learn from the new varieties of actions captured by the surveillance camera. Different varieties of anomalies can also be included for detection in future.

3.7 Conclusion

A crowd counting model and a violence detection model is built using Convolutional Neural Network. The violence is detected at an average accuracy of 95%. The crowd estimation model is built with Dilated Convolutional Neural Network using different custom layers and the MAE and the time taken for each is mentioned in the results section of this paper. This system is a prototype which can be further improvised and deployed as a real time violence detection and alert system along with crowd density estimation. The GUI of this system makes it easier for the user to obtain the desired result. Overall this paper is a study on usage of a system built upon image processing technique in public areas to monitor and control the crowd.

References

1. Muhammed Anees, V., Santhosh Kumar, G.: Direction estimation of crowd flow in surveillance videos. In: 2017 IEEE Region 10 Symposium (TENSYMP) (2017)
2. Alamri, S.S.A., Bin-Sama, A.S.A.: Developing an intelligent system for crowd density estimation. Int. J. Comput. Sci. Inf. Secur. (IJCSIS) **14**(4) (2016)
3. Alamri, R.J.: Al-Masjid An-Nabawi crowd advisor crowd level estimation using head detection. In: 2018 IEEE 1st International Conference on Computer Applications and Information Security (ICCAIS) (2018). 978-1-5386-4427-0/18/$31.00
4. Musa, A., Rahman, M.M., Sadi, M.S., Rahman, M.S.: Crowd reckoning towards preventing the repeat of '2015 Hajj pilgrims stampede'. In: 2017 IEEE 2nd International Conference on Electrical and Electronic Engineering (ICEEE) (2017). 978-1-5386-2303-9/17/$31.00
5. Muhammed Anees, V.: Deep learning framework for density estimation of crowd videos. In: 2018 Eighth International Symposium on Embedded Computing and System Design (2018)
6. Fu, J., Yang, H., Liu, P., Hu, Y., Wang, Y., et al. (eds.): A CNN-RNN neural network join long short-term memory for crowd counting and density estimation. In: IGTA 2017, CCIS 757, pp. 85–95 (2018)
7. Zhang, Y., Zhou, D., Chen, S., Gao, S., Ma, Y.: Single-image crowd counting via multi-column convolutional neural network. In: 2016 IEEE Conference on Computer Vision and Pattern Recognition (CVPR), Shanghaitech University (2016)
8. Li, Y., Zhang, X., Chen, D.: CSRNet: dilated convolutional neural networks for understanding the highly congested scenes. In: 2018 IEEE/CVF Conference on Computer Vision and Pattern Recognition (2018)
9. Tan, R., Atmosukarto, I., Han Lim, W.: Video analytics for indoor crowd estimation. In: 2018 IEEE International Conference on Service Operations and Logistics, and Informatics (SOLI) (2018). 978-1-5386-4987-0/18/$31.00
10. Ojha, N.: Spatio-temporal anomaly detection in crowd movement using SIFT. In: 2018 IEEE 2nd International Conference on Inventive Systems and Control (ICISC) (2018). 978-1-5386-0807-4/18/$31.00
11. Chibloun, A., El Fkihi, S., Mliki, H.: Abnormal crowd behavior detection using speed and direction models. In: 2018 IEEE 9th International Symposium on Signal, Image, Video and Communications (ISIVC) (2018). 978-1-5386-8173-2/18/$31.00
12. McGuinness, K., Little, S., Marsden, M., O'Connor, N.E.: Holistic features for real-time crowd behaviour anomaly detection. In: 2016 IEEE International Conference on Image Processing (ICIP) (2016)
13. Abdali, A.-M.R., Al-Tuma, R.F.: Robust real-time violence detection in video using CNN and LSTM. In: 2nd Scientific Conference of Computer Sciences (SCCS), University of Technology - Iraq (2019). 978-1-7281-0761-5/19/$31.00
14. Semwal, V.B., et al.: An optimized feature selection using bio-geography optimization technique for human walking activities recognition. Computing **103**, 2893–2914 (2021)
15. Semwal, V.B., et al. Pattern identification of different human joints for different human walking styles using inertial measurement unit (IMU) sensor. Artif. Intell. Rev. **55**(2), 1149–1169 (2021)
16. Dua, N., Singh, S.N., Semwal, V.B.: Multi-input CNN-GRU based human activity recognition using wearable sensors. Computing **103**, 1461–1478 (2021)
17. Zhang, Y., Zhou, D., Chen, S., Gao, S., Ma, Y.: Single-image crowd counting via multi-column convolutional neural network. In: 2016 IEEE Conference on Computer Vision and Pattern Recognition (CVPR), pp. 589–597 (2016). https://doi.org/10.1109/CVPR.2016.70

Chapter 4
Role of ML and DL in Detecting Fraudulent Transactions

Sindhu Rajendran, Alen Aji John, B. Suhas, and B. Sahana

Abstract Ever since the inception of online transactions, it has positively impacted the ease of business by making money transactions straightforward and secure, irrespective of location or amount of money. However, along with the increase in online transactions, the number of fraudulent transactions also increased. With the rapid growth in technology in the current environment, fraudsters are creating new methods to conduct these fraudulent transactions, which seems legitimate. Therefore, there is an ever-growing need to curb these incidents using real-time detection and reporting. This chapter explores the different techniques such as the *ANNs or Artificial Neural Networks, CNNs or Convolutional neural networks, Rule-based methods(RBM), Hidden Markov Models(HMM), Autoencoders*, and much more, in machine learning and deep learning. Most of the datasets for these models are not accessible to the public because of the privacy concerns of the financial institutes. We also assess various parameters like *precision, accuracy*, and *recall* of these solutions to make a comprehensive study. The chapter is concluded with the latest improvements and the future prospects to keep track of these fraudulent transactions.

Keywords Machine learning · Deep learning · Fraudulent transaction · Credit card · Net Banking

S. Rajendran (✉) · A. A. John · B. Suhas · B. Sahana
R. V. College of Engineering, Bangalore 560059, India
e-mail: sindhur@rvce.edu.in

A. A. John
e-mail: alenajijohn.cc19@rvcc.cdu.in

B. Suhas
e-mail: suhasb.ec19@rvce.edu.in

B. Sahana
e-mail: sahanab@rvce.edu.in

A. Biswas et al. (eds.), *Artificial Intelligence for Societal Issues*, Intelligent Systems Reference Library 231, https://doi.org/10.1007/978-3-031-12419-8_4

4.1 Introduction

4.1.1 Introduction to Fraudulent Transaction

A fraudulent transaction is the unauthorized use of an individual's accounts or payment information. Fraudulent transactions will lead to the victim's loss of money, personal property, or personal information. Fraudsters carry out fraudulent transactions without the knowledge of the account holder. They withdraw large sums of money from the bank account by pretending to be the account holder and make transactions from the bank to themselves. The banking system will assume the fraudsters as authentic account holders as they will have the information and the authentication codes as the account holders. Therefore, the banks will not get to know that a fraudulent transaction has taken place until the account holder finds discrepancies in their bank account and lodges a complaint to the bank stating that there had been transactions without the account holder's knowledge.

The banks provide various types of services, among which credit cards and debit cards are some of the popular services. The banks grant a line of credit to the credit card holder using which the cardholder can make payments and money transactions. The banks love the credit cardholders to build a balance of credit which the cardholder has to repay along with interest for borrowing the amount. On the other hand, the debit card will transfer the card holder's money to a personal account when the car holder initiates a money transaction. Here, the money which will be transferred belongs to the cardholder, unlike in the case of debit cards, where the money to be transferred belongs to the bank. The banks have sophisticated systems that will track the money and the transactions carried out by the cardholder so that there are no discrepancies. The cardholder can also track how much money he/she is crediting or debiting from the bank.

There is a requirement where the credit/debit cardholders have to come across a Personal Identification Number (PIN) authentication for the cardholders to carry out money transactions. If the fraudsters can get the card details and the pin authentication number, they can do transactions with the bank posing as the cardholder [1]. Fraudsters prefer to carry out fraudulent transactions on credit cards rather than debit cards as there will be a risk of the amount requested for transfer being more than the amount in the card holder's account.

In the past couple of years, a significant increase in complaints registered against fraudulent transactions is prevalent. More and more people are trying to find loopholes in the banking system to conduct fraudulent transactions. Thus, there have been significant attempts to make fraudulent transaction detection and prevention algorithms. Nevertheless, the fraudulent transaction prevention algorithms have mainly been inaccurate and are not feasible to use. The fraudulent transaction detection algorithms have also faced many obstacles, among which getting reliable data sets to train the models as the data sets contain sensitive information regarding money transactions related to the banks.

For deep learning models, the accuracy will increase with the size of the data sets. So getting large data sets is very important. Online/Internet banking's electronic payment systems enable customers of financial institutions such as banks to conduct different kinds of finance-related transactions with the help of the bank's website or the latest bank-backed mobile applications. The Internet banking service is connected to, or a part of the banking administration operated to provide the customers of the bank access to avail their services through the internet without the customers reaching the regional bank branch.

4.1.2 Influence of Online Banking on Fraudulent Transaction

The Nottingham Building Society provided the initial online banking service in Great Britain. This online banking service was first opened on a limited basis in 1982 and publicly expanded later in 1984. By the year 2000, almost 80% of the banks in the United States had already started to provide online internet service. ICICI Bank introduced Internet banking service in 1998, and 23 years down, India is one of the leading global financial markets with Internet banking transaction processings crossing 2 Billion for the year 2020 itself with a whopping money movement of Rs 4.5 lakh Crore with companies like Google, SBI, PhonePe, HDFC, etc.

Before the introduction of the online banking system, fraudsters could make a fraudulent transaction only after getting stolen credit cards and relevant information regarding the credit cardholders. The cardholders whose cards were stolen could have gone to the bank and deactivated the credit cards to minimize the damage. However, due to the online banking system, there is a greater risk of fraudulent transactions. The fraudsters can collect information about the cardholders using various ways, such as data breaching accompany and collecting its customers' card details, hacking into the bank's database and collecting the card holder's details, using phishing sites and pharming, trojan horse, cross-site scripting, keylogger and collect the information of the victims [2]. The fraudsters will use this information to conduct fraudulent transactions. The bank will only get to know about the fraudulent transaction after the customers check their transaction history and raise a complaint to the bank about it.

4.1.3 Statistics of Fraudulent Transactions

The online banking system is improving over time. Newer technologies are being integrated with the online banking system to make it more secure. However, along with the technology, even the fraudsters are improving and are coming up with newer and innovative ways to breach the security of the banking system and conduct fraudulent digital transactions. This is evident in the steady increase in the number of attacks and the number of attempted fraudulent transactions. The fraudsters have targeted the pandemic as there was an increase in digital transactions due to lockdown and social distancing.

There was around a 200% increase in customers using mobile banking. The number of suspected fraudulent digital transactions between march 2020 and march 2021 increased by 40% compared to the number of suspected fraudulent transactions between march 2019 and march 2020(March 2020 was when the pandemic was declared by the world health organization). The increase of suspected fraudulent transactions In the same period was 22%. This was because of the increase in the online transactions during the pandemic due to lock-downs and social distancing. These fraudulent transactions mainly affected the telecommunication, travel, and leisure industries.

Table 4.1 consists of the statistical data of various fraudulent transactions.

The Table 4.1 shows the number of cases reported during the third quarter of the financial year of 2020. In a short period of three months, around 22 thousand cases have been reported involving Rs 128 crores, which were lost in such transactions.

The graph in Fig. 4.1 shows the increase in the number of fraudulent transactions reported and the amount stolen by the fraudsters. Globally around 20 billion dollars is said to be lost due to fraudulent transactions. It is estimated that there has been a 35% increase in attempted fraudulent transactions, and the trend is expected to continue.

The above pie chart in Fig. 4.2 shows the cybercrime complaints received by the police. Around 60% of them are directly related to Fraudulent digital transactions. The police informed that the complaints regarding fraudulent transactions have tripled during the pandemic.

4.1.4 Current Preventive Systems

Ever since the fraudulent transactions started, more and more fraudsters have tried to breach the bank security system and steal the money. The banking sector and other

Table 4.1 The statistical data [27]

Month	October		November		December	
Type	Cases	Amount (in crores)	Cases	Amount (in crores)	Cases	Amount (in crores)
ATM/debit	3376	73.65	3533	10.55	4149	10.33
Credit card	1641	4.04	1711	4.77	2765	10.87
Net bank	360	7.01	2256	4.31	1250	2.27
Total	5377	84.7	7500	19.63	8164	23.47

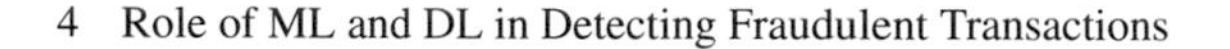

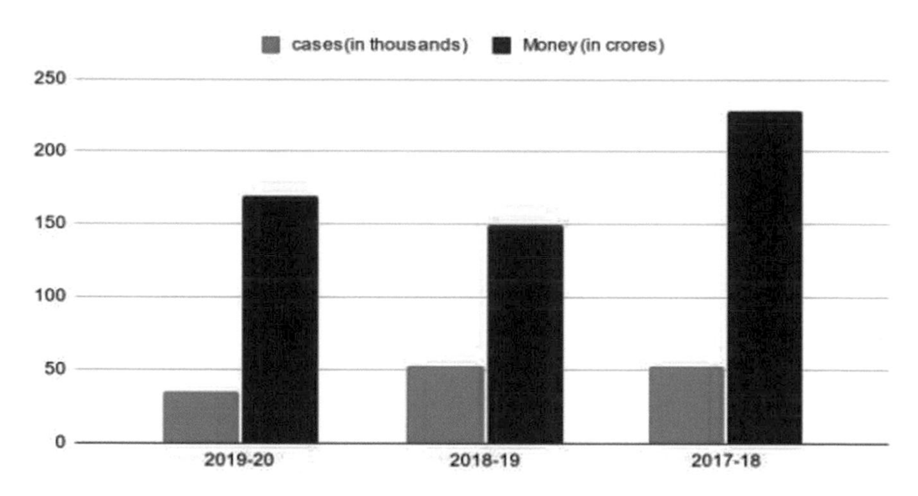

Fig. 4.1 Trend of fraudulent transactions in India

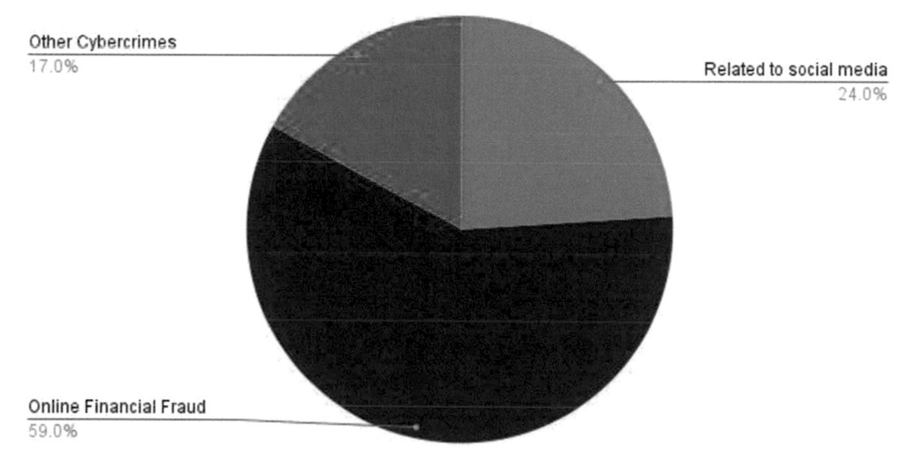

Fig. 4.2 Classification of cyber security complaints in India [26]

financial institutions actively tried to Preview it by creating various fraud detection and prevention systems. The banks are using systems such as Primary Account Number(PAN), Temporary Authentication Number(TAN), Personal Identification Number(PIN), One Time Password(OTP) to authenticate the transaction and ensure that the cardholder is genuine. Fraud detection and prevention software systems have been in the market since the early 1990 s. The first company to introduce them in the market was Falcon. Other popular companies which are the emerging leaders in cyber-security systems are BAE Systems, IBM, SAS technologies, Actimize, which also provides solutions in the same field. This software analyses the usual behavior of regular credit card transactions to track and flag fraudulent transactions. Changing the data to be read is very large; the software generally uses artificial intelligence, machine learning, or deep learning [3, 4]. This software tracks various information of the user to determine if the transaction is valid or fraudulent. They check if the transaction is requested near the usual area where the cardholder does the transaction, if the amount is unusually high if multiple transactions are requested from the same IP address, and other anomalous behavior to see if the transaction is legitimate. Huge data sets are given to the software models to analyze the data and generate an algorithm that can track the transaction and get to know if the transaction requested should be flagged as fraudulent or not [5].

4.1.5 Introduction to Artificial Intelligence

Artificial Intelligence(AI) is another synonym to a much-talked topic in the technological industry known as Machine Intelligence, and this is the thinking capacity showcased by the running machines, converse to the intelligence naturally exhibited by us humans. AI has an unparalleled capability of being able to design intelligent working tools to leverage higher efficiency, to create and develop software that can mimic human traits and aid such tools in fields such as reasoning,problem-solving, planning, etc., to an extent where researchers rely on such AI tools to make decisions pertaining to optimal research results and prospects of the project they have given the AI to decode. The main idea behind artificial intelligence is to exceed human action in the same field and make engagement more efficient. The introduction of AI into different algorithms and its expanded utilization potential has caught the attention of businesses and research communities worldwide [6].

AI is a growing field in terms of the innovation going behind and its prospects in areas under computer science. Leading AI developers have also showcased and implemented ideas such as how an AI can run a machine without humans in the same manner as a human would run the machine. The latest example where we can relate to AI working in our everyday lives is based on an app that most of us use every day or occasionally use our very own YouTube. Every month close to 2 billion people log into the YouTube platform and over a billion hours of watch time every day. With this amount of data and user base, it makes sense for YouTube to employ the power of AI to ease operations. Some of the ways how AI helps YouTube run smoothly are as follows.

The Up-Next feature pops up every time you complete or pause for the following video. YouTube has a thriving community presence, and its dataset is constantly with its users uploading hours of video content every passing minute. To control and maintain the audience's attention on the platform, the AI must upkeep a recommendation engine that works differently than that of Netflix or Spotify. The AI is developed to handle real-time recommendations while its users are adding new content every minute. Automatically remove objectable content since anyone from anywhere can upload content irrespective of the platform's rules on YouTube.

Therefore, it is YouTube's work to remove such content that can promote or showcase violence and explicit content. This is where AI comes into scan and removes such content before it reaches a broader audience. AI has enhanced YouTube's capability to identify and immediately flag the explicit content present on the platform. Before introducing AI into its service, only 8% of such videos were identified and removed before it had crossed 20 views; now, more than 75% of such content is removed even before the video hits ten views [7].

4.1.6 Introduction to Deep Learning

Deep learning or DL is an advanced section of ML developed on the core concepts of Artificial Neural Networks. This branch of ML will rely on models which are similar to the functioning of the human brain. It is a subset working within Artificial intelligence frameworks; under this branch, a computer can act itself without being assertively programmed for the action. Deep learning is so engraved in our life that we use it even without noticing it. Deep learning is not new. In the last couple of years, it has been on hype due to the increasing number of ways it can collect and generate data.

Our brain has approx 100 billion neurons altogether, and we need to create individual neurons, and each of these neurons is connected to thousands of their neighbors. We create the same in a computer using an artificial structure termed Artificial Neural Net, which consists of Nodes and Neurons. Some of these neurons are assigned for input values and some for output value, while most of them are present in the hidden layer, that is in between the initial and final layer where the neurons are interconnected to each other, forming a sewed network channel [8].

Deep learning is a well-devised branch of Machine learning. They mainly differ in the way each algorithm tries to learn the data and what amount of data is utilized by the algorithm to finally conclude on the targeted or the required data. Most of the feature extracting process, which requires manual human intervention, is eliminated when deep learning is involved. DL requires a mammoth amount of datasets Fig. 4.3, therefore, being a large, scalable machine learning system. The capability with which DL automates much of its work is exciting because much of the business/research organization's data is considered unstructured and unorganized. This added feature can help ease human interference.

On the other hand, ML is comparatively more reliant on human inputs to learn. The development of the hierarchy of features through which the ML can identify the difference between the data, usually requiring more structured data to learn, is developed under human supervision [9]. Many of the applied features of ML on the platform need to be identified by an expert, and each of the variations should be hand-coded based on the domain and data type, mostly regarded as low-level feature detailing. In contrast, High-level features are extracted from the dataset and Deep learning tools try to learn from it with far more accuracy with increasing dataset size in Fig. 4.4. This is a step way ahead in terms of the traditional Machine learning

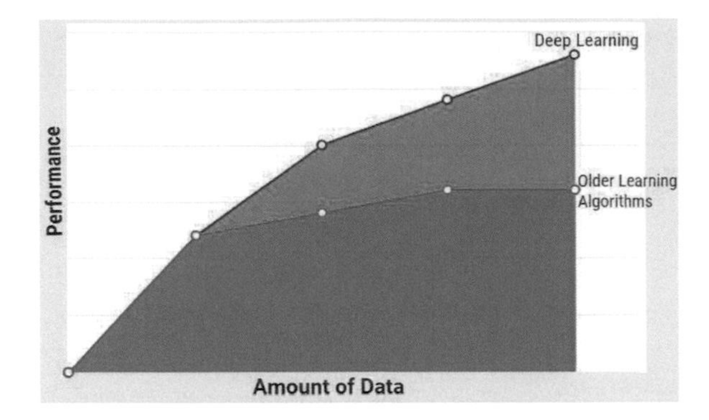

Fig. 4.3 Performance of deep learning

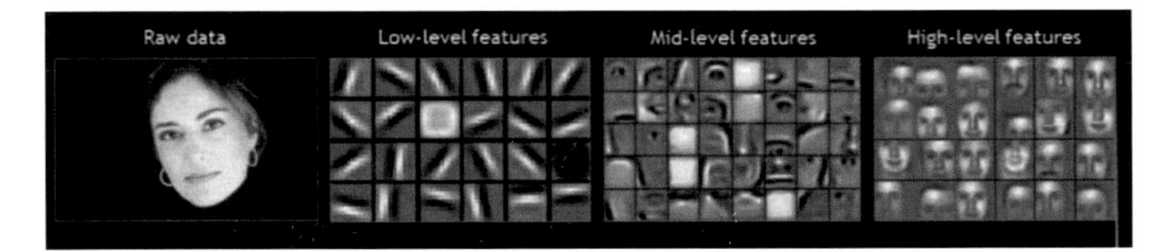

Fig. 4.4 Features of deep learning

techniques, therefore DL reduces the work of creating a new feature extraction for every difficulty and detailed attachments, which can help fast-track real-time fraud detection and implode newer fraud methods in real-time.

4.2 Different Detection Systems for Fraud

There are numerous algorithms and models which are being tried and tested to successfully curb such unwanted instances, some of the implemented and in-test phase-detection systems are mentioned below

4.2.1 Hidden Markov Model

The Markov chain is a simple model which identifies the probabilities of a sequence of random variable states, and every state can take a value from some set. This model makes strong assumptions that if the user wants to predict the future, the prediction is based only on the present state. Therefore, as we are considering only the present state, any states prior to the present state are disregarded or have no impact in predicting the future.

Since the Markov chain is valid when we need to compute the probability for a sequence of observable events, it can not be used when we are interested in hidden events or for events where we do not observe directly. For example, in a text, we usually do not observe or emphasize the part of different speech tags on the text; rather, we see the words and must infer the tags from the word sequence. In this case, we call the tags a hidden event because they are not observed [10]. Using HMM (Abbreviated from Hidden Markov Model), we can learn about the observed and the hidden events that we consider as causal factors in our probability model.

For our study of the HMM, we employ the following payment range:

- Large Amount as High
- Median related Amount as Medium
- Lower amount as Low.

During the credit card transaction dispensation process, we need to record the transaction in real-time in the preferred condition for the HMM; therefore, it is created through inspection symbols in our representation. We group the purchase value 'A' into X price ranges Y1, Y2, Y3, ...Yx from the analysis symbol issued from the side of the bank. Depending on the expenditure trait of the user, a price variety is configured for each symbol arising from the transaction taking place.

HMM learns from data and forms the price range. In a running context, it uses a clustering algorithm like the "K clustering algorithm" [11] to price the values of every user transaction. It uses cluster Vk for clustering algorithms such as 1, 2, 3, ...M, which can represent the observation of the price value symbol and price value range.

Under the Fraud detection approach, we have considered three price ranges as above, High, Medium, low { **H, M, L**} as a set of possible observations, let's say **L** = (0, Rs 5000), **M** = (Rs 5000, Rs 25,000), **H** = (Rs 25,000, Card Limit). If the cardholder makes a transaction of Rs 10,000 it would come under the category of medium and resulting symbol as M. These symbols would later be assigned to each transaction ID for the HMM to create an expenditure pattern depending on the amount and the given parameter.

The Hidden states of HMM are created from the collection of all plausible types of purchase and the collection of all plausible lines of business of a merchant as the user utilizes the card more. The HMM learns more about the spending pattern and the business the user is more interested in; this helps narrow down the chances of unrelated sums of money transactions happening outside the spending pattern, terming them as fraud until the user acknowledges the transaction. HMM-based credit card FDS transactions do not need fraud indicators but, it can detect fraudulent transactions by studying the customer's expenditure pattern [12].

4.2.2 Artificial Neural Network (ANN)

This is one of the leading tools that help classify and find hidden patterns between different parameters and characteristics. Artificial Neural networks consist of many discrete layers, among which the initial layer is the input layer, and the final layer is the output layer, and between them, they can have different numbers of hidden and non-hidden layers for better data calibration as shown in Fig. 4.5. It is considered to be a deep learning system if it has more than one hidden layer [11]. Different numbers of neurons are present in each of these layers, and each neuron is individually connected to a weighted edge, and the output of these neurons is a function of the unit. This is called the activation function, and some examples of these are linear function, the sigmoid function, which is one the most used function, and the threshold function.

The classification label has the same number of neurons as the output layer, and each of the neurons present in the final layer gives the probability of a given weight being in a specified class. In a typical transaction, the initial verification is conducted in the first layer along with the screening; here, the entered credit card detail is cross-verified. Generally, a standard cardholder carries out similar amount transactions, and a cluster can be formed under the types of the products purchased; however, when a fraudster impersonates the cardholder, the fraudster behavior is expected to change from the usual purchasing or transactions pattern of the cardholder as the fraudster is not aware of the card-holder's profile. This leads to the transaction being observed as an exception and this cluster being termed as an outlier. The doubtful transactions are moved to the suspicious table so that they can be made to undergo deeper analysis

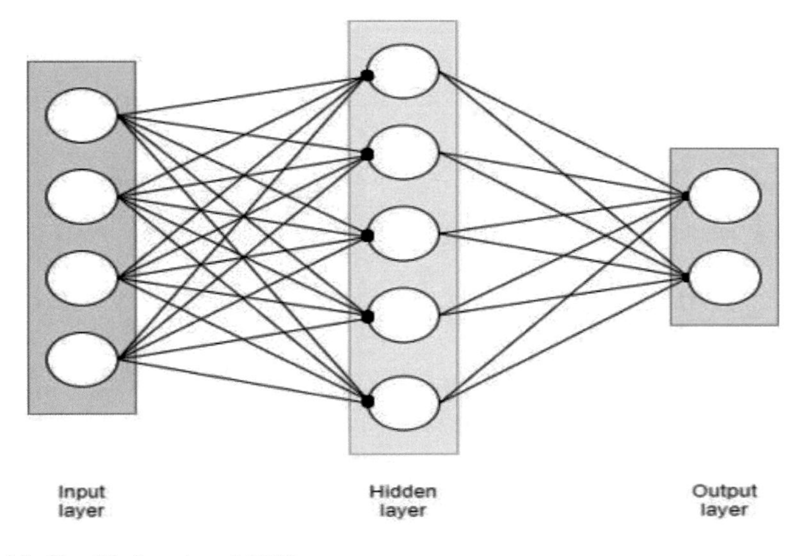

Fig. 4.5 Simplified version of ANN

or test classifications which are conducted using a feed-forward neural network of back propagation (FFNNBP).

A learning algorithm that is well supervised is based on the conjugate gradient method, also called scale conjugate gradient (SCG), used for learning purposes. Due to its faster execution, the standard backpropagation algorithm(BP) has set an excellent benchmark for its performance and accuracy. SCG further removes some drawbacks of using other backpropagation algorithms, such as low convergence rate and bad performance when used on the datasets which are large.

In the given working condition we have utilized parameters such as the articles bought by the user and the time of purchase. Therefore, based on the mentioned parameters, fraudulent transactions are classified and cross-examined statistically to understand the relationship between the input data value and the values for specific prime attributes; this is done to deeply understand how the fraud was conducted and decoded its pattern. The data that we get out of this algorithm is continuously fed into the feed-forward neural networks, thereby we are training a model set for future use.

With this knowledge, we can classify a fraud or a suspicious transaction when there is a similarity with the trained pattern [13].

With the further addition of components and software, we can reduce the percentage of miscalculation to give a close to the accurate foolproof system. The Banking major VISA was the initial major to incorporate Neural network technology into their system to fight excessive card/transaction fraud.

Table 4.2 consists of the latest Methods used with Neural Networks in increasing positive fraud detection rate.

4.2.3 Autoencoder

Sometimes fraud detection models are fed with partial data which contains irrelevant features in the input source. The presence of such uncorrelated features in the data can hinder the learning capability of the classifier when it comes across large amounts of data. A two-stage approach to sort this issue is introduced here wherein the initial stage, the extraction of lower dimension features from the input data takes place. In the consequent stages, the model tries to resolve whether the transaction that took place was fraudulent or not. This is where autoencoders are utilized, input data with the help of autoencoders the essential features are extracted which is followed by a categorizing algorithm. When a transaction occurs, a large file of the user's data is cross verified, such as mode of transaction, amount transacted, age, location of the ATM/bank, time of the transaction, etc. Having such unnecessary features can render the model behave poorly. This means having a higher number of characteristics, hence a very high dimension, and working with such massive data can lead the analysis to be expensive. Therefore the primary objective of the autoencoder is to find the useful parameters for the detection, and the autoencoder can efficiently handle the

Table 4.2 Latest Methods used with Neural Networks in increasing positive fraud detection rate

Techniques used	Datasets used	Results
Neural network	For this training dataset, it was obtained from the initial credit card traffic for the year 1994–95	The credit card operation achieved an average rating time of 60ms
Decision tree and Logistic Regression utilizing Neural network	A real-time dataset was created using real-time transaction history	This combination of different algorithms together worked better than decision tree in detecting frauds quickly
Deep Neural Network	European Cardholder history was considered which contained more than 285,780 transactions	Accuracy close to 96.37% was achieved
Deep Neural Network	An anonymized dataset containing more than 900 million card transactions during the year 2014–15 was clubbed with another dataset containing fraud transactions	ROC results were quite impressive and the accuracy rate was close to 98%
BP Neural Network optimized with whale algorithm	Data is a mixture of 498 fraudulent cases and around 290,00 genuine cases	F-measure was close to 98.04%

recognition of Non-linearly correlated features while creating a lower-dimensional representation of the input data [14].

Autoencoder comes under the Feedforward Neural network, which is utilized to learn and clean encodings of the training data. Autoencoder encodes the input data into hidden representation and then recreates the input data from the hidden representation; the input and output dimension of an autoencoder is the same. Autoencoder tries to reciprocate an identity function by setting curbs on the network by restricting the number of hidden units (Fig. 4.7).

The architecture of the Undercomplete Autoencoder is displayed above in Fig. 4.6, and it is the most popular type of autoencoder. In this kind of autoencoder, the input dimension is more than the hidden dimension. It is observed from the diagram that it consists of 6 input dimensions, whereas the dimensions of the hidden layer are three. The autoencoder contains two components, an Encoder, and a Decoder, which are modeled based on deep neural networks. The initial stage of the autoencoder is trained using the selected transaction characteristics; therefore, it can produce a newly encoded representation of the parameters. Since the newly encoded features have fewer parameters than the initial input, it becomes easier for the learning of the classifier in the next stage [15] (Fig. 4.8).

In the next stage, the deep neural network is connected and often used with the SoftMax cross-entropy as the loss function in a classification problem; this can lead to higher accuracy in predicting fraud transact by utilizing the SoftMax function in

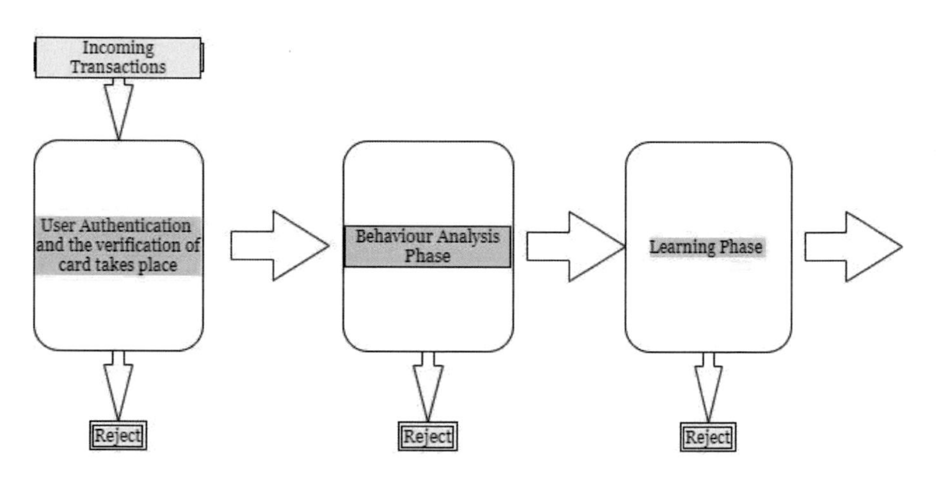

Fig. 4.6 Working of a neural network with its output

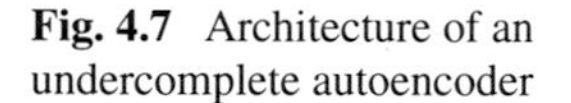

Fig. 4.7 Architecture of an undercomplete autoencoder

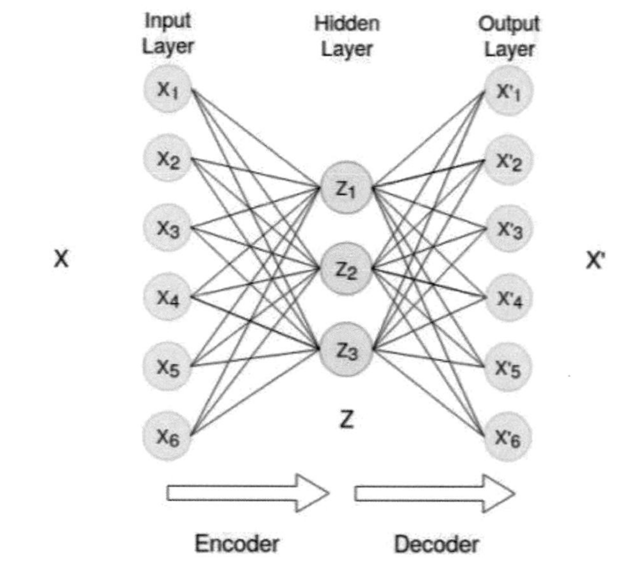

the final stage of the neural network-based classifier. The exponential value for each output is calculated, and then it is normalized. Using SoftMax and autoencoder, it is possible to increase the classification accuracy performance to 97.94% by controlling the threshold value around 0.6. This shows that we can detect fraud transactions much quicker and cleaner fashion with very few false-positive test cases.

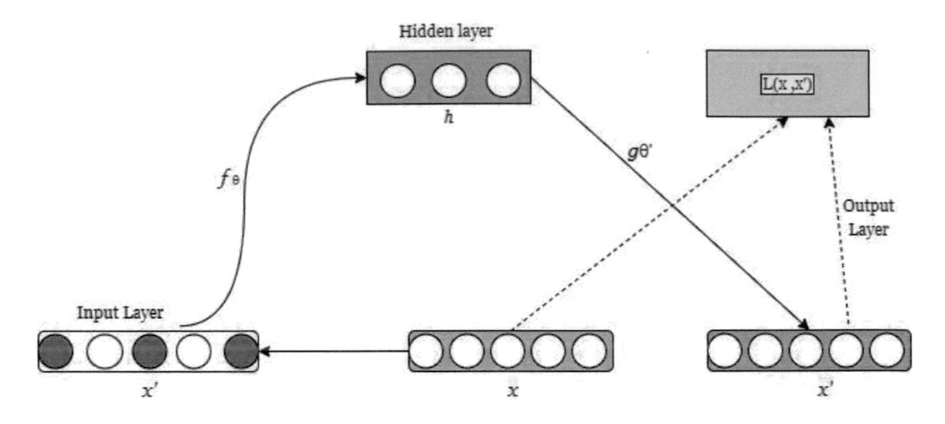

Fig. 4.8 Flowchart showcasing SoftMax function

4.2.4 Convolutional Neural Network

Convolutional Neural Network or CNN is a class of deep learning that is often utilized in image analysis. CNNs are also known as Shift invariant or space invariant artificial neural networks (SIANN). They are a type of artificial neural network designed dependent on the shared weight architecture of the convolution kernels or filters that analyze the input features using convolution to give translation equivariant outputs called feature maps [14]. The convolution neural networks are mainly used in applications based on image recognition, video recognition, image classification, Image segmentation, image analysis, NLP, and brain-computer.

The biological processes influenced the convolutional neural networks. The connectivity pattern of the neural network's neurons mirrored the visual cortex of the brain. Every single cortical neuron only responds to stimuli in a narrow section of the visual field, which is referred to as the receptive field. Each neuron's reception areas partially overlap, allowing the whole visual field to be covered. When compared to other classification algorithms, convolutional neural networks require less preprocessing. The filters in the primitive technique are hard designed, but with enough training, CNNs can learn these filters/characteristics. A standard CNN comprises three types of layers: one input layer, multiple hidden layers, and one output layer (Fig. 4.9).

A convolutional neural network-based framework can detect patterns for fraudulent transactions in credit card-based transactions, and it can convert the transaction information into a feature matrix for every transaction done by a cardholder [16]. Therefore, the inherent relation and transaction concerning time will be given as input to the CNN model combining the cost-dependent sampling method within a characteristic period to alleviate the data sets which are highly imbalanced to yield good performing fraud detection. The introduction of trading entropy, which is a novel trading feature. We are also introduced to a novel trading feature known as

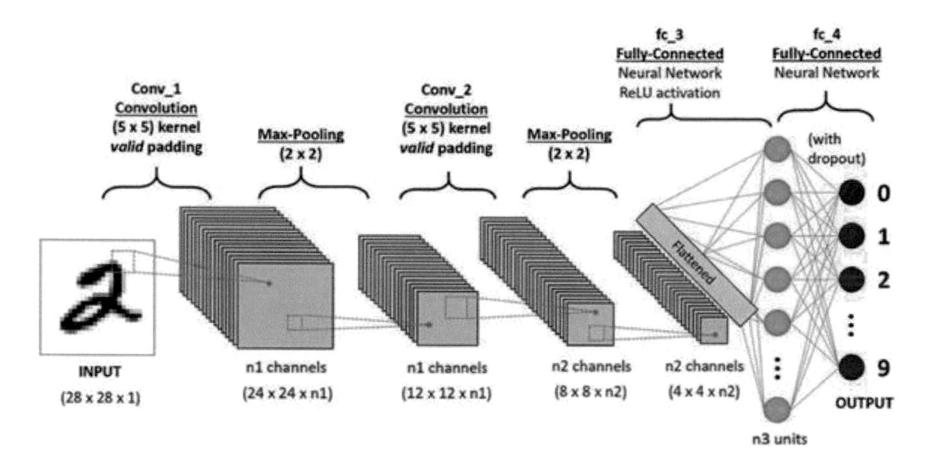

Fig. 4.9 Working of a CNN model

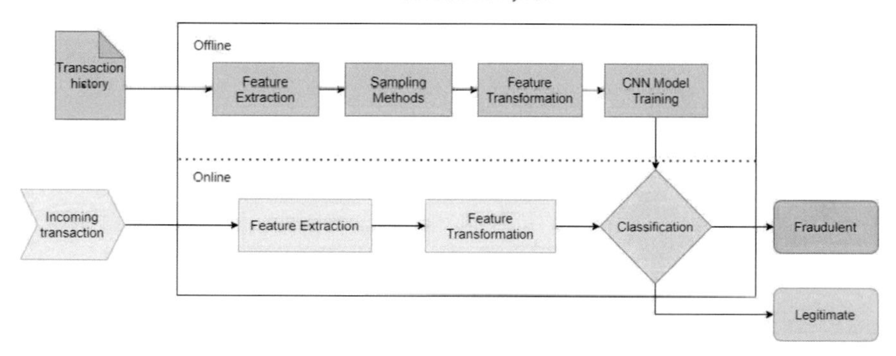

Fig. 4.10 Block diagram of the fraudulent detection model

trading entropy to deduce more complex fraud designs. The trading feature is present to identify complex ways of expenditure of the cardholder [17] (Fig. 4.10).

The fraud detection model is separated into two sections: the training section and the detection section. The first section comprises modules such as feature extraction, sampling methods, feature transformation, and a Convolutional Neural Network training model. The training phase takes place offline, where the data set used to train the CNN model will be delivered. When a transaction occurs, the data is made available online. The CNN module will automatically forecast and label the transaction as fraud or real in the Detection section. The detection section is made up of three modules: feature extraction, feature transformation, and classification [18] (Fig. 4.11).

Figure 4.7 shows how the transactions, which are one-dimensional features, are converted into feature matrices. The process that takes place during feature transformation (Fig. 4.12).

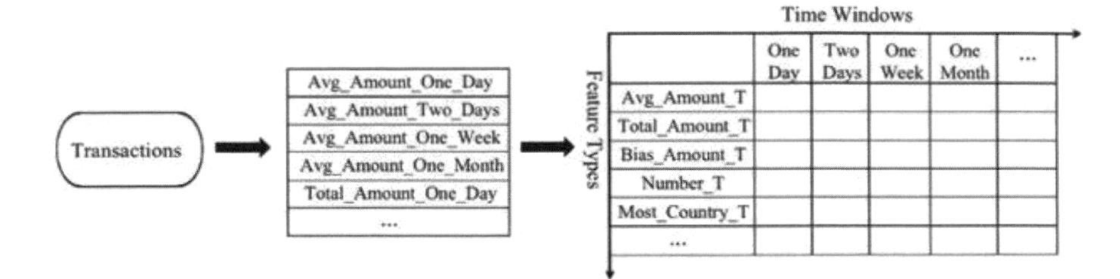

Fig. 4.11 Feature transformation

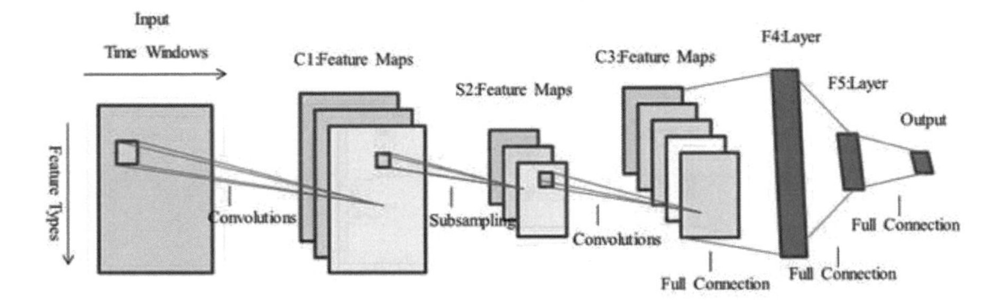

Fig. 4.12 The process that takes place during feature transformation

4.2.5 Rule-Based Method

Rule-based ML (RBML) is a machine learning approach that is based on the discovery and application of a set of relational rules that altogether consist of the intelligence gathered by the model. In comparison, other methods detect, learn, and evolve the rules to store, manipulate, or apply [19]. Machine learning methods frequently discover a single model that can be applied uniformly to various cases to decide classification. Learning classifier systems, association rule learning, artificial immune systems, and any other method that relies on a set of rules, each covering circumstantial knowledge, are examples of rule-based ML methodologies. Rules should be included in the form if: then expression. Rules-based fraud detection will identify fraudulent transactions based on rules that consider unusual attributes like unusual time stamps, account numbers, transaction types, and amounts, among others [20]. The transaction will be predicted as fraudulent or legitimate based on different criteria. Some of the commonly used rules are:

- The location of the transaction will be considered. If the transactions are made from too far away locations, they will be flagged as fraudulent transactions.
- The frequency of all the transactions will determine the legitimacy of the transaction. If a user suddenly starts making a large number of transactions in a very small instance of time, it will be flagged as a fraudulent transaction.

- If the user receives large amounts of money from multiple newly created accounts, it will be considered a fraudulent transaction.
- If there are multiple newly created accounts from the same IP address, it will be considered fraudulent transactions.
- If there is an unusually high amount requested, it will be considered a fraudulent transaction.

Consider the working of a Rule-based method with the following example[23].

S1: When the amount transacted is low, it is **legitimate**.

S2: When the amount transacted is high, and with this, the CVV is verified, it is **legitimate**.

S3: When the amount transacted is high, and the CVV is not verified, but the income of the user is high, it is **legitimate**.

S4: When the amount transacted is high, the CVV is not verified, and the income is low, it would be termed **fraudulent**.

4.2.6 Generative Adversarial Network

GAN or generative adversarial networks are a Subset of ML frameworks introduced by Ian Goodfellow and his coworkers in the year 2014. GANs are an intelligent method of preparing a generative model by outlining the issue as a directed training issue with two sub-parts: the generator model, which would be trained to generate new data, and the discriminator model, which will try to classify the generated data as either real (a part of original datasets) or fake (generated). The two models are trained in an adversarial zero-sum game until the discriminator model is tricked roughly half of the time, implying that the generator model is creating believable datasets [21]. Zero-sum game refers to the concept where there are no net losses or gains.

Thus, if a player faces some losses, it means one or more of the other players will gain the profit, which is precisely equal to the loss faced by the former player. The novel idea of this GAN system is that it uses two systems that compete with each other such that they have the opposite goals so that whenever a situation is faced by the system, either of the two will be benefited. Since the two systems or sub-models compete with each other, it is called adversarial [25] (Fig. 4.13).

The generator model will take a random vector as input feed and generate a sample from the domain. This random vector is drawn from the gaussian distribution, which was generated during the training of the model. This place is called latent space, and the variables are known as latent or hidden variables as they will not be available to the user. The variables are taken in from a specific part of the latent space, which is determined by the training dataset such that it remains as similar as possible to the real data. The data which is used to train the generated data will be passed into the discriminator model. It is similar to a classifier system. Its job is to determine if the data provided belongs to the dataset. In contrast, the job of the generator network is to trick the discriminator network into thinking that the fake data is generated as the real

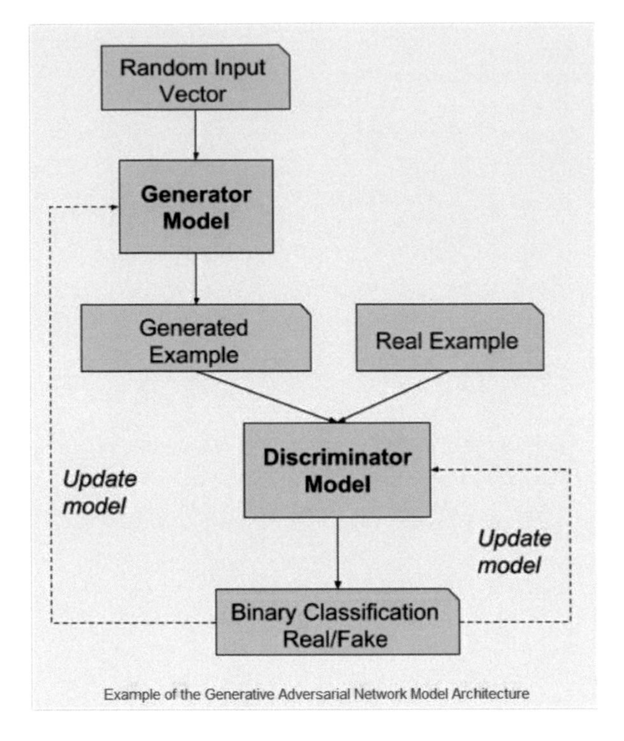

Fig. 4.13 GAN model architecture

data. Thus, together they generate data similar to the original dataset as the generated data that is not very close to the original data will be filtered by the discriminator model. This process will go on until each model keeps on improving based on the other system's mistake and they achieve maximum efficiency such that each system succeeds alternatively such that the discriminator model will accept every alternative data generated (Fig. 4.14).

The main reason for the success of this model is that it continuously takes feedback from its output and keeps improving. Also, when one of the sub-models improves,

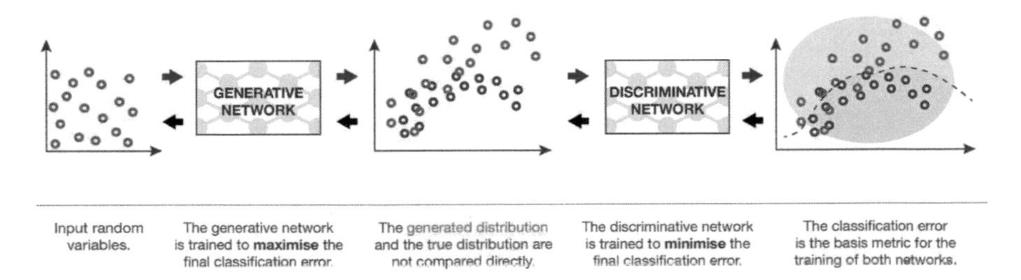

Fig. 4.14 GAN model working

it automatically improves the other model, reducing the margin of error and forcing the other model to make more accurate predictions. The above picture shows how the model improves and generates an accurate dataset. The generative network will generate a random data set from the lattice space created by the initial dataset.

The discriminator network will accept values that are close to the original datasets. Now the discriminator model will improve its accuracy by considering new classification parameters based on the new datasets, which consists of the old datasets and the newly accepted data. The accuracy will increase as the sub-model considers larger data.

Similarly, based on the data accepted and rejected by the discriminant model, the generative model will improve its accuracy by concentrating more on a smaller region in the lattice space where more data has been accepted [24]. Now, the accuracy of the discriminative model will increase if more data is accepted as it gets more extensive data to design classification. Due to this increased inaccuracy, it will reject more generated data. Due to an increase in the rejected data, the generated data will generate more data near the actual dataset and increase its accuracy.

The Paper [25] aims to design an efficient method to detect fraudulent transactions at the receiving bank. The information received at the receiving bank is significantly less, and thus, many fraud detection systems have been largely inefficient. The receiving banks only depend on the transaction history and the receiving account holder's information as they can't access the call records or the sender's details. Due to this inconvenience, the difficulty in supervised learning increases exponentially. Also, since the fraud transactions are relatively lower in number, less accurate fraud detection results in significantly less undetected fraud but a large number of false positives. Because of this, the banks may withhold many legitimate transactions, leading to customer dissatisfaction. The paper [25] proposes an innovative solution using adversarial learning to attain high accuracy and low misclassification.

Whenever a large amount of transfer is requested, the bank verifies if the receiving account has a high income. If it is true, the transfer is initiated. Else the details will be provided to the GAN model to calculate the probability of fraud. Based on that calculation, that particular transaction will be flagged and stopped until a delay period during which the customer can contact the bank and answer the question and carry out the transaction if the bank employee is convinced. In the meantime, the bank will notify the sending bank to confirm the senders' details. Until the bank receives credible information, the bank will stop the transaction till the delay period is over and conclude the transaction after it is over.

The basic idea of the paper[25] is to use a deep neural network to take out the latent response representation to build an effective classification system derived from a deep denoising model based on autoencoder to categorize the fraudulent transaction efficiently. The model uses adversarial learning made up of 2 sub-models to increase the accuracy of the model (Fig. 4.15).

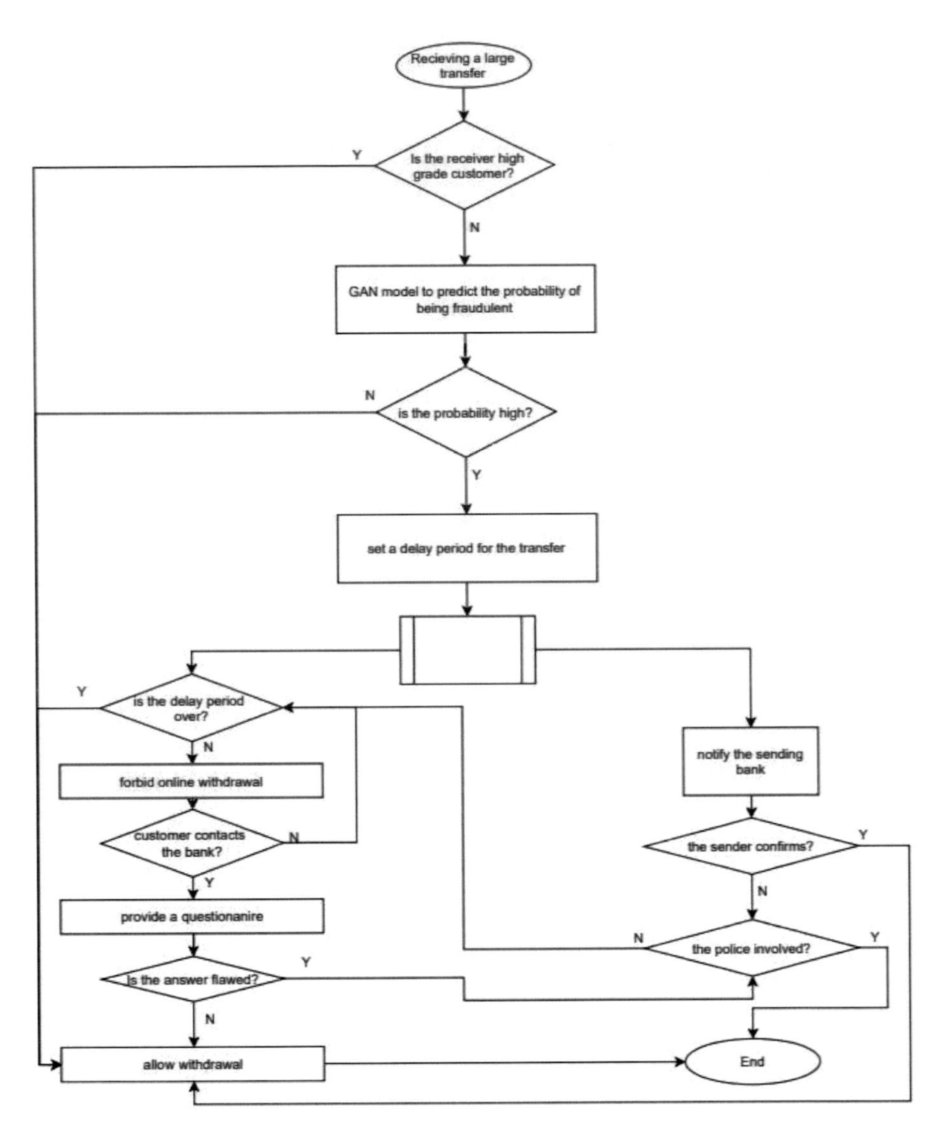

Fig. 4.15 Working model of the bank with the proposed GAN model

4.3 Future Scope

The usage of deep learning to avoid fraud transactions is an emerging technological development in the financial and banking sector. With this emergence, billions of dollars can be saved from losing in fraud ways. Fraud detection and its prevention is a dynamic field of research as it needs a constant upgrade frequently to cope with the advancement fraudsters could make with the advent of new methods of fraud. As we

Table 4.3 The statistical data [27]

Year	Techniques used	Dataset used	Results
2016	Autoencoder	Data generated from Brazilian Goods and products from 2014	The performance in increased by 20 times
2017	Autoencoder	German Credit Dataset	Dedection rate of Frauduelnt transaction is more than 84%
2018	Autoencoder	SAP ERP Dataset A and B	(Dataset A) f1 score is 32.9 and (Dataset B f1 score is
2018	Autoencoder	Transaction dataset from credit card registry and user interactions with related ads	(Stacked autoencoder's performance in better than the single and variational varients
2018	Autoencoder	European banking dataset, where 700 legitimate transactions were given to train the model whereas 980 transactions, out of which half of them were legitimate and other half were fraudulent, were used to test the model	Fraud detection accuracy of more than 89%
2018	Autoencoder	German dataset containing more than 1000 transactions, Australian Dataset having more than 700 transactions and European dataset containing more than 290,000 transactions	AUC's score based on AE is 0.9603 AUC's score based on RBM is 0.9505(for the European Dataset)

step ahead for better accuracy in the prediction of fraud transactions, this move can be further accelerated using Auto-encoder, Recurrent Neural networks, Generative Adversarial networks. Under Auto-encoder using different algorithms, it can cross 90% accuracy result in verifying legitimate transactions.

Table 4.3 consists of the statistical data comparing the performance of various Autoencoders.

Furthermore, in a study of different Generative Adversarial Networks (GAN), we will be able to cross the benchmark of 99% in test control dataset conditions, moving a step closer to a real-time scenario [24].

Table 4.4 consists of the latest Improvements in GAN.

Table 4.4 Latest Improvements in GAN for better precision detections

Year	Techniques used	Dataset used	Results
2017	Generative Adversarial Networks	Dataset consists of 492 Fake transactions and more than 284300 legitimate transactions	The sensitivity of the model is increased with a small increase in increase in the number of true positives
2019	Generative Adversarial Networks	Dataset consists of 492 Fake transactions and approximately 284315 legitimate transactions	Area under ROC(AUROC) is 99.1 Average Precision(AP): 99.3

With this rise in improvement ratio in keeping a favorable working condition for such algorithms, we can expect these tools to come into our near markets in the coming months. The research studies in the above tools were quite limited for a comprehensive understanding due to constrained access to the required datasets and the banks' hesitancy to share such info with the public. Hopes are high with the concept of blockchain technology as it can ease and bring more transparency in payment gateways, thereby giving a round secure transaction path in the coming future.

4.4 Conclusion

Fraudulent transactions are an act of criminal dishonesty which is increasing every day. In this chapter, we have discussed fraudulent transactions, how fraudsters steal the money, and the preventive measures banks and other businesses take. It is evident that the fraudulent transactions have increased over time and that the fraudsters are using newer technology to breach the bank's security system. Innovative and robust fraudulent transaction detection systems are the need of the hour. This chapter has gone through the various models and successful deep learning algorithms that the current banking systems can use to counter fraud transactions. Every year with up-gradation and development in this field, we are able to reduce the gap for a foolproof transaction system. It is pretty evident from the research data available in the public domain that most of the datasets required for future studies and research are not publicly available, or most of the details are hidden because of the privacy laws of the financial institutions. Such hurdles create hindrances in future research in this domain.

References

1. Sohony, I., Pratap, R., Nambiar, U.: Ensemble learning for credit card fraud detection. In: Proceedings of the ACM India Joint International Conference on Data Science and Management of Data (CoDS-COMAD'18). Association for Computing Machinery, New York, NY, USA, pp. 289–294 (2018). https://doi.org/10.1145/3152494.3156815
2. Singh, P., Singh, M.: Fraud detection by monitoring customer behavior and activities. Int. J. Comput. Appl. **111**(11) (2015). https://doi.org/10.1.1.695.5814
3. Maniraj, S.P., Saini, A., Ahmed, S., Sarkar, S.: Credit card fraud detection using machine learning and data science. Int. J. Eng. Res. Technol. (IJERT) **8**(9) (2019)
4. Fawcett, T., Haimowitz, I., Provost, F., Stolfo, S.: Ai approaches to fraud detection and risk management. AI Mag. **19**(2), 107–107 (1998)
5. Author, A.-B.: Contribution title. In: 9th International Proceedings on Proceedings, pp. 1–2. Publisher, Location (2010)
6. Annie Brown: AI Changing the Entertainment World. www.forbes.com/sites/anniebrown/2021/07/13/making-the-youtube-algorithm-less-elusive-with-the-help-of-Gregory-chase-a-creator-with-10m-subscribers/?sh=ac0b3bcd681f. Accessed 4 Sept 2021
7. Marr, B.: The amazing ways YouTube uses artificial intelligence and machine learning. Sept 5, 2021. www.forbes.com/sites/bernardmarr/2019/08/23/the-amazing-ways-youtube-uses-artificial-intelligence-and-machine-learning?sh=47b3720858522
8. Chase, M.: Introduction to deep learning (2021). www.geeksforgeeks.org/introduction-deep-learning. Accessed 7 Sept 2021
9. IBM Cloud R&D: Introduction to AI, Deep Learning and AI. http://www.ibm.com/cloud/learn/deep-learning. Accessed 12 Sept 2021
10. Srivastava, A., Kundu, A., Sural, S., Majumdar, A.: Credit card fraud detection using hidden markov model. IEEE Trans. Dependable Secur. Comput. **5**(1) (2008)
11. Behera, T.K., Panigrahi, S.: Credit card fraud detection: a hybrid approach using fuzzy clustering & neural network. In: 2015 Second International Conference on Advances in Computing and Communication Engineering (2015)
12. Modi, K., Dayma, R.: Review on fraud detection methods in credit card transactions. In: 2017 International Conference on Intelligent Computing and Control (I2C2) (2017)
13. Singla, J.: A survey of deep learning-based online transactions fraud detection systems. In: 2020 International Conference on Intelligent Engineering and Management (ICIEM), 2020, pp. 130–136 (2020). https://doi.org/10.1109/ICIEM48762.2020.9160200.
14. Chen, J., Shen, Y., Ali, R.: Credit card fraud detection using sparse autoencoder and generative adversarial network 1054–1059 (2018). https://doi.org/10.1109/IEMCON.2018.8614815
15. Misra, S., Thakur, S., Ghosh, M., Saha, S.K.: An autoencoder based model for detecting fraudulent credit card transaction. In: Procedia Computer Science (2020)
16. Saha, S.: Towards data science. In: A Comprehensive Guide to Convolutional Neural Networks - the ELI5 way (2021). https://towardsdatascience.com/a-comprehensive-guide-to-convolutional-neural-networks-the-eli5-way-3bd2b1164a53. Accessed 14 Sept 2021
17. Singh, P., Singh, M.: Fraud detection by monitoring customer behavior and activities. Int. J. Comput. Appl. **111**(11) (2015). https://doi.org/10.1.1.695.5814
18. Prusti, D., Rath, S.K.: Fraudulent transaction detection in credit card by applying ensemble machine learning techniques. In: 2019 10th International Conference on Computing, Communication and Networking Technologies (ICCCNT), pp. 1–6 (2019). https://doi.org/10.1109/ICCCNT45670.2019.8944867
19. Ghattamaneni, S., Portilla, R., Gupta, N.: Combining rules-based and AI models to combat financial fraud. Eng Blog. https://databricks.com/blog/2021/01/19/combining-rules-based-and-ai-models-to-combat-financial-fraud.html. Accessed 23rd Sept 2021
20. Leonard, K.J.: The development of a rule-based expert system model for fraud alert in consumer credit. Eur. J. Oper. Res. **80**(2), 350–356 (1995). ISSN 0377-2217, https://doi.org/10.1016/0377-2217(93)E0249-W, https://www.sciencedirect.com/science/article/pii/0377221793E0249W

21. Brownlee, J.: A gentle introduction to Generative Adversarial Networks (GANs), machine learning mastery (2021). https://machinelearningmastery.com/what-are-generative-adversarial-networks-gans/. Accessed 25 Sept 2021
22. Saha, S.: Towards data science. In: A Comprehensive Guide to Convolutional Neural Networks - the ELI5 way. https://towardsdatascience.com/a-comprehensive-guide-to-convolutional-neural-networks-the-eli5-way-3bd2b1164a53. Accessed 14 Sept 2021
23. Vatsa, V., Sural, S., Majumdar, A.: A rule-based and game-theoretic approach to online credit card fraud detection. IJISP **1**, 26–46 (2007). https://doi.org/10.4018/jisp.2007070103
24. Maniraj, S.P., Saini, A., Ahmed, S., Sarkar, S.: Credit card fraud detection using machine learning and data science. Int. J. Eng. Res. Technol. (IJERT) **8**(9) (2019)
25. Zheng, Y.J., Zhou, X.H., Sheng, W.G., Xue, Y., Chen, S.Y.: Generative adversarial network-based telecom fraud detection at the receiving bank. Neural Netw. **102**, 78–86 (2018). ISSN 0893-6080, https://doi.org/10.1016/j.neunet.2018.02.015
26. Times of India: https://timesofindia.indiatimes.com/city/delhi/virus-of-cybercrime-over-3000-cases-every-month/articleshow/77967994.cms. Accessed 20 Jan 2021
27. Times of India: https://timesofindia.indiatimes.com/business/india-business/in-92-days-india-lost-rs-128-crore-in-card-online-fraud/articleshow/74571025.cms. Accessed 20 Jan 2021

Part II
Agriculture and Education

Chapter 5
Employing Image Processing and Deep Learning in Gradation and Classification of Paddy Grain

Sudhanshu Ranjan, Anurag Sinha, and Susmita Ranjan

Abstract Agriculture has profound influence on the economy of India. In India Agriculture is a major source from where the collective amount of turnover is responsible on the crop, quality and Weather. In major part of the northern India rice grain agriculture is done and to ensure the quality of grains Traditional method is usually used. With the rise in technology Artificial intelligence is being proven the boon for bio medical and other industry. But due to leverage of technology in agriculture it's not being used on a larger scale. Thus, motivation of this chapter is to show how Ai can be used for society and for agricultural community. In this paper we are using Deep learning method and algorithm to detect and classify paddy grain seeds based on its morphological quality after detecting size, shading, surfaces, and thickness. For better precision of classification model we are using several images processing technique for data Pre-processing like Image Pre-preparing, Feature Extraction, Acquisition, Image Filtering, Image Data Linearization and so forth. The data set contains overall 570 set of images contains binary scaled image of paddy grain based on utilizing RGB shading division and separating its mathematical highlights. We have also utilized Region of interest boundaries edge detection for in depth feature extraction to train the model of our multi class CNN which accuracy of results is tested based on overall model accuracy.

Keywords Grain quality · Identification · Deep learning · Image-processing · CNN

S. Ranjan
Department of Information Technology, Amity University Jharkhand, Ranchi, India

A. Sinha (✉)
Department of Computer Science Engineering, Amity University Jharkhand, Ranchi, India
e-mail: anuragsinha257@gmail.com

S. Ranjan
Department of Food Technology, Warner College of Dairy Technology, Allahabad, Uttar Pradesh, India

A. Biswas et al. (eds.), *Artificial Intelligence for Societal Issues*, Intelligent Systems Reference Library 231, https://doi.org/10.1007/978-3-031-12419-8_5

5.1 Introduction: State of Agriculture Sector in India

For many years, the main source of livelihood in India is from the agriculture sector. As a substantial fact, it is considered as one of the important sectors in the country as it offers abundant job and employment opportunities for the locals. Currently, India ranks second in the world in terms of production of agricultural goods. Thus, this sector subsidizes a major share in the Gross Domestic Product (GDP) of the country [1].

5.1.1 Problems and Challenges Faced by the Agriculture Segment of India

In India, the agribusiness area is as of now confronting certain troubles and difficulties. A portion of this happened quite a while past while some are increasing a direct result of the present horticultural standards and practices. These issues incorporate, yet not restricted to the accompanying:

Lessening arable domain: Arable land turns out to be less and less due to nonstop strain from the quickly developing populace and advancing urbanization and industrialization. Truth is told, the number of inhabitants in India, floods speedier than its ability to yield wheat and rice. That squeezes the agribusiness area.

Sluggishness in significant harvest creation: It is inconvenient that some significant yields, for example, wheat are getting stale underway. This made a huge hole between the interest and supply of rising populace and creation.

Soil Exhaustion: Besides the positive effects, there are some adverse consequences of Green Revolution; one of it is soil depletion. It is because of the utilization of compound composts. Likewise, the redundancy of same yield debases the supplements in the dirt.

Decline in Fresh Ground Water: Another adverse consequence of Green Revolution is the declining measure of ground water. In dry areas cultivating is refined with the assistance of water system offices did by the ground water utilization. The constant act of such rural exercises has prompted a disturbing state in setting of ground water circumstance.

Increase cost in Farm Inputs: The increment in the costs of homestead information sources like pesticides, manures, ranch work, apparatuses utilized in cultivating and so on, put the low and medium land-holding ranchers in a difficult spot.

Effect of Global Climate Change: Increase in temperature influences the agrarian practices in India too.

Farmer Suicides: Farmers are ending it all; it's anything but a significant portion of submitted suicides in India. It is a significant issue looking by farming area of India. The higher self destruction rate is accounted for in regions where there is higher commercialization and privatization of horticulture and higher worker obligation.

Refrain from cultivating: Farmer's kids stopping from their calling is additionally a significant issue. Regardless of to the arduous and monotonous work, the procuring is less in contrast with its expensive homestead inputs that are making ranchers to head towards different alternatives [1].

5.1.2 Problem Statement and Paper Organization

The significant problem with agriculture is to detect and classify the granolas of paddy, rice and wheat with accuracy for quality assurance using conventional approach. Thus in this chapter we are proposing deep learning for classification and detection of grains based on various quality parameter. Throughout the most recent ten years, computer vision has been generally utilized in different areas. A few strategies in the field of PC vision have been changed from factual techniques to profound learning techniques since it offers more prominent precision for undertakings like item identification and picture acknowledgment. The innovation can help AI researchers to foster undertakings in different fields quickly.

Authors Contribution: The corresponding author of this has contributed the methodology and literature survey and second author of this paper contributed the problem definition and background study of the paper. The third author of this paper have implemented the model study of deep learning and machine learning with relevant case study and helped in paper organization.

5.2 Background: The Role of Artificial Intelligence in Agriculture Sector

5.2.1 Usability of Artificial Intelligence and Machine Learning in Agriculture

Man-made reasoning uncovers the effect of human insight on machines intended to think like people and shape their conduct, for example, learning and critical thinking. Designing is essential for computerized reasoning as displayed in the Fig. 5.1. AI is a device used to recognize, comprehend and dissect designs in information. Quite possibly the main spaces of examination on the planet is present day natural innovation. This innovation is getting increasingly more troublesome because of the headway of innovation and its viability for issues, and the majority of them can't be successfully addressed by conventional numerical conditions just as by people. A comparative region is especially significant for horticulture where around 30.7% of the total populace is chiefly occupied with development of 2781 million hectares. So ranchers need to confront numerous difficulties from seed to reap. The significant weight of cultivating is the insurance of the collect, which isn't sufficient

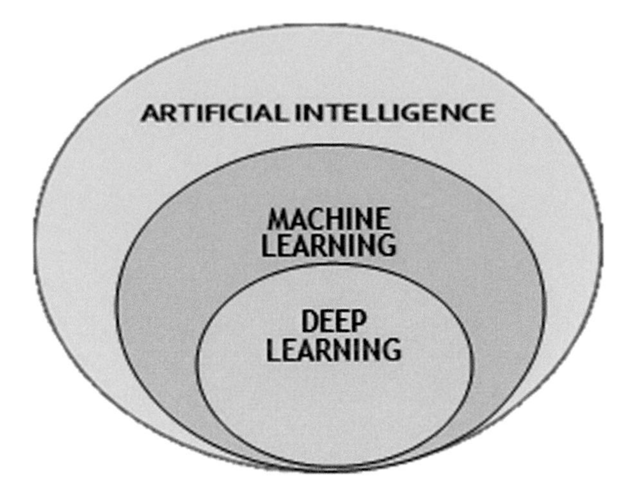

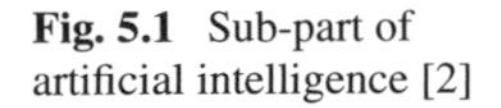
Fig. 5.1 Sub-part of artificial intelligence [2]

utilization of synthetic substances, bugs and irresistible sicknesses, insufficient water system and water system, timberland the board, and the sky is the limit from there. Agribusiness is an incredible land and conditions can't end with the installment of the normal e xp. Computerized reasoning Techniques have enabled us to catch explicit subtleties, everything being equal, and send the most ideal reaction to that specific issue. Fitting joining issues have been opened up through the advancement of numerous AI procedures [2].

Artificial Intelligcncc and AI arc turning into onc of the immense supported and progressed field in the area of software engineering whose application and significance are expanding step by step, among them the grain quality acknowledgment is one of the conspicuous uses of AI where the prepared framework can without much of a stretch perceive the nature of grain and the framework additionally delivers the exact outcomes according to our assumptions. Horticulture is the biggest monetary area and it assumes a significant part in the financial advancement of any country or state. The interest for nature of food item we tend to devour is expanding step by step in light of the fact that the accomplishment rate is expanding in India in this manner is that might want for nature of food creation is expanding. Our nation is that the second-biggest maker of paddy grains first being China. Since the creation of paddy is expanding subsequently is that the interest for its quality. This interest in the nature of food grains is expanding because some of the merchants cheat the businesspeople by trade low-quality food grains. Paddy crops are high burned-through and ideal grain in Asian nations. It very well may be found in the entire world. Numerous significant items are additionally made utilizing the paddy grain. The estimation of its quality likewise turns out to be vital because of the popularity of imports and fares. Paddy tests may contain husk and dead paddy without rice inside it. Paddy quality additionally fluctuates likewise these pollutant substances. The grain quality identification framework assumes a vital part for the ranchers just as merchants and government.

This framework assesses the seed quality in a fitting way. It is additionally significant for the seed fabricating industry where they can utilize it on their creation line to decide their creation. The customary grain discovery and reviewing methods are completely reliant upon human assessment with the master judgment which is dependent upon some sort of slip-ups. To decrease the human exertion and increment the precision we can take help from AI strategies and make a robotization framework. These robotization frameworks help in the identification of the nature of grain and lessen the time devoured. AI is an arising innovation for quick distinguishing proof of grain quality. It is likewise a computerized, non-dangerous, and savvy procedure. To make a computerization framework, machine vision is required. A machine vision framework is an elective choice to manage the quality testing of agrarian items. Machine vision is generally utilized for natural item grouping and assessment. For machine vision and AI execution, we need countless advanced pictures identified with the item and store them on PCs for an additional errand. The dataset of paddy crop is utilized in this task which is an advanced picture. Advanced pictures are a vision for PC vision AI frameworks. The dataset pictures are prepared by picture handling techniques like widening, otsu, disintegration, and so forth Utilizing these techniques and calculations on computerized pictures, a few highlights are removed and designs are created. Further, this information is utilized on AI calculations to characterize the pictures into their individual request. In this framework, the paddy seeds are tried dependent on their surface, shading, size, and shape highlights. This undertaking presents a plan to recognize the paddy seeds quality utilizing machine vision and Artificial neural organization techniques dependent on their mathematical and morphological highlights. Picture of various sorts of paddy are taken utilizing the camera and their highlights are removed utilizing diverse AI calculations. The acquired highlights of the pictures are essentially used to prepare the man-made consciousness model. The proposed neural organization model is tried for grouping of paddy crop as per their quality as great, medium, and poor. The field of AI and computerized reasoning is developing step by step where we can see various organizations working on their venture's dependent on AI or ML, as the improvement of self-driving vehicles, food-serving robots, and a lot more farming item. What's more, for the fruitful improvement of these sorts of robots or vehicles which need to play out a progression of activities dependent on object identification, here it is a need to recognize paddy grain quality by the assistance of which the framework can successfully react to the distinctive situation [3–7].

5.3 Literature Review

Before this we tend to acclimated acknowledge for shading extraction and orders by the Support vector machine by this, we tend to can't say at the satisfactory granules square measure blessing and what sort of value paddy. The grain quality recognition framework assumes a vital part for the ranchers. This framework assesses the seed quality in a proper way. It is additionally significant for the seed fabricating industry where they can utilize it on their creation line to decide their creation. A decade ago, the greater part of the specialists proposed such undertaking for identifying the grain quality. A portion of the new methodology for grain location are summed up as follows; In [8] introduced ANN and BPNN methods for grain identification utilizing ANN classifier calculations to accomplish a 94% exactness rate. They centre on ordering rice grain quality utilizing counterfeit neural organization (ANN) approach with the assistance of picture preparing to distinguish the contaminations contained in the rice grains. They utilized the ANN procedures as a classifier for rice grain arrangement and examine the rice quality and rice morphological highlights. In the morphological area, they utilized the grain's size, shape to remove the highlights from the computerized picture. The space of the grains is additionally determined in this segment utilizing the number of locale pixels inside the seed edge. Further, they played out the picture division to remove picture data which helps in the concealment of undesirable contortion or commotion in pictures. They have utilized GLCM which is a Gray-level co-event framework that gives the Grayscale picture and helps in force investigation. They utilized shading highlight extraction by tracking down the mean, change, standard deviation, and scope of each picture. It helped them in picture handling for deciding the rice test and aided in acknowledgment. They had performed Binarization to isolate the foundation and item in which the picture diminished to grayscale. This picture handling strategy brings about the improvement of the classifier execution and highlight choice. After Image preparation, they had played out the distinctive characterization strategies, for example, Back Propagation Neural Network (BPNN) which had the option to order precisely helpless pictures of assortments of rice grain. Another procedure they had utilized was Support Vector Machine (SVM) yet it didn't give them much exactness on the grounds that SVM strategies need more highlights. Thus, in the end, they had inferred that BPNN characterization strategies give them great exactness that is 94% and considered better compared to SVM and k-NN classifiers. In paper [3] introduced a financially savvy picture handling method for the acknowledgment of corruption and characterization of debasement levels (%) from the pictures of tainted mass paddy seeds utilizing cutting edge tone and surface highlights. In this paper they had worked upon seven corrupted mass paddy tests that were thought of and every one of the examples was set up by blending a superior paddy assortment with an indistinguishable looking and economically best assortment. In this investigation they had performed pre-handling techniques on the wheat pictures for highlights extraction then the presentation of the model on three classifiers were BPNN, SVM, and k-NN. In [4] proposed a grain location framework in which they utilized picture procurement utilizing a CCD camera. Greyscale transformation, clamour decrease, Linearization, morphological activities have been used on the obtained picture. Under pre-preparing steps, they changed the picture over to Grayscale and compute the limit worth of the picture under Linearization, and utilized the information to figure the histogram. In picture preparing, they had performed highlight extraction work in which the trademark highlights of grain-like tone and surface are separated and utilized in ID and characterization. In shading extraction, the mean, fluctuation, and standard deviation of each shading channel of pictures are removed. Shapes of the articles have been assessed by utilizing the utilization of form discovery. For division of covering and contacting rice parts watershed calculation was utilized. Binary Pattern (LBP) surface's capacity and shade perspectives separated from fragmented pictures. After picture preparing and include extraction, a Linear Kernel-based help vector machine was utilized to group the rice tests and to characterize the examples. These were assessed utilizing two strategies that are accuracy and review rate. The exactness is the small amount of accurately distinguished examples through every chosen case. The order had an 87% normal accuracy rate utilizing straight bit-based SVM strategies. In [5] proposed another component for the grouping of rice grains is introduced, features dependent on both spatial and recurrence is utilized for order of 9 assortments of economically available grains in the South Indian district. They had gathered the datasets from the power source across Karnataka. The pictures were pre-handled utilizing distinctive shading ranges and highlights were separated. The calculations utilized for extricating the recurrence includes in this paper are as per the following: Algorithm (Fast Fourier Transform), Input: Scanned picture of a rice grain, Output: Feature Vector. Characterization was done on two levels, first in the 2d level and the classifier of the NB Tree in the classifier of the SMO. They performed WEKA instruments for the grouping on the grain information. This examination used the multi-facet perceptron, Naive Bayes Tree, and SMO classifier for the best presentation of the framework. The outcomes acquired utilizing NB Tree groupings were 95.78% exact and with SMO classifier were 87% precise. In [6] introduced machine vision, comprising of AI capacities and a very much prepared complex neural local area classifier that was achievable, engineering of unpredictable rice grains set up neuro-local area filled in various agrarian and natural districts of the nation, can be utilized as a gadget for acquiring a superior, extra-reasonable, good appraisal of rice. They had utilized computerized pictures and Weka devices for the order of seeds. The highlights of the seeds utilized in this paper were border, conservativeness, length of the piece, width of part, lopsided coefficient, and length of portion groove. In WEKA order apparatus they had utilized capacities like Bayes, Meta, and Lazy strategies. For seeds arrangement the classifiers, for example, Multilayer Perceptron, Logistics, SMO, Naive Bayes and LWL are utilized. After then they utilized 10-overlap, 5-crease, 2-overlay cross approval to gauge their model presentation. The Sequential Minimal Optimization (SMO) calculation was utilized to take care of the improvement issue. The outcome got utilizing Weka calculations and capacities were 97% precise and with MLP 99.5% exactness. In [7] introduced another element for the order of rice grain utilizing multi-class SVM. This paper proposed AI calculation to review the rice pieces utilizing Multi-Class SVM. They had utilized computerized pictures of rice as a test and afterward performed picture smoothing, division, Binarization to extricate the highlights. Further, they had played out the investigation of rice virtue and its shape utilizing grain part. The order was finished utilizing the support vector machine and preparing the model. This model acquired a decent precision rate that is about 86%. They had additionally referenced a few disadvantages of the SVM model that this model is created for double grouping issues. The SVM upheld them in evaluating and grouping rice pieces precisely 83.9% in subgroups 1 and 2 and the aggregate gathering, separately, while the best exactness on the profound learning procedures was at 95.15% from InceptionResNetV2 models.

5.4 Proposed Approach: Image Processing

5.4.1 Involved Steps

In Fig. 5.2 first step is to save the image and remove noise from the image using a filter. The second step is to use the removal algorithm to separate the sensitive seeds. In the third step, we conduct extremist intelligence to find the border area. In the fourth step, the grain is measured and the same length, the width and length are also measured. In the fifth step of the algorithm, the rice is classified according to its size and shape [9].

A. A filter is used to remove any noise that may occur during image recovery. The filter emphasizes the image. The threshold algorithm is used to separate the rice grains from the black background.
B. Morphological retraction operation: It is used to distinguish sensitive components of rice without damaging the characteristics of the grain. The process of expansion is the same as the process of erosion. The purpose of expansion is to grow them back to their original shape without reusing the distinctive features [10].
C. Extreme intelligence: Help to understand the limits of the limits of rice. We use a sophisticated algorithm to find the edges.
D. Object measurement: The measurements indicate the number of grains. After counting the grains, the final values of each grain are obtained by the image detection method and the application of extreme algorithms. We measure the ends of the caliper as well as the length and width of each grain. If you know the value of the length and the width, you can calculate the length and the ratio.
E. Classification of objects: Discrimination is standard; all measured and calculated results are required. The standard database for measuring the size and shape of rice is mentioned in the Rice Quality Testing Laboratory Manual. According to the standard database, the types of rice are shown in the following table. Table 5.1 classifies rice grains according to their length/width ratio.

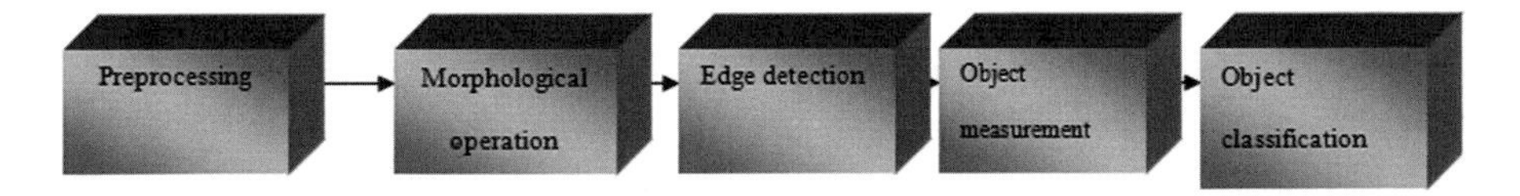

Fig. 5.2 Steps involved in image processing method

Table 5.1 Model result summary

Model	Accuracy (%)	Statistics of accuracy
SEP	87	0.990993
CNN	96.2	0.965702
k-nearest neighbor	91	0.973102

5.4.2 *Materials and Tools*

Rice sample: Based on grain size, three well-known American rice varieties were selected for testing. Medium and short paddy varieties, respectively The size and weight of the kernels are measurement. After purchasing the rice samples, 56 grains of each variety were spread out on a greenish gray surface. We chose a resolution of 58 and a balanced image capture ratio. Too many seeds in an image will reduce the resolution of individual seeds, and a smaller number per image will increase the number of images. Each core is integrated into a separate grid. Broken fish seeds were also selected and added to the perch. Seed length and width were accurately measured on the caliper at 0.001 mm.

Preparing the dataset: In Fig. 5.3 we need to search for grains of rice. We can do this with the canny edge detection algorithm. We then save these seeds from the training images. This code will take the guide from the dataset that contains the seeds that match the name of the crop type, such as the code name below, and then create another guide with a set of information about the crops that can be used to make changes, then step by step. The stages are [11]:

- Use edge detection for a complete image
- Look for curves on the edges to find seeds
- After removing the seeds, create a mask to completely darken the base
- Cut the seeds using the opencv boundingRect () function.
- Cut images to precise size (maintain proportions)
- Save the padded image to a specific directory.

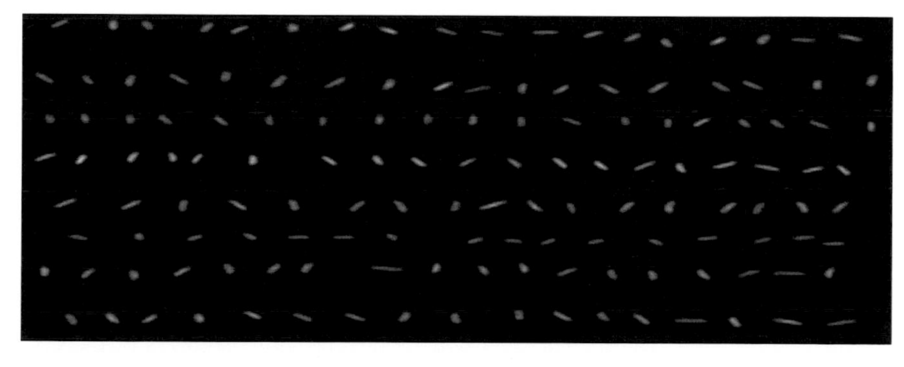

Fig. 5.3 Dataset sample

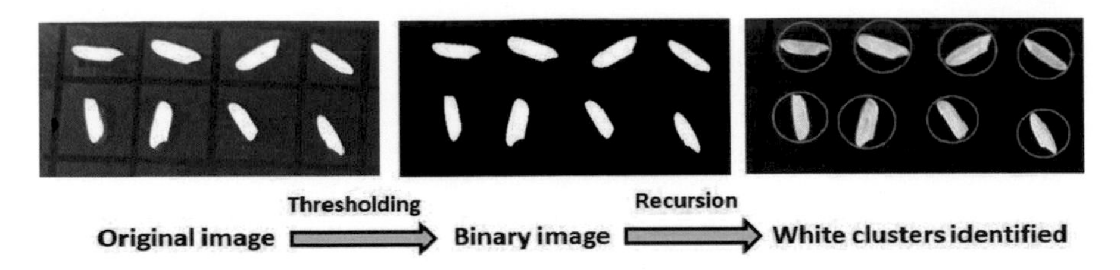

Fig. 5.4 Conversion to binary scale

Image transformation algorithm: The grains are placed on a gray surface to balance the background and the white grains. The grain (rectangular plate 13 × 7 mm) was also placed with the grain on a photo takes a photo of a grain of rice about 0.2 m from the screen and turns off the camera flash. Fixed size camera the original 3553 × 2013 pixel camera image includes nothing but grains of rice and reference paper. The cropped image is converted to a grayscale image with the average r, and b pixel values. Grayscale an image allows you to use a single threshold value to create a binary image. As the name suggests, the binary image contains black (r = 0, g = 0, b = 0) or white (r: 243, g: 164 and b: 278) pixels. In Fig. 5.4, adjusting the pixels of the image contained in a white sheet of paper (13 × 7 mm) contained in an image converts the image to pixels. The size of the white cord More than any other grain, it is stored with great purpose. We used the code to easily classify the white pods as the larger white pistons and to refine the kernels. To calculate the size of a kernel, the distance between all the pixels of the other pixels in the cluster is calculated. The two pixels with the greatest distances are considered. The cropped image is converted to a grayscale image with the average r, g, and b pixel values in Eq. 5.1. Below the pixel level, any pixel with a grayscale value is defined as a black background or pixel. Anything equal to or greater than this is considered a pixel representing a grain of rice or a white pixel (Eq. 5.2) [12].

$$image\ pixel = \frac{RGB\ pixel\ values}{2} \tag{5.1}$$

$$binary \frac{pixel\ 1}{pixel\ 2} [greysacle\ thresholding] \tag{5.2}$$

Stacked Assembly Model (SEM): To improve the grain size and the estimate of the mass of each image, we used different ML formats to make the estimates more precise. Feeds the ML shapes of the first layer. Each ML model of the first layer estimates the output variable as a function of the vector of input characteristics. The individual estimates made from each ML model in the first layer N1 are the number of ML models in the first layer. The feature vector V1 is used as input for each ML model on the second layer. The second layer ML models provide estimates of the kernel output (size and mass). Based on estimates from N1 ML models, this step improves the accuracy of the predictions made from the first layer. For our SEM, we only used the second layer defines an output layer. The performance of any degenerate form is usually measured by the Root Mean Square Error (RMSE), Mean Absolute Percent Error (MAPE), and Correlation Coefficient (R2) such as

The equation is shown in Eqs. 5.3 and 5.4. The actual value and the estimated amount are represented by "y" and "x" for "n" [13]

$$RMSE = \sqrt{1/N \sum_{I-1}^{N} (Y^2 - \mathrm{X}^2} \tag{5.3}$$

$$CV = \{CV(X) - CY(Y) \frac{1}{2\sqrt{CV(X, X)} + CY(Y, Y)} \tag{5.4}$$

K-neighbors (knn): k-The nearest neighbor ML method is based on estimation as its name suggests K For the input feature vector (mean, 2017) from the nearest training data point Our study used the neighborhood factor (k = 3) for the final estimate (Fig. 5.5). We are using For the characteristics given the Euclidean distance (Eq. 5.3) to find the three nearest neighbors The vector algorithm (A_N) calculates its Euclidean distance from all the training data On a two-dimensional domain, the mass of the grain is estimated on average The three grains closely correspond to his An and his training set. This Its simple approach is one of the fastest ML methods and works just as well. Disseminate big data, in particular in a special area of the investigation area [14, 15].

$$AM = \frac{(1 - N)A^2}{A^N} \tag{5.5}$$

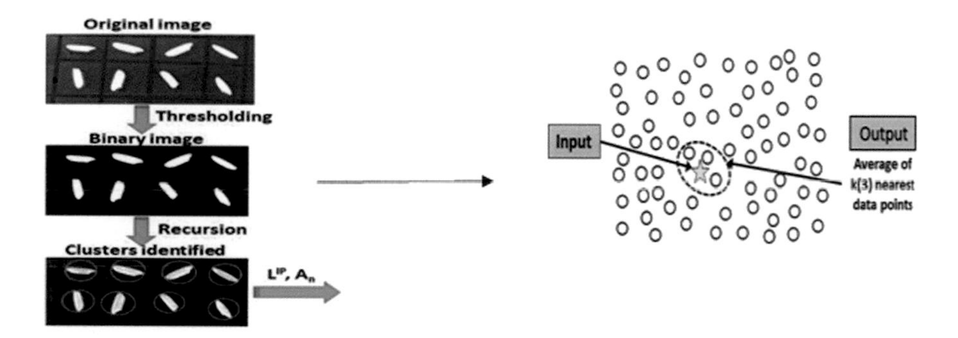

Fig. 5.5 KNN working mechanism

5.5 Methodology and Implementation

5.5.1 *Plan and Proposed Architecture*

The cycle selects by the information set of pictures it portions the picture by the totally various parts and applies the component extraction to determine the information of the granules and subsequently the condition of position to recognize the extent connection in and this calculation square measure training and testing. The items in the info picture are for sectioning and arranging. Inside the instructing segment, properties of the picture alternatives square measure isolated and, upheld these, a selective explanation of each order class. Inside the testing segment, picture choices square measure ordered by exploitation the component region allotments. Further, we play out the investigation of paddy immaculateness and its shape utilizing grain part. The characterization is finished utilizing the convolutional neural organization (CNN) of the model [16].

Plan and Proposed Architecture: In this Fig. 5.6, we will examine the plan and engineering of the paddy grain quality discovery framework/machines. Also, the design of the proposed Grain recognition frameworks can be extensively ordered into various stages:

i. *Pre-Processing of the sources of info*
ii. *Feature Extraction*
iii. *Recognition by CNN*

The given picture's portrayal shows the proposed framework engineering.

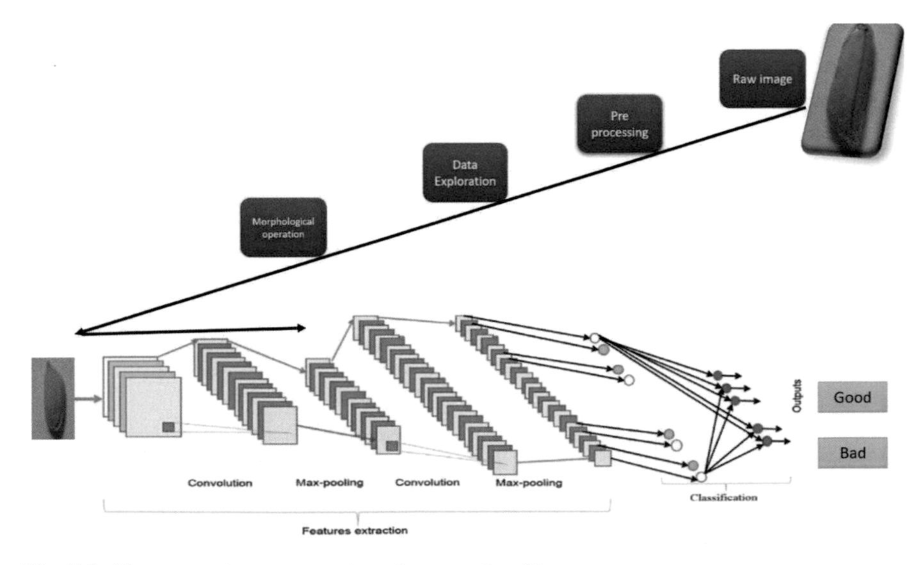

Fig. 5.6 Diagrammatic representation of proposed architecture

As examined in the above Fig. 5.6, the grain location framework takes input pictures which should be pre-prepared for acknowledgment. Furthermore, it further aids in diminishing the information by thresholding the given information pictures to a paired picture. After the undertaking of pre-preparing, the framework plays out the element extraction by the different divisions like every paddy division and edge discovery with the assistance of different picture handling calculations out of the paddy from the info picture.

According to the proposed design, the pre-handling assignment of grain quality recognition can be in everyday ordered into 6 sections [17]:

(a) **Image acquisition**—The first step of the interaction is picture procurement. The pictures of paddy tests and foundation were obtained with an advanced camera with a fixed gap and controlled climate for gaining great outcomes as demonstrated in Fig. 5.7. When catching pictures, a steady distance keeps up between the paddy test and the camera. Likewise, some uniform lightning frameworks are utilized. Picture of paddy tests and foundation is utilized to eliminate the non-uniform enlightenment impact. The goal of gained pictures is 250 × 250.

(b) **Pre-handling**—First step is changing over RGB shading computerized pictures to grayscale pictures. The primary objective of this cycle is to simplify the picture and lessen the intricacy of code. Grayscale pictures appear in Fig. 5.4. At that point the non-comparability of the pictures computes by taking away the foundation picture from the forefront picture. The primary objective of this progression is to forestall the impact of non-uniform enlightenment and light spots. Imaging measure utilizes uniform lightning conditions to procure pictures. Yet, a little distinction of enlightenment influences the division cycle just as of the little size of paddy bits. Consequently, the evacuation of light spots on the picture is finished by taking away the foundation picture from the forefront picture.

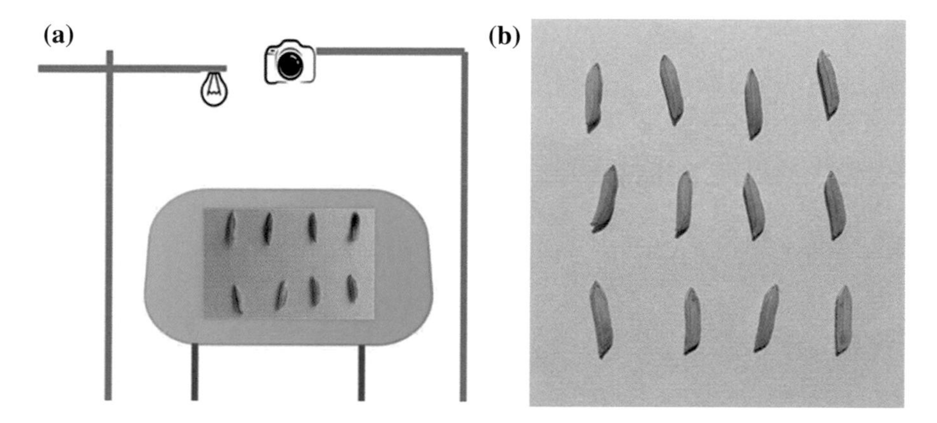

Fig. 5.7 **a** Input image capturing for data, **b** input image of paddy sample

(c) **Otsu's binarization**—The principal objective of performing parallel thresholding on the picture is to check the pixel worth of the picture. On the off chance that pixel esteem is more noteworthy than edge esteem, it is appointed one worth (might be white), else it is relegated another worth (might be dark). The capacity utilized is cv2.threshold Eq. (5.5). The first contention is the source picture, which ought to be a grayscale picture. The second contention is the edge esteem which is utilized to characterize the pixel esteems. The third contention is the maximal which addresses the worth to be given if pixel esteem is more than (once in a while not exactly) the limit esteem. Since we are working with bimodal pictures, Otsu's calculation attempts to discover limit esteem (t) which limits the weighted inside class fluctuation given by the connection: It really discovers a worth of t which lies in the middle of two pinnacles to such an extent that differences to the two classes are least given in Eqs. (5.6) and (5.7) [18].

$$q1(t) = \sum_{i=0}^{t} \mathrm{p(i)} \text{ and } \mathrm{q1(t)} \sum_{\mathrm{i=t+1}}^{\mathrm{i}} \mathrm{p(i)} \tag{5.6}$$

$$\mu 1(t) = \sum_{i=1}^{t} \frac{ip(i)}{q1(i)} \text{ and } \mu 2(t) = \sum_{i=1}^{t} \frac{ip(i)}{q2(i)} \tag{5.7}$$

It really discovers a worth of t which lies in the middle of two pinnacles to such an extent that differences to the two classes are least.

(d) **Smoothing pictures**—To diminish the haze impact from the picture we by and large utilize this Open cv capacity. Concerning one-dimensional signs, pictures additionally can be separated with different low-pass channels (LPF), high-pass channels (HPF), and so on A LPF helps in eliminating commotion, or obscuring the picture. A HPF channels helps in discovering edges in a picture. For discovering the averaging the picture is convolving with a standardized box channel. It basically takes the normal of the relative multitude of pixels under part territory and replaces the focal component with this normal. This is finished by the capacity cv2.blur() or cv2.boxFilter() cv2.filter2D(). Separating with the above shown piece gives yield in coming up next being performed: for every pixel, a 5 × 5 window is fixated on this pixel, all pixels falling inside this window are summarized, the outcome is then partitioned by 25. This likens to figuring the normal of the pixel esteems inside that window. **cv2.blur() or cv2.boxFilter() cv2.filter2D()** [19].

$$\boldsymbol{k} - \frac{1}{25} \begin{matrix} 1\,0\,0 \\ 0\,1\,0 \\ 0\,0\,1 \end{matrix} \tag{5.8}$$

In Eq. 5.8 Separating with the above shown piece gives yield in coming up next being performed: for every pixel, a 5 × 5 window is fixated on this pixel, all pixels falling inside this window are summarized, the outcome is then partitioned by 25. This likens to figuring the normal of the pixel esteems inside that window.

(e) **Edge detection**—cv2.canny() strategy is utilized for recognizing the edges of the image. Watchful edge recognition is a mainstream edge identification calculation. Since edge location is powerless to clamour in the picture, the initial step is to eliminate the commotion in the picture with a 5 × 5 Gaussian channel. The smoothened picture is then sifted with a Sobel piece both level and a vertical way to get the first subsidiary even way () and vertical course (). From Eq. (5.9) these two pictures, we can discover edge angle and bearing for every pixel as follows [20].

$$Edge\_gradient\ G = G2X + G2Y \tag{5.9}$$

Nearby greatest in its neighborhood toward angle as shown in Fig. 5.8a Point A is on the edge (in a vertical course). A slope course is typical to the edge. Points B and C are in angle headings. Thus, point A is checked with guides B and C to check whether it frames a neighborhood greatest. Provided that this is true, it is considered for the next stage, else, it is stifled (put to nothing). In Fig. 5.8b it is shown that how the edges are distinguished from the picture.

(f) **Feature Extraction**—This includes extraction from the paddy pictures that are considered for getting the shading, shape, and surface. The extricated features are put away in the model. Features are separated from new pictures and those highlights are utilized in example recognizable proof and grouping measure utilizing portion based Convolutional Neural Network (CNN). Items are named GOOD_PADDY and BAD_PADDY, The Fig. 5.9 representing the block diagram of the proposed system. A picture includes lattice infers a reflection of the picture which is utilized to portray and mathematically evaluate the substance of the picture. Regularly genuine, Integer or paired qualities are framed. Essentially, include network is a rundown of numbers used to address a picture. For better characterization precision it is essential to choose the most proper element extraction technique. The proposed calculation is utilized to extricate five-tone and picture highlights from the preparation and test dataset. The shading means, change, and standard deviation descriptor of RGB tones are utilized as shading descriptors. The surface highlights are removed by utilizing Local Binary Pattern (LBP). The pictures test comprises various surfaces. They give insights concerning the power variety of a surface by assessing properties like routineness and perfection [21, 22].

5.5.2 The CNN Architecture

Man-made reasoning has been seeing fantastic development in overcoming any issues between the abilities of people and machines. Analysts and devotees the same, work on various parts of the field to get astounding things going. One of numerous such

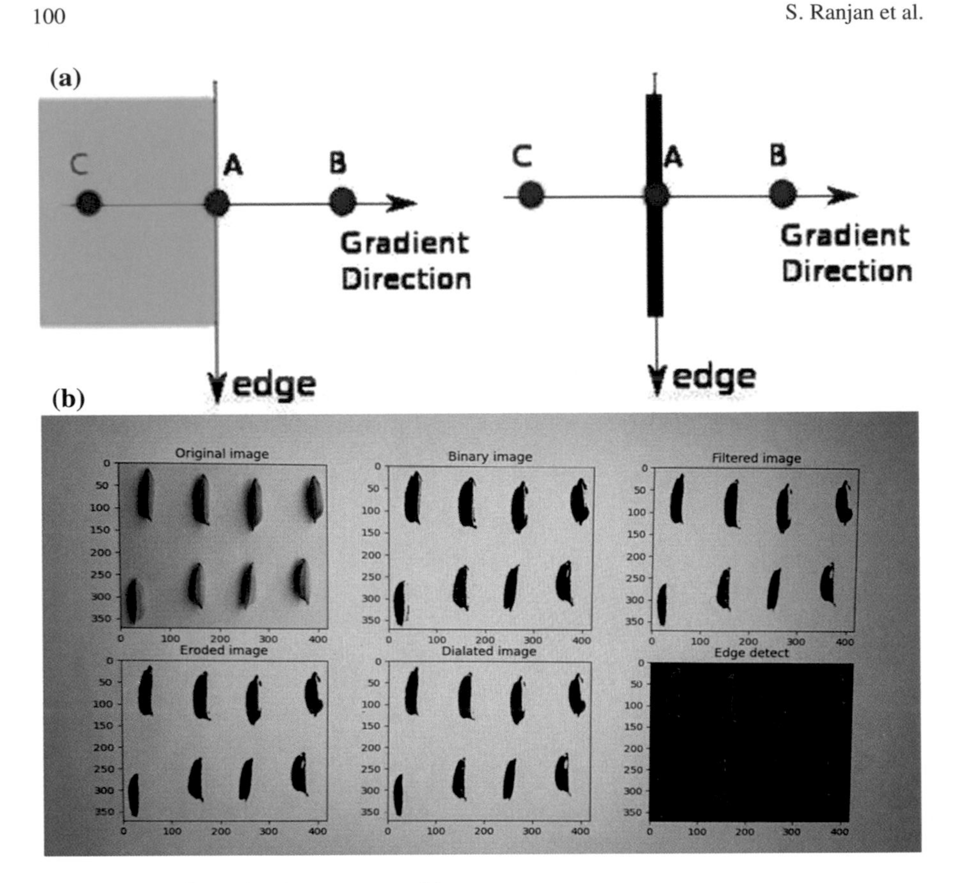

Fig. 5.8 **a** Gradient edge; **b** pre-processed image

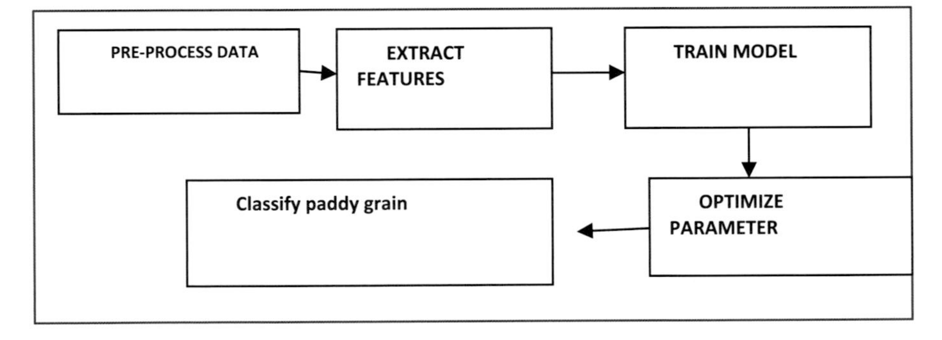

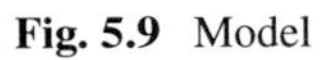

Fig. 5.9 Model

territories is the space of Computer Vision. The plan for this field is to empower machines to see the world as people do, see it likewise and even utilize the information for a large number of undertakings like Image and Video acknowledgment, Image Analysis and Classification, Media Recreation, Recommendation Systems, Natural Language Processing, and so forth. The headways in Computer Vision with Deep Learning has been built and idealized with time, principally more than one specific calculation—a Convolutional Neural Network. It is a Deep Learning calculation that can take in an information picture, dole out significance (learnable loads and inclinations) to different viewpoints/objects in the picture, and have the option to separate one from the other. The pre-handling needed in a Conv-Net is a lot lower when contrasted with other grouping calculations. While in crude strategies channels are hand-designed, with enough preparation, Conv-Nets can get familiar with these channels/attributes. The CNN engineering examined in Fig. 5.8, shows the profound neural organization which is utilized for Paddy grain from the information pictures, the main layer of the above design is the info layer, in the middle the convolutional layer and at definite it is the yield layer. Here the distinctive layer of the design is momentarily talked about [23, 24].

a. *Input Layer*—The information layer of CNN design allows to that layer which is utilized to pass input pictures to the CNN. The information layer is associated with the first convolution layer which plays out the cushioning task.
b. *Convolution Layer*—It alludes to the layer of CNN which measures the paddy pictures by the cycle of the division of the info pictures. It contains a bunch of boundaries which is learned by the model. Additionally, the weight and tallness of the channels in this layer are more modest than those of the information layers. In this layer, the numerical activity of convolution is performed between the info picture and a channel of a specific size M × M. By sliding the channel over the info picture, the spot item is taken between the channel and the pieces of the information picture concerning the size of the channel (M × M). The yield is named as the Feature map which gives us data about the picture like the corners and edges. Afterward, this component map is taken care of two different layers to become familiar with a few different highlights of the info picture.
c. *Pooling Layer*—The convolution layers are trailed by the pooling layers which targets decreasing the size of channels to reduce the number of boundaries. The essential point of this layer is to diminish the size of the convolved include guide to lessen the computational expenses. This is performed by diminishing the associations among layers and freely works on each component map. Contingent on the strategy utilized, there are a few kinds of Pooling tasks. In Max Pooling, the biggest component is taken from the highlighted map. Normal Pooling ascertains the normal of the components in a predefined measured Image area. The all-out amount of the components in the predefined segment is processed in Sum Pooling. The Pooling Layer normally fills in as a scaffold between the Convolutional Layer and the FC Layer [25].
d. *Fully Connected Layer*—The pollable centre of the CNN design is trailed by the completely joined layer that measures the yields of the pool capable layer

and convolution layer and uses it for investigating them and produces the end-product. In this, the information picture from the past layers is straightened and taken care of to the FC layer. The smoothed vector at that point goes through hardly any more FC layers where the numerical capacities activities ordinarily occur. In this stage, the characterization cycle starts to occur.

e. *Dropout*—Usually, when every one of the highlights is associated with the FC layer, it can cause over fitting in the preparation dataset. Overfitting happens when a specific model functions admirably on the preparation information causing an adverse consequence in the model's exhibition when utilized on information. To beat this issue, a dropout layer is used wherein a couple of neurons are dropped from the neural organization during preparing measure bringing about the diminished size of the model. On passing a dropout of 0.3, 30% of the hubs are exited haphazardly from the neural organization.
f. *Activation*—Finally, quite possibly the main boundary of the CNN model is the actuation work. They are utilized to learn and inexact any sort of consistent and complex connection between factors of the organization. In basic words, it chooses which data of the model should fire the forward way and which ones ought not toward the finish of the organization.
g. *Output*—After all the arrangement the yield layer examines them and produces the end-product.

5.5.3 Implementation

Execution: This part will talk about the execution interaction of transcribed image data acknowledgment as examined in the prior segment.

Task Deployment: With the end goal of sending this task, we will utilize PyCharm for picture and dataset handling and Anaconda manager where the Notebook interface is being utilized in the venture improvement. Also, as we realize that the Notebook editorial manager is one that is the most broadly utilized editors for empowering logical-based programming just as exploration-based application. Here we have likewise utilized the NumPy, Open cv, TensorFlow, Keras, Pickle, OS, Matplotlib libraries of python which is utilized to give mathematical information controls just as plotting and preparing to test [26].

Dataset-Creation: A dataset is a vital piece of any AI—profound learning project. It assumes a critical part in propelling PC vision and man-made consciousness. There are a few destinations like Kaggle, MNIST, and so on give the handled and all-around rearranged dataset. Yet, in my task "Grain quality recognition" I need paddy grain information that isn't accessible at any spot thus it turns out to be hard for me. We as a whole realize that the greater part of the datasets are made by broad human exertion and are very costly proportional up. Thus, to execute my model as a matter of first importance I need to make a dataset of in any event a great many pictures with the goal that I can prepare and test my model. I had gone through various exploration papers that depended on dataset creation, where I got the possibility of dataset creation.

I have utilized my telephone which has a 16MP camera to catch every one of the pictures as shown in Fig. 5.8b. At that point, the pictures were trimmed in square pixel design. The size of the pictures is too huge to even think about handling into AI. Along these lines, I have utilized Ms. Paint to resize it and save into various document as pr there quality. The picture's names are arbitrary as they were clicked utilizing telephone so our next work is to rename every one of them and for that, I have utilized python code given underneath, it renames every one of the pictures of the large number quickly.

Initialisation: The statement of the proposed calculation is finished by playing out the picture pre- handling and by bringing in every one of the modules which are required for the preparation of the framework. The made dataset is kept in their correct registry. Subsequently, the dataset can be brought effectively and we start chipping away at it. Our essential work is to play out the pre-preparing of the pictures with the goal that the highlights can be separated without any problem. The underneath code blocks show the introduction of the pre-preparing and profound learning library.

Method Invocation and establish the activation mapping: The accompanying code is utilized to send the information for preparation. It ought to be noticed that here we can't matter the information straightforwardly to the model. Along these lines, here the element of the preparation information is (256, (3, 3)). The CNN model will require one more measurement with the goal that we update the lattice to shape ((256, (3, 3),1). So appropriately, we will apply the information to the framework. At the point when the prepared information is being given to the model then we need to set up the planning of the yield information. The CNN model fundamentally comprises of convolutional and pooling layers. Along these lines, it generally performs better for information that is established as network structures; here the dropout layer is utilized to stop a portion of the neurons, and keeping in mind that preparation, decreases present connections of the model. Here to accumulate the particular model we will utilize the Ad delta enhancer which is generally used to empower the planning likewise once the code gets assembled.

Training the Model: This stage discusses the preparation of the proposed model when the program had aggregated and the planning has been set up. As a rule, we will empower it after the planning. The preparation set of code is planned for the model, where the model.fit() technique for Keras will begin the arrangement of the model. It takes the preparation subtleties, approval subtleties, ages, and bunch size as boundaries. It will require some investment for the preparation of models. In the wake of preparing, we save the loads and model definition in the 'pickle.wb' record. This document is fundamental for the infusion of the code during the runtime. At the point when the informational index prepare then the assessment is to be finished. The assessment of the model should be possible utilizing the Kera code as appeared underneath.

model.fit(X, y, batch_size = 32, epochs = 3, validation_split = 0.3) [27]

5.5.4 GUI Creation and Testing

GUI (Graphical User Interface) allowed to the client characterized layer which can be utilized by any client for distinguishing the grain quality and in like manner the framework will decide the outcomes. Here we need to plan another record that will assist with embellishment a cooperative window to draw picture design on a level surface and with a catch. Here we may likewise take the grain close by, clicked it, and use it as part which will be utilized to recognize the particular grain picture and in like manner the pixels will be initiated and we can recognize the grain. The pickle library's foresee picture () strategy is essentially utilized which assists us with snapping the photo as information and afterward utilized the prepared model to expect the grain quality.

Testing: The testing area examines the different testing and approvals which are performed for estimating the exactness of the proposed framework. As we probably are aware the testing of any product machine can be sorted into various parts, here we have talked about the unit testing, joining testing, approval testing, and finally the GUI testing through its yield and conversation [28].

Unit Testing: As the name proposes, in-unit testing we have attempted to try out the distinctive little square of codes and strategies which are being utilized. The unit testing was focused on testing the capacity which was utilized for picture collection and highlight extraction, pre-handling with the assistance of disintegration, parallel, edge recognition, after that component evacuating and acknowledgment of grain information.

Integration Testing: As we as a whole realize reconciliation testing is the joining of various units of the product, which is to be complete and check the fidelity of the product. The equivalent is the situation with this proposed framework, where the various modules like the pre-handling module are being coordinated with the division module, and the division module is being benefitted with the element extraction module.

Validation Testing: The approval testing for the created program or set of rules was performed to see whether the created framework meets the necessary particular or not. The primary objective of performing approval testing for the grain quality discovery machine was to discover absconds in the framework. It additionally helps us in the comprehension of different functionalities of the framework. The approval of datasets and the models likewise assume a significant part in making the framework more precise [8].

GUI Testing: As individuals of this framework are for the most part going to simply interface with the GUI part, subsequently the testing of GUI is fundamental and has a significant job. This testing helps in distinguishing the relapse blunders, helps in the recognizable proof of any lethargy in the GUI part. The output of the model after all testing is attached below in Fig. 5.10. The GUI testing of this venture incorporates a few stages like:

(a) Testing of the measurement and covering of grain.
(b) Verification of grain and picture body arrangement.

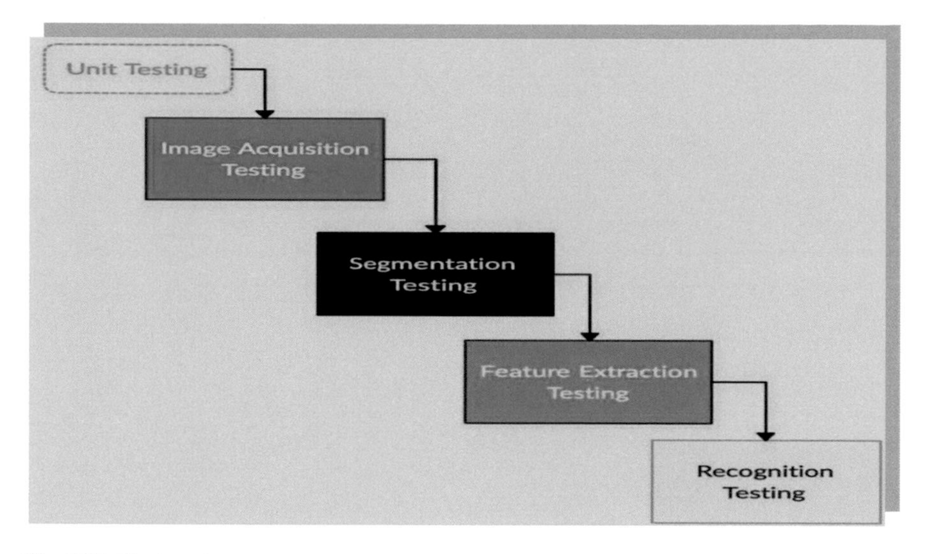

Fig. 5.10 Testing phases

5.6 Results and Discussion

The RMSE values for all three species were 0.251 mm, 0.176 mm and 0.198 mm; MAPE values were 3.69%, 3.87% and 7.65%, respectively. The quantitative indicators for medium and long seeds (Jupiter and CL153) were slightly higher than for small seeds. There are two main reasons for this difference. This is due to measurement errors and image resolution errors. Adjustment of the source of image distortion errors; this can be attributed to grading errors and image sharpness. The error is the result of a measurement error in the reference material which causes errors in the estimation of the grain size. During the evaluation process, the image is converted to a gate value and a duplicate image. The pixel associated with the grain can be adjusted relative to the background pixel (rgb = 2, 0, 0.1). It is not possible to estimate the number of pixels in a kernel. The sharpness of the image also plays an important role. An image with a higher resolution has a lower resolution because each grain has fewer pixels [29–31] (Fig. 5.11).

In Fig. 5.12 the classified sections of broken an accurate RGB image is shown. Manually characterize the component for order. For this situation, essentially binarizing the picture, extricating the region size of each picture and afterward characterizing a limit will get along nicely. All of the picture change and region estimating should be possible with the elements of skimage and scipy.

The hierarchical various leveled characterization approach yielded a few benefits to separate rice fields. The overall classes were arranged by applying edges and afterward eliminated from additional handling. The limit condition for any class was changed in accordance with accomplishes higher characterization precision without influencing the condition for different classes. Along with the edge condition, the

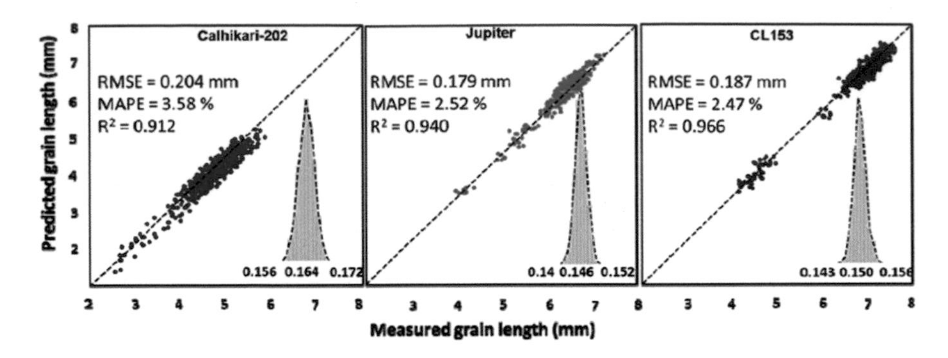

Fig. 5.11 Measured kernel length

closest neighbor grouping has likewise been applied for the underlying arrangement of rice. The closest neighbor arrangement considered the factual appropriation of the classes. It assisted with distinguishing the classes that couldn't be grouped by the edge condition. Also, the applied hierarchical progressive methodology was legitimate to separate the rice class.

Extend the information base of rice and lay out a thorough apparatus for arrangement system. The information increase strategy will be utilized for building a decent classifier when the quantity of tests is insufficient. Study other profound brain organization designs and make the most of profound learning calculations to further develop arrangement accuracy, and improve the dependability and heartiness of rice order framework. In a few different cases, too observe the stones looking like rice grains. They can likewise be identified alongside the grains and displayed in the result. To stay away from this issue, need to group the picture all the more precisely.

In Fig. 5.13 the total loss accuracy is shown for proposed model.

In Fig. 5.13a the total accuracy graph is shown based on the predicted samples by the model in Fig. 5.13b total multi-class accuracy is shown based on the output predicted by model, in Fig. 5.13c the loss epoch graph is shown based on total training and testing loss of the model. In Fig. 5.14 the good paddy and bad paddy is predicted based on morphological features of the image. In Table 5.1 cnn gives 96% overall accuracy, k-nearest neighbor 91% 0.973102 and SEP 87% 0.990993 [32].

5.7 Future Work

The grain detection system is improving day by day as the algorithms for image processing and pre-processing of large datasets can be done easily. The proposed grain quality detection system can be improved in many was which will make it more accurate and reliable. Talking about the improvement which can be done in future includes, the enhancement of the pre-processing system focusing on keeping the least number of incorrect noise detection which will make the system faster. Further the

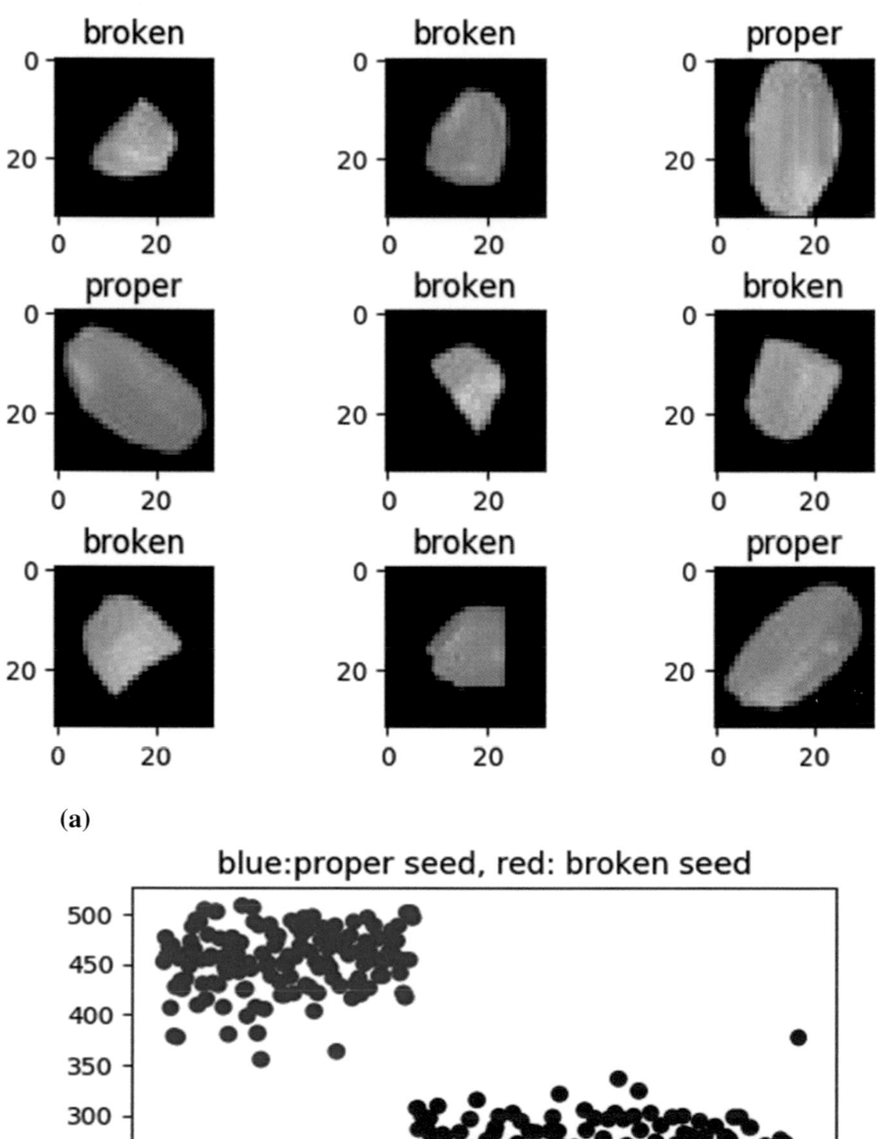

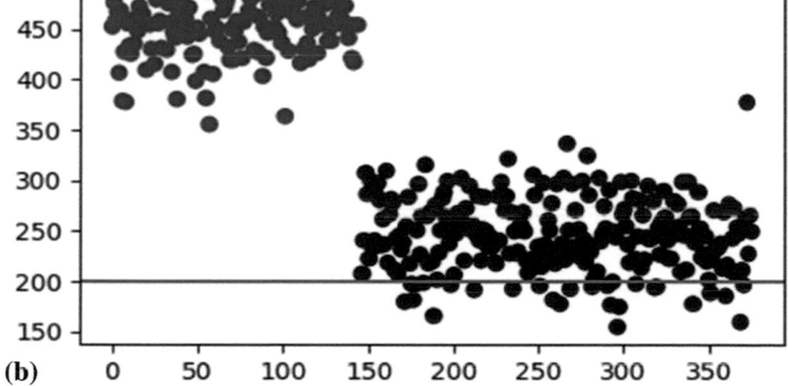

Fig. 5.12 Classified sections

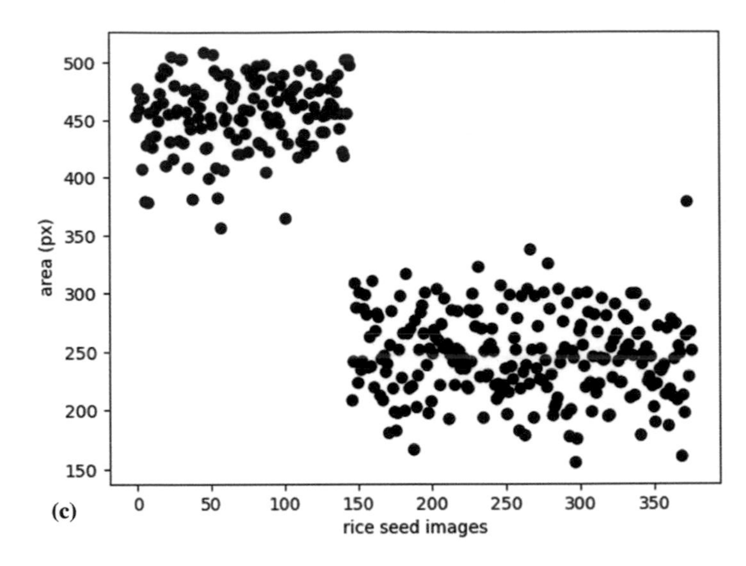

Fig. 5.12 (continued)

detection of straw and broken paddy grains needs to be upgraded with pre-processing of the input image into a high-quality binary image and image classification [33]. With the help of utilizing CNN, training of the model can be improved which will improve the capacity of the system to recognize the continuous grain or even the detection of a set of multigrain. The developed model can be extended to use it further in industries for getting the results between good and bad quality paddy grain automatically. In the Convolutional Neural Network, the neural components can be improved to get more concise extraction information from convolution layers. Along with this, the different datasets, including numbers of paddy grain with husk and broken grain can be used for the training purpose. After performing the entire above-discussed task, a whole grain detection system can be developed with more accuracy and efficiency [34].

5.8 Conclusion

In this chapter, many computer and artificial intelligence techniques mention are utilized with different algorithms to classify between the good and bad paddy. The classification of the paddy seeds is implemented using machine learning with the help of kernel-based CNN which has proved its capability and results. I can say that this system can beat the traditional approach of paddy grain quality detection and provides better results from them. The proposed system is much more effective than the traditional approaches. So, the proposed system can be extensively used in agriculture-industry, seed manufacturing industry, government PACS and to the

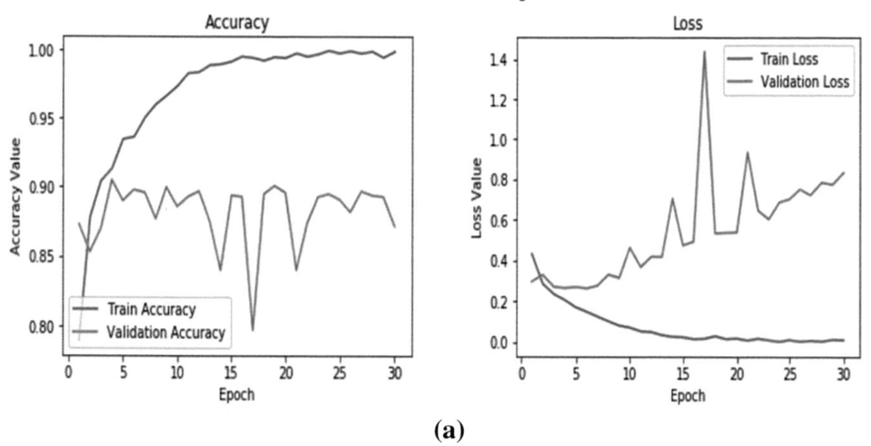

(a)

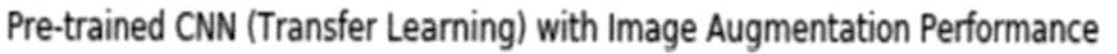

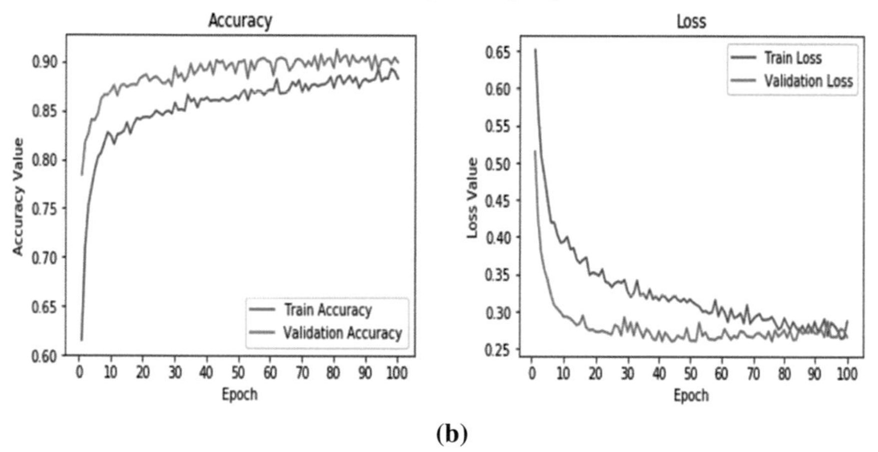

(b)

Fig. 5.13 **a** Epoch accuracy graph; **b** graph using tensor-board for creation of data model, on multi looped accuracy epoch data augmentation

farmer, although we have seen enormous developments in technology which helps in the recognition of grains. Paddy grain detection are not extensively used in industry as of now, but with the pace of evolution, the machine learning models are trained with numerous datasets to improve its accuracy, the application or usage of the proposed recognition system will definitely be going to increase in the future which will result in faster processing in the relevant industries where it can be used.

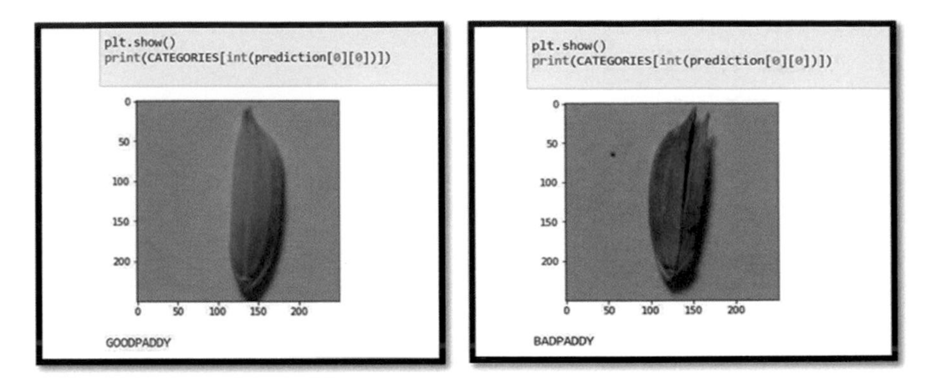

Fig. 5.14 Prediction by model

References

1. State of Rajasthan agriculture. December 2012
2. Sharma, R.: Artificial intelligence in agriculture: a review. In: Proceedings of International Conference on Intelligent Computing and Control Systems (ICICCS 2021), pp. 937–942 (2021). https://doi.org/10.1109/ICICCS51141.2021.9432187
3. Zhang, W., Liu, H., Wu, W., Zhan, L., Wei, J.: Mapping rice paddy based on machine learning with Sentinel-2 multi-temporal data: model comparison and transferability. Remote Sens. **12**, 1–21 (2020). https://doi.org/10.3390/rs12101620
4. Urva, P.: Detection and classification of grain crops and legumes disease: a survey. Sparkling-light Trans. Artif. Intell. Quantum Comput. (STAIQC) **1**, 41–55 (2021)
5. Yamaguchi, T., Tanaka, Y., Imachi, Y., Yamashita, M., Katsura, K.: Feasibility of combining deep learning and RGB images obtained by unmanned aerial vehicle for leaf area index estimation in rice (2021)
6. Aznan, A., Viejo, C.G., Pang, A., Fuentes, S.: Computer vision and machine learning analysis of commercial rice grains: a potential digital approach for consumer perception studies (2021)
7. Li, D., Wang, R., Xie, C., Liu, L., Zhang, J., Li, R.: A recognition method for rice plant diseases and pests video detection based on deep convolutional neural network
8. Hamzah, A.S., Mohamed, A.: Classification of white rice grain quality using ANN : a review. IJAI **9**(4), 600–608 (2020). https://doi.org/10.11591/ijai.v9.i4.pp600-608
9. Zhao, X., Yuan, Y., Song, M., Ding, Y., Lin, F., Liang, D.: Use of unmanned aerial vehicle imagery and deep learning UNet to extract rice lodging. Sensors **19**(18), 3859 (2019)
10. Jalaleddin, S., Rad, M.: Classification of rice varieties using optimal color and texture features and BP neural networks classification of rice varieties using optimal color and texture features and BP neural networks. November 2011. https://doi.org/10.13140/2.1.1571.4568
11. Zhou, C., Ye, H., Hu, J., Shi, X.: Automated counting of rice panicle by applying deep learning model to images from unmanned aerial vehicle platform (2019)
12. Sonawane, S., Awasthy, M., Choubey, N.: A literature review on image processing and classification techniques for agriculture produce and modeling of quality assessment system for soybean industry sample. Int. J. Innov. Res. Electron. Commun. **6**(2), 8–16 (2019)
13. Zhang, M., Lin, H., Wang, G., Sun, H.: Mapping paddy rice using a convolutional neural network (CNN) with Landsat 8 datasets in the Dongting Lake area, China. https://doi.org/10.3390/rs10111840
14. Gudipalli, A., Prabha, A.: A review on analysis and grading of rice using image processing. J. Eng. Appl. Sci. **11**(23), 13550–13555 (2016)

15. Qing, Y.A.O., Qing-jie, L.I.U., Guang-qiang, D., Bao-jun, Y., Hong-ming, C.: An insect imaging system to automate rice light-trap pest identification. J. Integr. Agric. **11**(6), 978–985 (2012). https://doi.org/10.1016/S2095-3119(12)60089-6
16. Narendra, V.G., Hareesha, K.S.: Prospects of computer vision automated grading and sorting systems in agricultural and food products for quality evaluation. Int. J. Comput. Appl. **1**(4), 1–9 (2010)
17. Kong, H., Chen, P.: Mask R-CNN-based feature extraction and three-dimensional recognition of rice panicle CT images. Plant Direct **5**(5), 1–8 (2021). https://doi.org/10.1002/pld3.323
18. Ajaz, R.H., Hussain, L.: Seed classification using machine learning techniques. J. Multidiscip. Eng. Sci. Technol. **2**(5), 1098–1102 (2015)
19. Prakhar, K., Kumar, A., Dwivedi, S.: Automatic wheat grain quality estimation. EE604 Project Report, IIT Kanpur
20. Teoh, C.C., Bakar, B.H.A.: Immature paddy quantity determination using image processing and analysis techniques. J. Trop. Agric. Food Sci. **37**(2), 241–248 (2009)
21. Kaur, H., Singh, B.: Classification and grading rice using multi-class SVM. Int. J. Sci. Res. Publ. **3**(4), 1–5 (2013)
22. Silva, J., et al.: Method for the recovery of images in databases of rice grains from visual content. Procedia Comput. Sci. **170**, 983–988 (2020). https://doi.org/10.1016/j.procs.2020.03.097
23. Tab, F.A.: Design of an expert system for rice kernel identification using optimal morphological features and back propagation neural network. May 2014 (2012)
24. Inácio, D., Rieder, R.: Computer vision and artificial intelligence in precision agriculture for grain crops: a systematic review. Comput. Electron. Agric. **153**, 69–81 (2018). https://doi.org/10.1016/j.compag.2018.08.001
25. Golpour, I., Parian, J.A., Chayjan, R.A.: Identification and classification of bulk paddy, brown, and white rice cultivars with colour features extraction using image analysis and neural network. Czech J. Food Sci. **32**(3), 280–287 (2014)
26. Nagoda, N., Ranathunga, L.: Rice sample segmentation and classification using image processing and support vector machine. In: 2018 IEEE 13th International Conference on Industrial and Information Systems, no. 978, pp. 179–184 (2018). 10.1109/ICIINFS.2018.8721312
27. Wang, R., Han, F., Wu, W.: Estimation of paddy rice maturity using digital imaging. Int. J. Food Prop. **24**(1), 1403–1415 (2021). https://doi.org/10.1080/10942912.2021.1970581
28. Sinha, A.: Dimension analysis and gradation of rice grain using image processing technique. Int. J. Mod. Trends Sci. Technol. **6**(10), 69–73 (2020)
29. Ali, T., Jhandhir, Z., Ahmad, A., Khan, M., Ali Khan, A., Choi, G.S.: Detecting fraudulent labeling of rice samples using computer vision and fuzzy knowledge. February 2017 (2020). https://doi.org/10.1007/s11042-017-4472-9
30. Abbaspour-Gilandeh, Y., Molaee, A., Sabzi, S., Nabipur, N.: A combined method of image processing and artificial neural network for the identification of 13 Iranian Rice Cultivars (2020)
31. Philip, T.M., Anita, H.B.: Rice grain classification using Fourier transform and morphological features. Indian J. Sci. Technol. **10**, 1–6 (2017). https://doi.org/10.17485/ijst/2017/v10i14/110468
32. Song, X., Huang, W., Shi, M., Zhu, M., Lin, H.: A QTL for rice grain width and weight encodes a previously unknown RING-type E3 ubiquitin ligase. Nat. Genet. **39**(5), 623–631 (2007). https://doi.org/10.1038/ng2014
33. Kiratiratanapruk, K., Temniranrat, P., Sinthupinyo, W., Prempree, P., Chaitavon, K., Porntheeraphat, S.: Development of paddy rice seed classification process using machine learning techniques for automatic grading machine. J. Sens. **2020**, (2020)
34. Ege, T.: A new large-scale food image segmentation dataset and its application to food calorie estimation based on grains of rice. In: MADiMa '19: Proceedings of the 5th International Workshop on Multimedia Assisted Dietary Management, pp. 82–87 (2019)

Chapter 6
Role of Brand Love in Green Purchase Intention: Analytical Study from User's Perspective

B. Charan and S. Vasantha

Abstract In this research, Relationship of Brand Love in Green Purchase Intention is studied. Brand Love plays significant role between Brand Experience and Green Purchase Intention as Brand Experience paves way for Brand Love of Electric Vehicles. Even purchase intention comes because of Brand Experience and Brand Love. In this research, the investigator took 152 respondents as sample size in Bangalore City. Those 152 respondents are the users of electric two-wheeler scooters. The details of consumers were collected from different outlets selling electric vehicles. Brand Love is Mediating Variable while Brand Experience is Independent Variable and Green Purchase Intention is Dependent Variable.

Keywords Green vehicles · Fuel vehicles · Sustainable development · Brand love · Brand experience and green purchase intention

6.1 Introduction

Commercial activities are out of control in entire world as it generates not only healthy economy but also fetch huge revenue to the participating countries. As these activities are rapidly increasing, production of goods and supply of such goods form major chunk of commercial activities to be transported across each country. These commercial activities consume major natural resources and cause damage to environment. Certain cognizable consequence of environment spoilage comprises in reduction of natural resources. Many nations across the world have started figuring out this natural environmental threat and taken the responsibility towards decreasing the carbon foot prints towards protecting the environment. The delayed awareness & apprehension for atmosphere in particular towards society in general have paved way for appearance of 'sustainable development' which focused on enhancing sustainability and augment healthy life of people living all over the globe.

B. Charan · S. Vasantha (✉)
School of Management Studies, Vels Institute of Science, Technology and Advanced Studies (VISTAS), Chennai, India
e-mail: vasantha.sms@velsuniv.ac.in

A. Biswas et al. (eds.), *Artificial Intelligence for Societal Issues*, Intelligent Systems Reference Library 231, https://doi.org/10.1007/978-3-031-12419-8_6

6.1.1 Green Purchase Intention

As we know, Green Purchase Intention is a move to reduce environmental damage like greenhouse gases and air pollution. At present people have the mindset of buying electric two wheelers as it reduces carbon dioxide and nitrogen. They are the causing agents of air pollution. Consumers have realized the significance of procurement of electric two wheelers as it has less impact on environmental damage and at the same time it conserves the natural habitat. In this study, Brand Experience is regarded as Independent Variable while Brand Love is treated as Mediating Variable and Green Purchase Intention is known as Dependent Variable. Our study portrays that how Brand Experience influences the Green Purchase Intention through the mediating variable of brand love.

6.1.2 Brand Love

Brand love is a systematic approach which is highly significant in the brand management concept to build relationships with a particular brand which leads to be loyal toward the brand and become a influencer to such brand. Brand Love drives the customers to choose the product of their preference. In today's competitive world, loving a product by someone is to go through incredible factors like a product must have impressive quality, affordable price, and competitive advantage and so on. When it comes to electric two wheelers, customers have passion over this because of it replaces the fuel vehicles thereby preventing customers from spending a lot towards fuel.

6.1.2.1 Objectives of the Study

- To study the influence of Brand Experience on Green Purchase Intention.
- To analyze the relationship of Brand Love among Brand Experience and Green Purchase Intention.

6.1.2.2 Statement of the Problem

As far as this study is concerned, Green Purchase Intention has taken off in the right direction in other developed countries. In India, the patronage of Green Vehicles is bleak as it does not have more features when compared to conventional vehicles. Even now, people do not want to switch over to fuel vehicles as it fulfills the requirements of users. In case of electric two wheelers, pulling power is not up to the satisfaction of buyers. Similarly, inadequate electric vehicle charging stations prevent the consumers from purchasing electric vehicles. Though, electric vehicles mushroom everywhere across country. The rate of response among the buyers is very less as compared to internal-combustion (IC) engine vehicles. However, usage of Fuel Vehicles also causes serious environmental concerns over time thereby future generation suffer from many health issues like non-communicable diseases like cancer, heart ailments, skin ailments and lungs problems also continue to be unabated.

6.1.3 Significance and Scope of Study

In this study, usage of Green Vehicles is gaining momentum as it causes enormous benefits not only to the users but also to the society at large. People slowly give up using conventional vehicles as it produces serious health issues and severely damage the environment. Due to which, there is shortfall of rainfall or lack of rainfall which paralyses the farming activities. People have brand experience of green vehicles and pass it on to their near and dear thereby patronage for electric vehicles gradually scale up. People of today are more care for destruction of environment. That is why; they have started switching their purchase patter over to green vehicles. By 2030, nearly 50% of population will use the green vehicles and demand for green vehicles will go beyond its supply. In this study, researcher has enunciated the relationship of Brand Experience with Green Purchase Intention by taking into account the dominating role of Brand Love. Green Vehicle buying consumers have passion over it due to reasons known to them. This study has described three variables as they are more relevant and purchase intentions of consumers to Green Vehicles are influenced by Brand Love and Brand Experience.

6.2 Review of Literature

Kanchanapibul et al. [1] major finding of the study says people are more responsive to the green products due to having known the repercussions of environmental dilapidation. They have eliminated the preference of smoke emitting vehicles and have the only option of green vehicles thereby protecting the environment from getting destroyed. Consumers taken for survey in this study are very particular about environmental damage and their consequences for future generation.

Milonia (2017) focused on environmental concerns due to the diminution of natural resources, regular usage of smoke emitting vehicles, global warming, diminution of ozone layer etc. Present generations have become aware of adapting to green vehicles rather than using the conventional one. Totally 255 respondents were interviewed by the researcher in order to know how brand experience affects the green purchase intention through the mediating variable of Brand Love. The study was conducted in Tripura, and researcher finally concluded that brand love influenced the customers in buying electric vehicles.

Chin et al. [2] stated the significance of green products and the rising purchases of the eco-friendly products by considering brand love as mediating variable. The research was carried out in Johore, Malaysia from 162 respondents using online survey questionnaire and snowball technique. From the investigations it was found that green brand knowledge and green brand positioning has positive impact on green purchase intention. The study predicted on the knowledge of the consumers about the green purchase intention and sated that companies focused on protecting the environment by providing eco-friendly products are having higher demand in the market as the usage outcomes of the products and services have more advantageous over the traditional products and services.

Keni et al. [3] demonstrated that there were three variables which were tested whether those three variables had influenced the green purchase intention. The three variables namely Green Perceived Value, Green Perceived Risk and Green Trust that have impacted the green purchase intention, like normal purchase of vehicles. Consumers are influenced by certain factors.

6.3 Research Methodology

In this study, responses were evoked from Target Audiences. Totally 152 respondents have been taken for this research study. Of total respondents for this study, few have been interviewed directly while others have been interviewed via online. Their details were obtained from the customers who use electric scooters for their travel commute. Well-Structured Questionnaire were developed and screened by the experts concerned. Some glitches were found and simultaneous removed. After that, they have been sent to the mail id of respective respondents while some responses were collected physically by meeting the respondents at several places. Descriptive Research Design has been used all through this study while deliberate sampling techniques were adopted as it is more suitable than other types of sampling. Even Secondary sources of information have been gathered wherever required. The sample unit is Bangalore city where electric scooter showrooms are available.

6.3.1 Research Model

See Fig. 6.1.

6.3.2 Description of Variables

- Green Purchase Intention is treated as Dependent Variable
- Brand Experience is regarded as Independent Variable
- Brand Love is Mediating Variable.

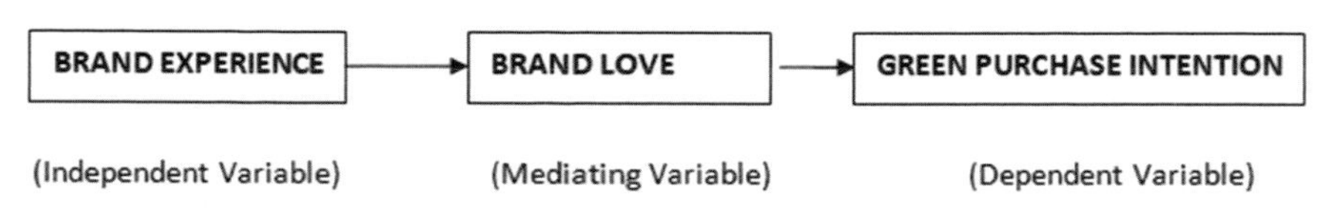

Fig. 6.1 Proposed research model

As far as this research model is concerned, Brand Experience, Brand Love and Green Purchase Intention are taken as 3 variables. As could be seen, Brand Experience is regarded as Independent Variable. The independent variable influences the dependent variable through mediating variable. The Mediating Variable is Brand Love while Green Purchase Intention is Dependent Variable. In this model, it is concluded that without brand love, green purchase intention is of least importance. Therefore, Brand Love is very important determinant while deciding about Purchase of Electric Vehicles.

6.3.3 Research Questions

The following research questions are designed in accordance with the framework of this research.

- Whether does Brand Love as Mediating Variable create relationship statistically between Brand Experience and Green Purchase Intention?
- Does Brand Experience drive the consumers to buy electric vehicles over the normal vehicles?

6.3.4 Hypothesis

- Brand Experience does not encompass difference over Green Purchase Intention through Brand Love.

6.4 Results and Discussion

6.4.1 Structural Equation Model

Structural Equation Model is one of the best tools that are applied to study co linearity of latent and manifest variables. The main purpose of using this technique is to measure and reduce the error present in the variables. In this study, researcher has showed the relationship of brand experience along with green purchase intention by means of considering Brand Love as mediating variable. The estimates of SEM also depict how independent variable (Brand Experience) is closely correlated with dependent variable (Green Purchase Intention) by imitating mediating variable (Brand Love). Both Standardized and un-standardized estimates have been obtained and presented along with this research model (Fig. 6.2 and Table 6.1).

Based on the test results, it is inferred that all the values are at permissible limit. The overall model for Role of Brand Love in Green Purchase Intention is ideally good fit. The test results have been compared with standard values thereby interpretation was put. Since p value is larger than 0.05, formulated null hypothesis i.e. Brand Experience does not have impact on Green Purchase Intention through Brand Love and it is accepted at 5% level of significance (Table 6.2).

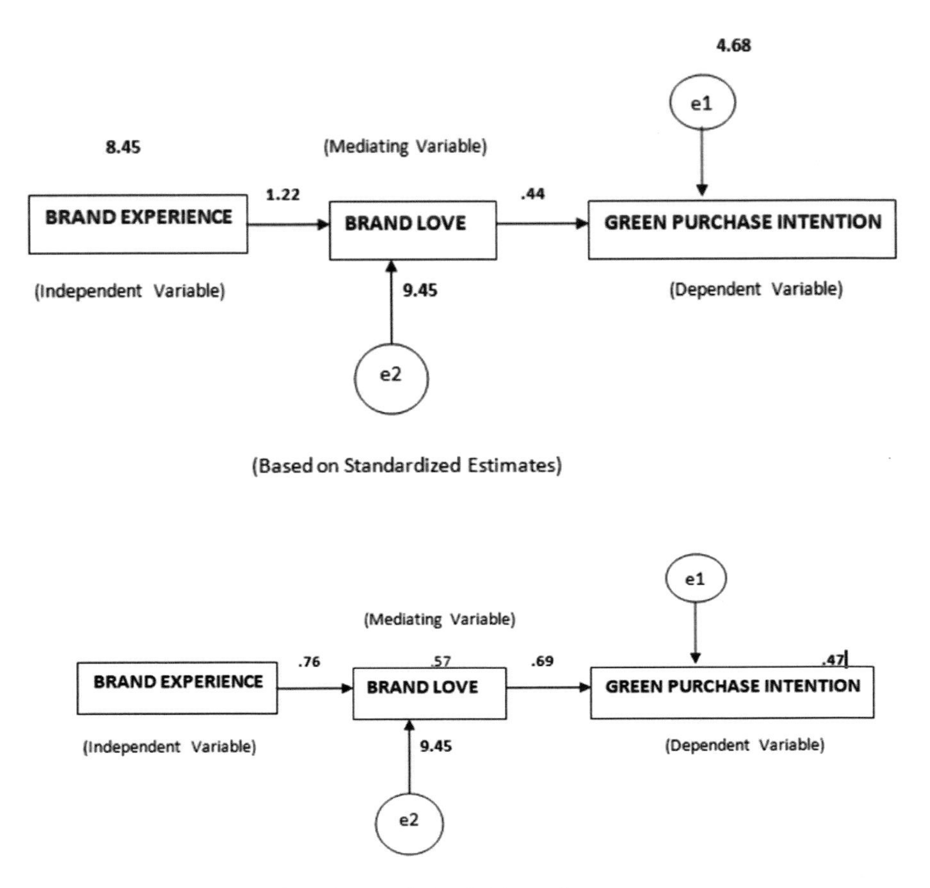

Fig. 6.2 SEM model

Table 6.1 Model fit summary for role of brand love in green purchase intention

Label	Chi-square	RMSEA	GFI	AGFI	CFI	TLI	NFI	IFI
Cut-off value	≤5.00	≤0.080	≥0.95	≥0.90	≥0.90	≥0.90	≥0.950	≥0.950
Test result	2.209	0.089	0.990	0.942	0.995	0.984	0.990	0.995

Table 6.2 Regression coefficients for the mediating effect of brand love among brand experience and green purchase intention

Estimate	S.E	C.R	P			
Brand love	<–	Brand Experience	1.220	0.086	14.177	***
Green purchase intention	<–	Brand Love	0.436	0.038	11.629	***

It is further stated that PLS was applied to keep a check on Mediating Effect among Brand Experience and Green Purchase Intention. As per the estimates of Regression, the p value is less than 0.0001 which significant and evidence of alternate hypothesis is strong and accepted i.e., there is mediating effect of Brand Love among Brand Experience and Green Purchase Intention.

6.4.2 Multi-group Analysis

Multi Group Analysis is used to examine the data group that has major difference in their group particular parameter estimates. Three groups are predefined namely Brand Experience, Brand Love and Green Purchase Intention. In Brand Experience, ten variables have been identified, Brand Love, thirteen variables have been identified namely and at last Green Purchase Intention that has nine variables. Brand Experience is of Independent Variable while Brand Love is of Mediating Variable and eventually Green Purchase Intention is of dependent variable (Fig. 6.3 and Table 6.3).

Based on the estimates of regression, it is believed that Latent Variable of Brand Experience are not Significantly different with specific observed variables such as BE2, BE3, BE4, BE6 and BE8. The null hypotheses i.e., Brand Experience is not significantly different with the above five variables namely BE2, BE3, BE4, BE6, and BE8 is accepted at 0.0001 level of significance. However, B5, B7 and B9 are significantly different with Brand Experience and null hypotheses i.e., Brand Experience and B5, B7and B9 are significantly different at 0.0001 level of significance.

However, in Green Purchase Intention, only two observed variables such as GPI5 and GPI9 are significantly different with Green Purchase Intention at 0.0001 significance level. Therefore, null hypothesis i.e., GPI5 and GPI9 are not significantly different with Green Purchase intention is rejected. Similarly, other observed variables called GPI2, GPI3, GPI4, GPI6, GPI7 and GPI 8 are not significantly different with Green Purchase Intention. Therefore, null hypotheses i.e., GPI2, GPI3, GPI4,

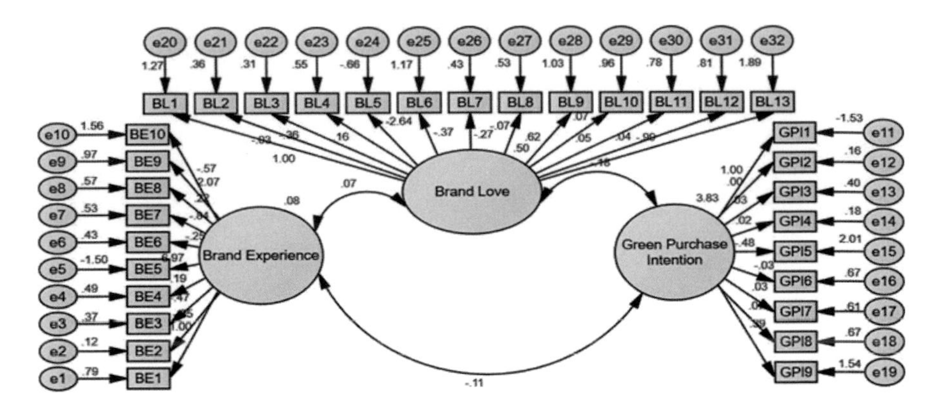

Fig. 6.3 Multi group smart PLS analysis of brand experience, brand love and green purchase intention

GPI6, GPI7 and GPI 8 and Green Purchase Intention are not significantly different is accepted at 0.0001 level.

Finally, in Brand Love, among thirteen variables only five observed variables such as BL5, BL7, BL9, BL3 and BL13 are significantly different with Brand Love (Latent Variable) at 0.0001 significance level. But, other eight observed variables such as BL2, BL3, BL4, BL6, BL8, BL10, BL11 and BL 12 are not significantly different with Brand Love (Unobserved Variable) at 0.0001 level of significance. Therefore, null hypotheses i.e. No significant relationship between BL2, BL3, BL4, BL6, BL8, BL10, BL11 and BL 12 (Observed Variables) and Brand Love (Unobserved Variable) is accepted.

6.4.3 C. Variances

The above table explains the residuals (error terms) that are present in the data. It means certain margin of error is prevalent in the model undertaken by the researcher. Residuals cannot be completely omitted due to conflict among the responses of respondents. They should be to certain extent and residual to observe variables are imminent but, these can be good to draw upon. In the above table, most of the residuals to observed variables are significantly different i.e., error is present to some extent in observed variables, but they prove healthy relationship even if they are present in data (Table 6.4).

Table 6.3 Multi-group analysis for brand experience, brand love and green purchase intention

			Estimate	S.E	C.R	P
BE2	<–	BEE	−0.053	0.064	−0.818	0.413
BE3	<–	BEE	−0.471	0.143	−3.286	0.001
BE4	<–	BEE	0.192	0.135	1.423	0.155
BE5	<–	BEE	−6.973	1.587	−4.394	***
BE6	<–	BEE	−0.250	0.130	−1.924	0.054
BE7	<–	BEE	−0.841	0.207	−4.071	***
BE8	<–	BEE	0.220	0.146	1.512	0.131
BE9	<–	BEE	2.069	0.424	4.877	***
GPI2	<–	GPIN	−0.003	0.010	−0.330	0.741
GPI3	<–	GPIN	0.030	0.017	1.797	0.072
GPI4	<–	GPIN	0.018	0.011	1.596	0.111
GPI5	<–	GPIN	−0.479	0.079	−6.054	***
GPI6	<–	GPIN	−0.025	0.021	−1.203	0.229
GPI7	<–	GPIN	0.031	0.020	1.512	0.131
GPI8	<–	GPIN	0.066	0.024	2.751	0.006
BL5	<–	BLE	−2.643	0.359	−7.367	***
BL6	<–	BLE	−0.369	0.115	−3.194	0.001
BL7	<–	BLE	−0.266	0.072	−3.717	***
BL8	<–	BLE	−0.066	0.072	−0.909	0.363
BL9	<–	BLE	0.623	0.124	5.011	***
BE10	<–	BEE	−0.565	0.255	−2.220	0.026
BL4	<–	BLE	0.164	0.076	2.164	0.030
BL2	<–	BLE	−0.028	0.059	−0.473	0.637
BL3	<–	BLE	−0.358	0.069	−5.163	***
BL10	<–	BLE	0.073	0.097	0.750	0.453
BL11	<–	BLE	0.052	0.088	0.595	0.552
BL12	<–	BLE	0.041	0.089	0.466	0.642
BL13	<–	BLE	−0.988	0.179	−5.510	***
GPI9	<–	GPIN	0.393	0.067	5.832	***

***Significant at 0.0001 level

6.5 Findings

The overall model for Role of Brand Love in Green Purchase Intention is ideally good fit. The test results have been compared with standard values thereby interpretation was put. Since p value is larger than 0.05, formulated null hypothesis i.e., Brand Experience don't have major difference on Green Purchase Intention through Brand Love is accepted at 5% level of significance.

Table 6.4 Estimates of variances

	Estimate	S.E	C.R	P		Estimate	S.E	C.R	P
BEE	0.084	0.035	2.375	0.018	e16	0.666	0.076	8.709	***
GPIN	3.833	0.461	8.317	***	e17	0.615	0.070	8.721	***
BLE	0.498	0.137	3.635	***	e18	0.671	0.076	8.814	***
e1	0.790	0.088	9.023	***	e19	1.540	0.181	8.533	***
e2	0.118	0.014	8.699	***	e20	1.275	0.142	8.973	***
e3	0.373	0.042	8.891	***	e21	0.357	0.041	8.692	***
e4	0.491	0.056	8.719	***	e22	0.309	0.034	8.977	***
e5	−1.501	0.428	−3.507	***	e23	0.554	0.063	8.743	***
e6	0.427	0.049	8.747	***	e24	−0.662	0.220	−3.013	0.003
e7	0.530	0.059	9.031	***	e25	1.172	0.133	8.808	***
e8	0.567	0.065	8.723	***	e26	0.425	0.048	8.850	***
e9	0.966	0.109	8.831	***	e27	0.532	0.061	8.698	***
e10	1.561	0.178	8.768	***	e28	1.026	0.114	8.965	***
e11	−1.531	0.452	−3.387	***	e29	0.965	0.111	8.695	***
e12	0.158	0.018	8.691	***	e30	0.785	0.090	8.693	***
e13	0.403	0.046	8.736	***	e31	0.807	0.093	8.692	***
e14	0.182	0.021	8.725	***	e32	1.894	0.211	8.995	***
e15	2.008	0.240	8.376	***					

***Significant at 0.0001 level

Based on the estimates of regression, it is believed that Latent Variable of Brand Experience are not Significantly different with specific observed variables such as BE2, BE3, BE4, BE6 and BE8. Therefore, the null hypotheses i.e., "Brand Experience is not significantly different with the above five variables namely BE2, BE3, BE4, BE6, and BE8" is accepted at 0.0001 level of significance. However, B5, B7 and B9 are significantly different with Brand Experience and null hypotheses i.e., Brand Experience and B5, B7and B9 are significantly different at 0.0001 level of significance.

In Green Purchase Intention, only two observed variables such as GPI5 and GPI9 are significantly different with Green Purchase Intention at 0.0001 significance level. Therefore, null hypothesis i.e., GPI5 and GPI9 are not significantly different with Green Purchase intention is rejected. Similarly, other observed variables called GPI2, GPI3, GPI4, GPI6, GPI7 and GPI 8 are not significantly different with Green Purchase Intention. Therefore, null hypotheses i.e., GPI2, GPI3, GPI4, GPI6, GPI7 and GPI 8 and Green Purchase Intention are not significantly different is accepted at 0.0001 level.

In Brand Love, among thirteen variables only five observed variables such as BL5, BL7, BL9, BL3 and BL13 are significantly different with Brand Love (Latent Variable) at (0.0001) significance level. But, other eight observed variables such as

BL2, BL3, BL4, BL6, BL8, BL10, BL11 and BL 12 are not significantly different with Brand Love (Unobserved Variable) at 0.0001 level of significance. Therefore, null hypotheses i.e. No significant relationship between BL2, BL3, BL4, BL6, BL8, BL10, BL11 and BL 12 (Observed Variables) and Brand Love (Unobserved Variable) is accepted.

In variance analysis, residuals (error terms) that are present in the data. It means certain margin of error is prevalent in the model undertaken by the researcher. Residuals cannot be completely omitted due to conflict among the responses of respondents. They should be to certain extent and Residual to observe variables are imminent but, these can be good to draw upon. In the above table, most of the Residuals to observed variables are significantly different i.e., error is present to some extent in observed variables, but they prove healthy relationship even if they are present in data.

6.6 Suggestions

Brand Experience drives the Purchase Intention of Electric Scooters but country like India where lack of awareness of Electric Vehicles results in very less patronage among consumers even it looms large in India. People those who use electric vehicles should inculcate the essence of using electric vehicles to protect the environment in near future and leave some good to our future generation.

Those who constantly use the green vehicles should make word of mouth publicity so that people those who have the purchase intention will execute the purchase at once. Similarly, Brand Experience will promote the sales of electric vehicles. The reason is that durability of product depends on experience of Brand. Therefore, consumers should inform their friends, colleagues, family members to go for electric vehicles instead of convention vehicles and avoid environmental damage in the years to come. Brand Love is a mediating Variable that plays important role between Brand Experience and Green Purchase Intention because root cause of purchase of green vehicles is Brand Love and Brand Experience.

6.7 Conclusion

Purchase of Electric Vehicles is going to be out-numbered the conventional vehicles due to Brand Experience and Brand Love. Brand Experience through Brand Love drive the Green Purchase Intention. Even results of statistical analysis proved there is positive relationship among Brand Experience and Green Purchase Intention by considering the Brand Love. Therefore, Manufacturers of Green Vehicles must conduct the survey and redress the grievances of the consumers. They should increase the engine capacity from what is being now to the expectation of consumers to increase the high volume of sales in Future without compromising on conservation of Environment.

6.8 Questionnaire

Brand Experience

BE 1 I think this brand makes a strong impression on my visual due to environmental concerns.
BE 2 I feel the brand induces feelings and sentiments for the environment.
BE 3 I do have strong emotions for this brand,
BE 4 I engage in physical actions and behavior's when I use this brand.
BE 5 I engage in a lot of thinking when I encounter this brand.
BE 6 This brand stimulates my curiosity and problem-solving.
BE 7 This brand appeals to my senses.
BE 8 I have strong emotions for this brand.
BE 9 I depend on this brand and I find this brand interesting in a sensory way.

Brand Love

BL 1 I feel myself using this favorite brand.
BL 2 I feel this brand is a favorite for my mode of transportation.
BL 3 I feel that this brand is exciting.
BL 4 I feel this brand full fill my expectations and part of my life for a long time.
BL 5 I am passionate about using this brand.
BL 6 I am very attached to this brand.
BL 7 I feel emotionally connected with this brand.
BL 8 I feel I have a "bond" with this brand.
BL 9 I feel there is a natural "fit" between me and this brand.
BL 10 This brand seems to fit in my daily transportation mode.
BL 11 I make my life more meaningful when I use this brand.
BL 12 I am very attached to this Brand.
BL 13 I love this Brand.

Green Purchase Intention

GPI 1 I intend to buy products from green brands because it reduces pollution.
GPI 2 I intend to switch to eco-friendly brands for ecological reasons.
GPI 3 I prefer to purchase green brands which can reduce environmental pollution.
GPI 4 I prefer green brands, even if the price of the product is more expensive.
GPI 5 I intend to purchase green brands because they are zero-emission.
GPI 6 I consider green brands as a high priority in my purchase decision.
GPI 7 Government initiatives motivated me to purchase green brands.

GPI 8 I have time, available resources, multiple payment options & willingness to buy green brand products.

GPI 9 There are plenty of opportunities for me to purchase green brands

References

1. Kanchanapibul, M., Lacka, E., Wang, X., Chan, H.K.: An empirical investigation of green purchase behaviour among the young generation. J. Clean. Prod. **66**, 528–536 (2014)
2. Chin, T.A., Sulaiman, Z., Mas'od, A., Muharam, F. M., & Tat, H. H.: Effect of green brand positioning, knowledge, and attitude of customers on green purchase intention. J. Arts Soc. Sci. **3**(1), 23–33 (2019)
3. Keni, K., Asali, A., Ping, T., Muthuveloo, R.: Factors Influencing Green Purchase Intention (2020). https://doi.org/10.2991/assehr.k.201209.161

Chapter 7
Effect of Online Review Rating on Purchase Intention

A. Navitha Sulthana and S. Vasantha

Abstract The recent adoption of Web 2.0 technology has bought an enormous change in customer buying behavior. The study proposed to find the effect of online review rating on purchase intention. The relationship between online review rating and purchase intention was investigated using primary data. The primary data was gathered through an online survey among one hundred and ninety two online buyers as respondents. The study is based on descriptive analysis. The independent variable is the customer review rating and the dependent variable is Purchase Intention. The result of the study shows that review and rating has a significant effect on purchase intention. Rating or star numeric from 5 stars to 1 star are given to any product and service through the recommendation system which has an impact on purchase decision. The implication of the study provides detailed insight for the researchers, online marketers, web retailers and online buyers.

Keywords Customer reviews · Online buyer · Purchase intention · Rating and social media

7.1 Introduction

From past few years internet has been a necessity for day to day life. Social media and social media platform usage has provided a large scope for all types of business. No smaller scale business depends on big commercial advertisement. A small content on a blog, a picture on Facebook says more information about anything. Product and service achievement is reached based on customer review rating. Rating or the five stars given for any product or service determine the overview about it. The higher number of star or stars a product or service gets creates large goodwill for it. The Rating has a big impression among online user who is in search of any information about the product or service. User depends on digital rating and review before making buying decision. Rating creates accuracy and reliability about the

A. Navitha Sulthana · S. Vasantha (✉)
School of Management Studies, Vels Institute of Science, Technology and Advanced Studies, Pallavaram, Chennai, India
e-mail: vasantha.sms@velsuniv.ac.in

A. Biswas et al. (eds.), *Artificial Intelligence for Societal Issues*, Intelligent Systems Reference Library 231, https://doi.org/10.1007/978-3-031-12419-8_7

product and service. Based on the product purchased customer discuss or share the opinion about the product. The customer gives a rating from five stars to one star. The rating helps the other online user to review information about the product and service before they intent to buy anything. According to Prabha and Vasantha [1] The interaction with social network communities has an effect on consumer purchase intention.

The wide usage of social media via internet has initiated two way communications between buyer and seller. The growth of seller product has reached large customers doorstep through online shopping. The user generated content, electronic word of mouth has a vital role in promoting and creating brand awareness about product and services. People completely rely on review rating before making buying decision. A good product promotes by itself. Online products reach a large number of people all over the world through social network platforms. People are able to distinguish various products among many available alternatives and select a particular product according to their need. Rating in social media act as the one the source for online buyer before making a purchase decision. The star given by a customer after purchasing the product or obtaining the service helps the other people before making decision.

7.1.1 Role of Review Rating in Social Media

The Rating has a significant role in social media. From the perception of online user first the user searches the various alternative products available on social network. At the time of deciding a particular product from various alternatives, he or she look into positive and negative reviews or content shared by consumers who bought the product or obtained service. The next phase the customer looks into a rating that how much rating that a specific product has reached. The star rating can be termed as an overall judgment of the product purchased through online or service obtained. The major difference between rating and review are, rating are the numbers given to any product purchased online. It can be 5 stars or 10 stars rating. It is the feedback number given by customers. Review means the short or long summary given by consumers who have purchased the product or obtained any service. It can only be text content or it can be both text content with image and the number of stars. The review is one form of recommending the product or service to benefit others. Rating and review can be best derived from the movie, product, and any service by taxi, any customer care execution phone conversation or online chats clarifying any query [2]. The researcher combined three factors such as attributes types, user preference and Item relation as attribute boosting for review and rating. Rating prediction is determined by user preference [3]. There were speedy growth of different online platforms. Social media has become emerging trend to search and find needed information. As a result the consumer prefers these online platforms for making their buying decisions. Chen [4] found that individuals purchase decision is influenced by five star rating. The cognitive fit based on 191 subjects obtained help in buying decision of individuals.

7.1.2 *Effect of Review Rating on Purchase Intention*

There is the rapid shift from traditional buying to online buying. The online buyers use the maximum sources available on the internet on various social media platforms, electronic commerce websites, service websites more before making the intend to buy anything online or to get service. The numeric rating and review are the one of the factors they followed before buying anything through internet. Ling et al. [5] discussed the brand orientation, trust on online, purchase experience shared by prior customer, quality orientation and impulse buy intention are positively connected with purchase intention. According to the Thomas et al. [6] the credibility of online reviews influences to the positive intention of purchasing through online. The online review rating approach are widely available in online websites, social media platforms, products community pages. The reviews are over taken the offline buying habit to online purchase and creating a purchase intention. The numeric star ratings display the overall score about the products and services. customer reviews and ratings assist the customers to locate the products easily and get more information through post purchaser than the sellers.

7.1.3 *Objective of the Study*

The purpose of the study is to analysis the importance of review and rating posted by online customer. The online users purchase various need products through online portal and leaves their reviews and rating as bookmark for new and potential purchasers. Therefore, the main objective of the study is to find the effect of online review rating on purchase intention. To investigate the relationship between online review rating and purchase intent, primary data were gathered from online buyers. The primary data was gathered using an online survey method. The survey research gathered information from 192 online buyers who participated in the survey. The research is descriptive in nature. The Hypothesis of the study is (H_1) Customer review rating has positive influences on Purchase Intention.

7.2 Literature Review

7.2.1 *Review Rating on Purchase Intention*

Wang et al. [7] They studied how the product evaluation review is posted by users. Compared to rating reviews are most considered. They purpose the review rating forecast method rely on user context and product context. They integrate user and review information into review texts. They took their study in three models such as review rating prediction on worldwide, review rating prediction on product specific and review rating prediction of user specific. The study aimed at problem of sentiment word dependency by combining Review rating prediction (RRP) plus, User context (UC) plus and product context (PC). The triple word has a more important role in review rating prediction such as negative word, modifier and sentiment word. Web 2.0 and E-commerce increasing the number of online user reviews. Opinion seeking, quality monitoring becomes important before making purchase decision. Wang et al. [8] explained that the internet rich web 2.0 created rapid growth in review rating. The authors suggested personalized sentiment expression compared with an existing unified model of review rating forecast. They predict the user personalized information by interpreting user item rating matrix and review text. Yelp and Douban dataset are taken to validate the prediction method. Yelp consisting of restaurant reviews and Douban is a Chinese website consisting of movie review. The review rating explosion of information helps others in making purchase decision. The online review rating has wide content about opinion mining, quality tracking and personalized utilization decisions which influence in a purchase decision.

Ling et al. [9] The authors suggested unified model to content based filtering and collaborative filtering. By utilizing rating and review information they reach accuracy of the recommendation system. Suggestion system has become mandatory in every aspect to finalize anything regarding purchasing a product or obtaining a service. The humans suggested system makes a way for the other people to buy or select anything online. Wang et al. [10] The growth of web 2.0 technology allows users to join in various community group and share their opinion freely related to a product or service. The shared reviews are useful for the others in their decision making. Latent Aspect Rating Analysis LARA was proposed by authors to analyze the deep understanding about the review. They gather both aspect weights and aspect rating of individual reviews to find the user rating behavior. Online users join various communities in their personal interest to share their purchase experience and access knowledge in order to support purchase decision and purchase intention of upcoming buyer or purchaser.

McAuley and Leskovec [11], the authors presented Hidden Factors as Topics or HFT model. The Hidden Factors as Topics model for product recommendation by merging rating along with review text. The obtained forty two million reviews in the time of eighteen years. Total forty two million reviews from ten million users and three million items. This model works to analyses hidden factors and topics in product rating and review. The authors combine the rating and review text. By combing rating and review text the hidden rating for a product and hidden topic for a product are gathered. This help for the new product recommendation. According to the Sun [12] the informational role of the product and framed Hotelling 1929 linear city theoretical model. The model has two attributes as quality and mismatch cost. The customer like or dislike the product on average rating and variance. If the average rating is lower the demand for variance product is very high. People look for more information online about the product before intent to buy them. The customer review and collect information about product from http://www.Yelp.com and consumer reports posted by early patrons. They consider family and friends recommendation as a favorable suggestion about the product. He emphasizes based on report at least 68 percent of the online buyer look for four reviews and one quarter of buyer review at least eight reviews. The website like http://www.Yelp.com, http://www.Walmart.com and http://www.Amazon.com place the bar chart which often appears on the product page. The bar chart strengthens the percentage of reviews in connection with a height of rating. The marketer can be benefited if they obtain product variance of rating to create better demand forecast.

Kusgen and Kocher [13] discussed based on the price and brand the traditional buying was carried on. They also consider family and friends suggestion before making purchase decision. The technology based buying conversion practice leads to more than a hundred messages about a product and its feature. The widely shared review rating about each product or service posted by a previous buyer sharing their experience about purchased product helps the other user in making the purchase decision. The wide information available on web 2.0 makes a way to buy online. Li [14] examined how online review rating shared via social media for the restaurant has an impact on consumer to redeem the voucher. He examines the dispersion of review rating as moderating effect and the impact on discount entry to bulk buying deal. Consumers share and endorse about the purchase in Facebook to convey the product information to family and friends. He examined online review rating was antecedent of social media support. He concluded dispersion and discount threshold of review ratings have a moderated effect on average rating widely.

Bao et al. [15] proposed latent factor model as TopicMF. The study model will combine rating and unstructured review text. Recommendation has become an important element in today's online business. Ratings only predict the like and dislike preference of the customer. In order to get more information why the item is like and dislike, the rating can be associated with the review. The segment of review text explodes the different preference choice of user and item. Hung and Lai [16] proposed this study to find relationship between rating and comment post by online purchaser. The content posted by them have influence on other online users purchase decision. The Web 2.0 has transformed electronic commerce from product oriented environment to social environment. User generated content and word of mouth emerges for interaction and communication. Customer review is determined as an asset for business and this review has increased sales. The customer reviews help the online users purchase decision. Facebook like page benefit both seller and buyer. Seller promotes the brand and buyer gather information before intent to purchase anything from Facebook. Customer star rating and open comment both benefit the web shopper.

Park et al. [17] discussed how there is a rapid increase in the use of online platforms. Reviews with images posted on online platforms increase the purchase intent. The rating provided by previous customers combined with the image multiplies the likelihood of increasing purchase intent. From Smironva et al. [18], they explored user created rating perform as an important factor in decision making. Online review rating helps the customer in reducing uncertainty and increase the chance of buying decision from online platforms. Zhang et al. [19] with the advancement in usage of information technology social media users are more profited by sharing the information in social networks. The review sharing network assist in repeating the purchasing intention. Le et al. [20] investigated online reviews for purchase decision making by including three concepts such as text, image, and star as one of the dimensions under behavioral components. Ghimire et al. [21] exposed that online users reviews are more reliable posted on google reviews page.

Zhang et al. [22] discussed that the online buying trend of customer through the internet and electronic commerce is increasing day by day. They highlighted more than ninety percent of online buying customers read and refer the customer reviews about the relevant product which they have a buying intention in the future. The feedback posted by online purchaser have a strong impact on future buyers buying intent or behaviour. Product reviews posted in online networking sites are often a beneficial source of information for predicting consumer buying behavior. Dennis et al. [23] examined the factors like quality of information, richness of media, ratings and reviews influences millennial customers electronic purchase intent in Indonesia.

Mulyono [24] highlighted technology advancement in the internet era rapidly increased the purchase and sale transactions via online shopping mode. The author found from a study among four hundred respondents taking into consideration of particular online shop Lazada. The online shop contains a greater number of reviews and ratings influence positive effect on customer buying intention. More number of positive reviews and discussion build the reputation of the product and make the product popular.

According to Liu et al. [25] online media provides wide platforms to get more knowledge about various products and services available in the form of customer reviews. The reviews posted on online websites in the form of language style build the intention to purchase as the customer gets clear reviews in easily understandable language. The product type and social presence as moderator and mediator factors has a positive relationship between customer review rating and purchase intention.

Obieda [26] discussed the internet platforms are becoming more advanced to provide effective portal like shopping websites and forums to benefit online shoppers. With one hundred and twenty United Kingdom online customer the author examined and found customer review act as a smart method to create purchase intention. Further the high level of cognition provides more reviews about quality and quantity of the products rich available in online platforms. The purchase intention is very high when there is sound knowledge about the products. Elwalda and Lu [27] highlighted that there is a rapid increase in usage of online media. Numerous virtual communities engaged in electronic word of mouth sharing through online reviews about various products and services. The new form of electronic word of mouth recommendation has increased the customer buying decision. The customer is able to review and read more product related customer feedback online along with the star ratings.

Rahayu et al. [28] discussed the marketing method and strategy tends to change day by day with the growth of advanced social media. They conducted an empirical study among four hundred respondents to examine the reviews posted on Blibli shopping site. Online customers are benefitted through online reviews shared on internet platforms. The customer gets high impact to buy a product via customer reviews with dimension such as quality information shared, source reliability, timeliness, valence and volume. According to Filieri [29] the marketer should update themselves the new and trendy method to get more customer. The finding highlighted customer depend on online reviews more to gain more product knowledge and feature about it from post purchaser. The overall ranking and ratings act as a significant role in increasing the purchase decisions of customer along with customer reviews. The ranking system help the customers to differentiate various similar products available online. Engler et al. [30] discussed the importance of online rating which act as a key component for giving more insight about various products to online buyers, sellers and manufacture. The score of online rating was influenced by two sectors such as expectation before purchase of products and performance of a product after buying.

Kocher et al. [31] stated that in present scenario internet is accessed widely by the people and the companies. The companies use the connecting platforms to promote their products and services by reducing promotional expenses. The communication network opens an approach of creating and posting customer reviews in online electronic websites and platforms by increasing the sale and promoting the product through opinion sharing. They revealed from their finding negative reviews do not have an impact on rating when the product function is high. Aditya and Alversia [32] they conducted an empirical study among six hundred and fifty three respondents to analyse the customer reviews impact on purchase intention about cafe in Indonesia. The consumer was found to be loyal to the cafe business with six dimensions in relationship with online review and purchase intention. The online content posted in the form of reviews contain six factors such as effectiveness, timeliness, reviewer knowledge, comprehensiveness, favorable and unfavorable online reviews. Further Aditya and Alversia [33] discussed the digital era facilitate customers to share reviews in various social media platforms. Review platforms, customer reviews, profile of the customer and property features has an effect on customer buying decision.

Elwalda and Lu [34] highlighted that more study investigating the relationship of online customer reviews on customer purchase intention. The technology usage and advancement bring customer reviews as a main component to create an intention to buy from online platforms. The customers share and discuss through various platforms like online discussion forum, blogs and electronic websites. The non numeric approach will provide insight to sellers to understand the customer behaviour in online purchase decision. Schreck and Chin [35] discussed that in online shopping the customers are more dependent on post customers reviews as they are able to view the product picture online. The information shared by the post purchaser clear many questions in knowing the product better to purchase in the future. The requirement of knowledge and congruency for review rating has an important relationship on buying intention.

Based on the previously published articles few articles discuss about electronic word of mouth and user generated content are the two ways to post and retrieve online customer review rating. The content is the online reviews or feedback posted in virtual mode in various electronic websites and social media platforms about various products and services. When the products contain high number of reviews and numeric star rating create the intention to purchase the products. The score rating helps the customer to identify the best top product available in online platforms. Further the customer reviews ratings are shared with friends, family, online group, brand followers. Online members join with the interested community to get updates about particular products and services information. The customer review rating is recommended in order to increase sale as well to build the product reputation and make it popular among online buyers. Reviews and rating act as a powerful tool in creating intention to buy. When the products and services obtained high number of positive star ratings and reviews creates the high intention to buy the products in future.

7.3 Methodology

The chapter examined the effect of online review rating on purchase intention. The primary data have been collected from 192 respondents through the questionnaire survey. The construct items are adopted from previous research literature and modified according to the study. The scale for measuring customer rating was taken from Filieri [29] and Park et al. [36]. For measuring Purchase Intention the items have been adapted from Park et al. [36] and Celeste et al. [37]. Purchase intention construct questionnaire were developed using five point Likert scale with 1 for Strongly Disagree and 5 for Strongly Agree.

The survey respondents are customers who had purchased products from various online platforms. The purchaser also obtains the knowledge of customer reviews and rating. The post purchasers are the respondents who are familiar in posting various reviews and star rating in networking sites. The respondents completed the survey through the google form link shared. The chapter further examined the various electronic commerce websites to get insight about the customer reviews and rating.

Statistical Package SPSS version 21 is used to measure the percentage analysis of the demographic variables to find the gender wise and age wise respondents. Simple linear regression analysis is measured to find relationship between Customer review rating independent variable and purchase intention as dependent variable.

Figure 7.1 illustrate the block diagram about customer review and star rating display example. The five stars are the rating method given by the customer for the purchased product and services online. The text content in the form of reviews is the bookmark posted in the purchased online platforms by the post purchaser. The words customer review and rating are important two important terms for creating purchase intention. The post purchaser opinion about various products and obtained services are documented in social networking online sites. The reviews give in depth information about the product features by creating an intention to purchase in the future.

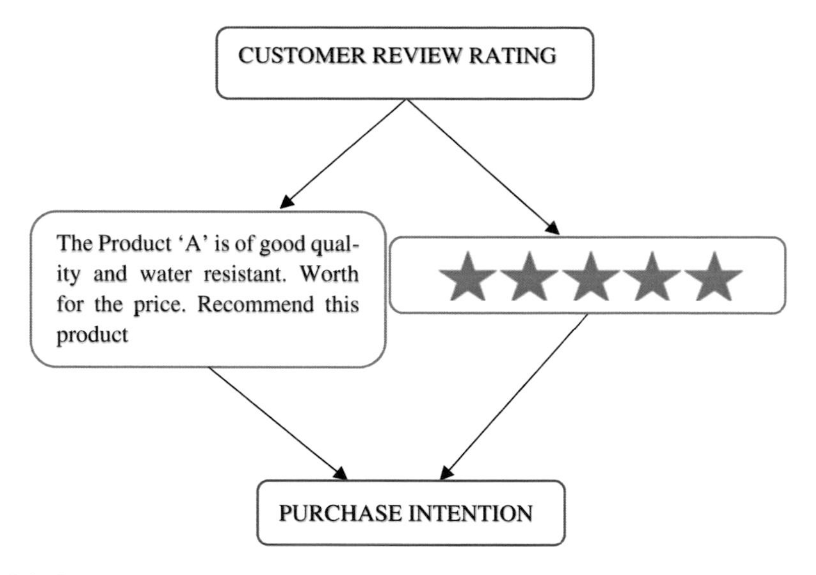

Fig. 7.1 Customer review rating on purchase intention. *Source* Authors

7.4 Analysis and Interpretation

The hypothesis of the study is measured by simple linear regression analysis to find the relationship between independent variable Customer review rating and dependent variable purchase intention. The Tables 7.1 and 7.2 display the percentage analysis of gender wise and age wise response of respondents. The Tables 7.3, 7.4 and 7.5 demonstrate the model summary, ANOVA and coefficients of regression.

Table 7.1 Gender wise distribution of respondents

Gender	Frequency	Percent
Male	91	47.4
Female	101	52.6
Total	192	100.0

Table 7.2 Age wise distribution of the respondents

Age	Frequency	Percent
Less than 20	6	3.1
21–30	69	36.0
31–40	101	52.6
More than 40	16	8.3
Total	192	100.0

Table 7.3 Regression analysis of customer review rating and purchase intention

Model	R	R square	Adjusted R square	Std. error of the estimate
1	0.819[a]	0.670	0.668	0.34583

[a]Predictors: (Constant), CRR—Customer review rating

Table 7.4 ANOVA for customer review rating and purchase intention

Model		Sum of squares	df	Mean square	F	Sig
1	Regression	46.126	1	46.126	385.677	0.000[b]
	Residual	22.724	190	0.120		
	Total	68.850	191			

[a]Dependent Variable: Purchase Intention
[b]Predictors: (Constant), Customer Review Rating

Table 7.5 Coefficient analysis of customer review rating and purchase intention

Model		Unstandardized coefficients		Standardized coefficients	T	Sig
		B	Std Error	Beta		
1	(Constant)	0.550	0.167	0.819	3.304	0.001
	Customer review rating	0.825	0.042		19.639	0.000

[a]Dependent Variable: Purchase Intention

Table 7.1 display the gender wise survey questionnaire distribution of the respondents. The results show that 47.4% of those surveyed are male and 52.6% are female. When compared to male responders, female respondents are higher.

Table 7.2 shows that 3.1% of respondents are less than 20 years, 36% of respondents are 20–30 years, 52.6% of respondents are 31–40 years and 8.3% of respondents are more than 40 years age group. The maximum respondents fall in the age group of 31–40 years and minimum respondents are in the age group of less than 20 years.

Table 7.3 shows that the R square value is more than 0.4 which is considered for further analysis. The study result shows the value of R square is (0.670) 67% of variation in Purchase intention.

Table 7.4 shows the ANOVA result indicate that the customer review rating has a positive effect on purchase intention ($F(1,190) = 385.677$, $p < 0.001$) where p value is less than 0.05 level of significance.

Table 7.5 inferred that the probability value for the variable Customer Review Rating is 0.000 where the p value is less than 0.05 level of significance. Customer Review and Rating positively influence the Purchase Intention with β values of (0.825). Therefore, Hypothesis (H_1) is supported.

7.5 Results and Discussion

The finding of the study reveals that the online customer review rating given by post or previous customers are the very important source to create purchase intention among online customer. By reviewing the star numeric and comments the new and potential customers have the intention to buy the products and obtain service through online mode. The finding of the study from 192 respondents reveals that when there is a higher number of online ratings for a product positively influence on purchase intention. This is similar with previous finding of Engler et al. [30] online rating given for a product integrate with purchase intention. Therefore, the research study concludes that there is a positive connection between online customer review, rating and purchase intention. Prastiwi et al. [38] They conducted a study among one hundred online user who purchased product online in Surakarta in Indonesia. The people are more interested to buy and sell in the media. They review online review and rating information which help them in a purchase decision. Buying online has increased the sales. Buyers are more conscious about buying decision and spending money. They have taken customer review, rating and perception into three factors to continue their study and found that, rating and review has a positive effect on customer perception. Customer perception shows affirmative and significant effect and influence on customer buying decision. The customer looks for quality of product and they prefer to get information online as it is less cost compared with conventional channels. Kiran and Vasantha [39] They highlighted that the online review has major impact on buying behavior of customers.

The study is carried forward how the transformation of social media has changed the purchasing pattern of an individual. People tend to believe in review rating than anything. The product and service with highest star have the potential to repeat the purchase or help in buying decision. The rating is the opinion suggested by buyer who have already purchased product or rendered service. Kiran and Vasantha [40] said that the social media transformation and the customers attitude direct to purchase decision. The digital media has bought wide changes in the lives of people. Social media turned out to be one of the trending and important pillars among online user. Ganu et al. [41] concluded that the advancement of web 2.0 technology and the user generated content and recommendation has enlarged people to post information about the product and service before they intent to buy or obtain any service. The author highlighted online review is the asset of the internet and help with business, movie review, restaurant review. The feedback given by previous customer helps in a purchase decision. Sometimes people ignore the review and look for rating alone. Tran [42] he validated and proved the study that online reviews are closely linked to customer purchasing intent from one thousand one hundred and twelve respondents.

7.6 Conclusion

This study focused on how the customer review and rating influence on purchase intention. The Study revealed that customer review and rating have a significant relation. The Hypotheses H_1 Customer review rating has positive influences on Purchase Intention is supported. The study helps the online marketer to strengthen the product. The increased numeric stars or rating will increase the sales. Consumers like to buy a higher price product, if there is a higher number of stars or rating given to the product. Consumers have less preference if the product has a lesser rating compared with much alternative availability. The rating or five stars are highly useful if the product is new launched. The Rating or star recommendation will be favorable to buy a newly launched product among online users or purchaser along with existing product. The user preferred to buy a product which has a greater number of ratings. Therefore the study concludes customer review rating in social media have a positive impact on purchase intention.

Further the research can be enlarged by implementing deep learning prediction of review rating for simplifier the learning process without human interaction. The bolts implement in webpage assist as supportive for interaction. The deep learning subset of artificial learning learns the keywords entered by the online users. This simplifies the job of the engineering process. According to the Ahmed and Ghabayen [43] they proposed and successfully experiment the deep learning framework into two phases. In second model they applied deep learning concept for predicting review rating from online users review text. The researcher found deep learning identify the words easily compared with traditional machine learning and had a significant impact on predicting review rating.

References

1. Kiran, P., Vasantha, S.: Analysing the role of user generated content on consumer purchase intention in the new era of social media and big data. Indian J. Sci. Technol. **9** (2016). https://doi.org/10.17485/ijst/2016/v9i43/104754
2. Zheng, L., Zhu, F., Mohammed, A.: Attribute and global boosting: a rating prediction method in context-aware recommendation. Comput. J. **bxw016** (2016). https://doi.org/10.1093/comjnl/bxw016
3. Kiran, P., Vasantha, S.: Car factors considered by consumers for making purchase intention—social media perspective. Aust. J. Basic Appl. Sci. **10**(1), 689–695 (2016). https://www.ssrn.com/abstract=2791728
4. Chen, C.-W.: Five-star or thumbs-up? The influence of rating system types on users' perceptions of information quality, cognitive effort, enjoyment and continuance intention. Internet Res. **27**(3), 478–494 (2017). https://doi.org/10.1108/IntR-08-2016-0243
5. Ling, K.C., Chai, L.T., Piew, T.H.: The effects of shopping orientations, online trust and prior online purchase experience towards customers' online purchase intention. Int. Bus. Res. **3**(3), 63–76 (2010)
6. Thomas, M.-J., Wisrtz, B.W., Weyerer, J.C.: Determinants of online review credibility and its impact on consumers' purchase intention. J. Electron. Commerce Res. **20**(1) (2019)

7. Wang, B., Xiong, S., Huang, Y., Li, X.: Review rating prediction based on user context and product context. Appl. Sci. **8**(10), 1849 (2018). https://doi.org/10.3390/app8101849
8. Wang, B., Chen, B., Ma, L., Zhou, G.: User-personalized review rating prediction method based on review text content and user-item rating matrix. Information **10**(1), 1 (2018). https://doi.org/10.3390/info10010001
9. Ling, G., Lyu, M.R., King, I.: Ratings meet reviews, a combined approach to recommend. In: Proceedings of the 8th ACM Conference on Recommender Systems—RecSys '14 (2014). https://doi.org/10.1145/2645710.2645728
10. Wang, H., Lu, Y., Zhai, C.: Latent aspect rating analysis on review text data: a rating regression approach. In: Proceedings of the 16th ACM SIGKDD International Conference on Knowledge Discovery and Data Mining, Washington, DC, USA, 25–28 July 2010, pp. 783–792
11. McAuley, J., Leskovec, J.: Hidden factors and hidden topics: understanding rating dimensions with review text. In: Proceedings of the 7th ACM conference on Recommender Systems, Hong Kong, China, 12–16 October 2013, pp. 165–172
12. Sun, M.: How does the variance of product ratings matter? Manage. Sci. **58**, 696–707 (2012). https://doi.org/10.1287/mnsc.1110.1458
13. Kusgen, S., Kocher, S.: The influence of customer product ratings on purchase decisions: an abstract. Dev. Market. Sci. Proc. Acad. Market. Sci. **953–954**,(2017). https://doi.org/10.1007/978-3-319-45596-9_176
14. Li, X.: Impact of average rating on social media endorsement: the moderating role of rating dispersion and discount threshold. Inf. Syst. Res. (2018). https://doi.org/10.1287/isre.2017.0728
15. Bao, Y., Fang, H., Zhang, J.T.: Simultaneously exploiting ratings and reviews for recommendation. In: Proceedings of the Twenty-Eighth AAAI Conference on Artificial Intelligence, Québec City, QC, Canada, 27–31 July 2014, pp. 2–8
16. Hung, Y.-H., Lai, H.-Y.: Effects of facebook like and conflicting aggregate rating and customer comment on purchase intentions. Universal Access in Human-Computer Interaction. Access to Today's Technologies, pp. 193–200 (2015). https://doi.org/10.1007/978-3-319-20678-3_19
17. Park, C.W., Sutherland, I., Lee, S.K.: Effects of online reviews, trust, and picture-superiority on intention to purchase restaurant services. J. Hosp. Tour. Manage. **47**, 228–236 (2021). https://doi.org/10.1016/j.jhtm.2021.03.007
18. Smironva, E., Kiatkawsin, K., Lee, S.K., Kim, J., Lee, C.-H.: Self-selection and non-response biases in customers' hotel ratings—a comparison of online and offline ratings. Curr. Issue Tour. **1–14**,(2019). https://doi.org/10.1080/13683500.2019.1599828
19. Zhang, N., Liu, R., Zhang, X.-Y., Pang, Z.-L.: The impact of consumer perceived value on repeat purchase intention based on online reviews: by the method of text mining. Data Sci. Manage. **3**, 22–32 (2021). https://doi.org/10.1016/j.dsm.2021.09.001
20. Le, L.T., Ly, P.T.M., Nguyen, N.T., Tran, L.T.T.: Online reviews as a pacifying decision-making assistant. J. Retail. Consum. Serv. **64**(1), 102805 (2022). https://doi.org/10.1016/j.jretconser.2021.102805
21. Ghimire, B., Shanaev, S., Lin, Z.: Effects of official versus online review ratings. Ann. Tour. Res. **103247**,(2021). https://doi.org/10.1016/j.annals.2021.103247
22. Zhang, J., Zheng, W., Wang, S.: The study of the effect of online review on purchase behavior. Int. J. Crowd Sci. **4**(1), 73–86 (2020). https://doi.org/10.1108/ijcs-10-2019-0027
23. Dennis, L., Ramdhana, F., Faustine, T.C.E., R.B., Hendijani: Influence of online reviews and ratings on the purchase intentions of gen Y consumers: the case of Tokopedia. Int. J. Manage. **11**(6), 26–40 (2020). https://doi.org/10.34218/IJM.11.6.2020.003
24. Mulyono, H.: Online customer review and online customer rating on purchase intention in online shop. Int. J. Res. Rev. **8**(1) (2021)
25. Liu, Z., Lei, S., Guo, Y., Zhou, Z.: The interaction effect of online review language style and product type on consumers' purchase intentions. Palgrave Commun. **6**, 11 (2020). https://doi.org/10.1057/s41599-020-0387-6
26. Obiedat, R.: Impact of online consumer reviews on buying intention of consumers in UK: need for cognition as the moderating role. Int. J. Adv. Corporate Learn. (iJAC) **6**(2), 16– (2013). https://doi.org/10.3991/ijac.v6i2.2910

27. Elwalda, A., Lu, K.: The impact of online customer reviews (OCRs) on customers' purchase decisions: an exploration of the main dimensions of OCRs. J. Cust. Behav. **15**(2), 123–152 (2016). https://doi.org/10.1362/147539216X14594362873695
28. Rahayu, A., Utama, D.H., Novianty, R.: Proceedings of the 5th Global Conference on Business, Management and Entrepreneurship (GCBME 2020), Advances in Economics, Business and Management Research, vol. 187, pp. 471–477 (2021). ISBN 978-94-6239-424-7,ISSN 2352-5428. https://doi.org/10.2991/aebmr.k.210831.094
29. Filieri, R.: What makes online reviews helpful? A diagnosticity-adoption framework to explain informational and normative influences in e-WOM. J. Bus. Res. **68**, 1261–1270 (2015)
30. Engler, T.H., Winter, P., Schulz, M.: Understanding online product ratings: a customer satisfaction model. J. Retail. Consum. Serv. **27**, 113–120 (2015). https://doi.org/10.1016/j.jretconser.2015.07.010
31. Soren, K., Stefanie, P., Sarah, K.: The recommendation bias: the effects of social influence on individual rating behavior. Dev. Market. Sci. Proc. Acad. Market. Sci. **353–354**,(2016). https://doi.org/10.1007/978-3-319-19428-8_93
32. Aditya, A.R, Alversia, Y.: The influence of online review on consumers' purchase intention. J. Manage. Market. Rev. **4**(3), 194–201 (2019). https://www.ssrn.com/abstract=3461000
33. Aditya, A.R., Alversia, Y.: The impact of online reviews on social media platform on consumer's purchase intention in choosing first visited cafe. In: International Conference on Rural Development and Entrepreneurship 2019: Enhancing Small Business and Rural Development Toward Industrial Revolution 4.0, vol. 5, no.1, pp. 1136–1144 (2019). ISBN: 978-623-7144-28-1
34. Elwalda, A., Lu, K.: The influence of online customer reviews on purchase intention: the role of non-numerical factors. In: European Marketing, Conference: European Marketing Conference LCBR Munich, Germany (2014). http://www.lcbr-archives.com/2014.html
35. Schreck, J.L., Chin, M.G.: Online product reviews: effects of star ratings and valence on review perception among those high and low in need for cognition. Proc. Human Factors Ergonom. Soc. Annual Meet. **63**(1), 401–405 (2019). https://doi.org/10.1177/1071181319631447
36. Park, D.-H., Lee, J., Han, I.: The effect of on-line consumer reviews on consumer purchasing intention: the moderating role of involvement. Int. J. Electron. Commer. **11**(4), 125–148 (2007). https://doi.org/10.2753/jec1086-4415110405
37. Celeste see-pui ng, intention to purchase on social commerce websites across cultures: a cross-regional study. Inf. Manage. **50**, 609–620 (2013). https://doi.org/10.1016/j.im.2013.08.002
38. Prastiwi, S., Umam, M., Auliya, Z.: Online costumer reviews (OTRs) dan rating: new era in Indonesia online marketing. **8**, 89–98 (2017)
39. Kiran, P., Vasantha, S.: Role of online reviews in building customers trust and its impact on online purchase behavior. Indian J. Appl. Res. **4**(12) (2014)
40. Kiran, P., Vasantha, S.: Transformation of consumer attitude through social media towards purchase intention of cars. Indian J. Sci. Technol. **9** (2016). https://doi.org/10.17485/ijst/2016/v9i21/92608
41. Ganu, G., Elhadad, N., Marian, A.: Beyond the stars: improving rating predictions using review text content. In: Proceedings of the Twelfth International Workshop on the Web and Databases, WebDB, Providence, RI,USA, 28 June 2009, vol. 9, pp. 1–6
42. Tran, L.T., Thuy,: Online reviews and purchase intention: a cosmopolitanism perspective. Tour. Manage. Perspect. **35**(10), 100722 (2020). https://doi.org/10.1016/j.tmp.2020.100722
43. Ahmed, B.H., Ghabayen, A.S.: Review rating prediction framework using deep learning. J. Amb. Intell. Human Comput. (2020). https://doi.org/10.1007/s12652-020-01807-4

Chapter 8
Artificial Intelligence: Paving the Way to a Smarter Education System

Neha Sharma

Abstract Once a figment of imagination, Artificial Intelligence today is a scientific reality establishing its presence in every industry and sector across the globe. The education sector is no exception as during the last few years AI has made its way deep into this field. The ubiquity of digital devices and the universal reach of the Internet have led to the new age of any time anywhere education. AI forms the backbone of this new education system and supports every stakeholder in the field of education in accomplishing their different goals. The inclusion of AI obliterates geographical boundaries making global education more affordable and accessible than ever before while automation of the formidable backend administrative tasks with AI, lends fluidity and transparency to the education system. AI is bringing profound changes to the field of education by providing endless learning opportunities, expanding access, modernizing the mode of dissemination of knowledge, and enabling round-the-clock support to students with the help of chatbots. The present book chapter aims to review the numerous benefits of adopting AI-powered solutions in education and their possible implications. The chapter outlines how AI is personalizing learning, creating virtual content, addressing learning needs of the differently abled leading to wider inclusion, deeper engagement, and enrichment of the teaching–learning process.

Keyword Artificial intelligence · Education · Inclusion · Chatbots · Engagement · Learning

8.1 Introduction

Artificial Intelligence (AI) is a thriving technological province capable of changing every facet of the way we interact socially. The reason why AI has gained such prominence in a short period is because of its ability to emulate human capabilities of understanding, planning, reasoning, perceiving, and communicating efficiently at

N. Sharma (✉)
Emaar Emerald Hills, School of Business studies (SBS), Sharda University, Greater Noida, India
e-mail: nehasharma161985@gmail.com

A. Biswas et al. (eds.), *Artificial Intelligence for Societal Issues*, Intelligent Systems Reference Library 231, https://doi.org/10.1007/978-3-031-12419-8_8

a low cost without the need for any human intervention. Most of the popular AI applications such as self-driving cars to the robots that play chess rely on Natural Language Processing (NLP) and deep learning [1]. Utilizing these technologies, machines can be trained to achieve tasks by processing huge volumes of data and recognizing data patterns. AI syndicates massive volumes of data with fast iterative processing and intelligent algorithms, letting the software learn automatically from features or patterns in the data. Machines that are AI-enabled can learn from experience, understand, and analyze new inputs and accomplish human-like tasks. These capabilities translate into numerous benefits for individuals, societies, and businesses. AI can be used across industries as it can considerably reduce the time taken to perform tasks, enable multi-tasking, execute complex tasks without substantial cost outlays, thereby easing the workload of the existing resources. 24 $\times$ 7 operating capability without interruption and downtime and faster and smarter decision-making are some of the key benefits of adopting AI and the reason for its exceptional popularity. AI today has made its way deep into every aspect of human existence as it has broad applicability in almost every industry and sector from manufacturing, transportation, entertainment, health care, agriculture, national defense, and cybersecurity, the list is endless [2]. Globally millions of dollars are being pumped into funding startups in the AI domain and a wide array of AI implementations. As the potential of AI is being realized, its adoption and use are proliferating. Above and beyond businesses, AI also can address social challenges in the real world and provide valued solutions to numerous societal issues. From wildlife protection to agriculture to modeling the spread of epidemics AI is being effectively used to address and resolve issues of social interest globally. One area of social and cultural importance that is vital for the development, growth, and prosperity of the entire humankind and can benefit immensely from the application of AI in education.

Education is the backbone of every economy; it is the foundation for development and progress and therefore the application of AI in the field of education has received huge attention in the recent past to overcome the challenges of the education sector and to ensure the dissemination of quality education to as many as possible. A report by markets.com projects the AI in the Education market to grow at a 47% CAGR to reach USD 3.68 billion by 2023 [3]. However, Education is a very complex domain, as the major stakeholders the learners and the educators belong to a wide range of age, capability, acumen, intelligence, and tech-savviness making it immensely difficult to create and implement AI-enabled learning systems that truly help serve the purpose of imparting quality education in a simplified way.

8.2 Education and Its Many Challenges

While the right to education is a fundamental right across most parts of the globe, imparting and disseminating quality education to everyone is not a simple task. Several challenges impact the dissemination of knowledge and the quality of education. From poverty to war zones to religious fanaticism, several roadblocks make it

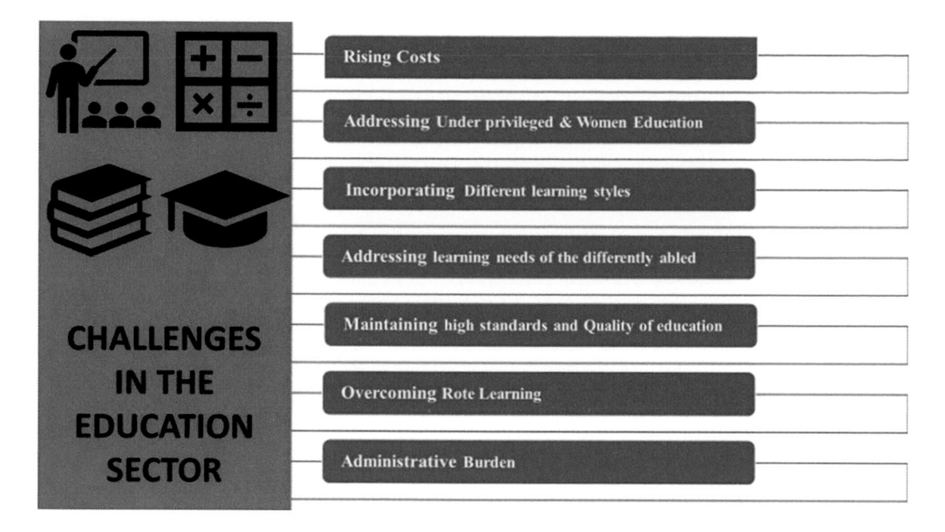

Fig. 8.1 Challenges in the education sector

almost impossible for children and adults to access even basic education. The figure below depicts the various issues that the current education system faces and should be overcome to maximize the reach and depth of quality education (Fig. 8.1).

8.2.1 *Rising Cost of Education Worldwide*

For centuries people have struggled to gain access to quality education, and it has remained to be the privilege of the affluent. The cost of education is on a constant rise and imposes a major challenge for the education sector. The Covid-19 pandemic has only added to the misery forcing millions of children to quit or take a break as they can't afford the basic technical infrastructure required to be able to attend the online mode of delivery or their education providers are not equipped to move from offline to online mode of education. A report from UNICEF claims that schools in many parts of the globe have been completely shut for more than a year impacting nearly 168 million children globally [4].

8.2.2 *Reaching the Less Privileged and Promoting Women's Education*

Inaccessibility to basic education is a reality even today. In the third-world countries, and even in the developed and developing economies, access to education is limited and, in some cases, it is outrightly denied to certain groups based on ethnicity or gender [5]. UNESCO's Global education monitoring report 2017 suggests that close to 264 million children do not go to school due to poverty and gender discrimination [6]. A report by United Nations published in the year 2020 claims that close to one-third of the world's Poorest girls, between 10 and 18 years of age have no access to education at all [7].

8.2.3 Addressing Different Learning Needs

Besides access and affordability, for learning to be effective it must be imparted to the learner in their preferred style of learning. As a learner, every individual is different and has a different learning style that enables deeper understanding and retention of knowledge. Visual, Auditory, Reading/writing, and Kinesthetics are the most common forms of learning [8]. While a classroom is generally a complex space where learners with different abilities acumen and learning styles come together, the classroom teaching and learning process generally follows a standardized method of teaching dominated by the one-way lecture delivery style, with minimal use of other methods to address different learning styles.

8.2.4 Learning Needs of the Differently-Abled

One of the biggest challenges of the education system is to address the learning needs of the differently-abled. More often than not, the differently-abled have found it difficult to gel with the regular classroom mode of teaching and learning with the regular learner, mostly dropping out mid-way due to issues in coping up or being neglected, and not feeling like an integral part of the generic system [9]. Even today 32 million children with disabilities from low and middle-income nations are out of school [10]. Over the past few decades, there has been a focus on addressing the needs of the differently-abled, with special schools being set up for the same and teachers being trained especially for educating them, however, a deep understanding of the learning process and personalization of education is what is needed to adequately address and effectively impart education to people with special needs.

8.2.5 Setting High Standards and Maintaining Quality of Education

If the quality of learning does not match up to the international standards, then the benefits are questionable as in the end the reason for undertaking education is to be able to earn a living from the knowledge skills and attitude acquired through the learning process. While with globalization the world has become one big marketplace for job seekers to pitch their candidature, the global skill market holds the expectation that is difficult to meet without achieving and ensuring standardization of learning globally [11].

8.2.6 Overcoming the Age-Old Problem of Rote Learning

Education has long been focused on memory retention rather than understanding leading to rote learning. While Rote learning helps in developing foundational knowledge and augments the ability to swiftly recall basic facts, it does not promote the acquisition of a deeper understanding of the subject matter and use of analytical and logical skills leading to very surface level or superficial learning [12]. As a result, Rote learning adversely impacts the quality of education and the employability potential of students that pass through this system. The current global unemployment rate stands at 13.6 percent with the active workforce witnessing declining participation from youth in the age group of 15 to 24 years owing to many reasons, rote learning leading to unemployability being one [13].

8.2.7 The Ever-Increasing Burden on the Education System

Quality education also mandates quality teachers, effective assessments, updated curriculum, and quality study material. Unfortunately, in most cases schools suffer from a very high student-to-teacher ratio putting an excessive burden on teachers and administrators, the assessment systems are ineffective outdated curriculum takes years to get updated and the study material is either inaccessible or very poor quality. Over the past few decades, education has been recognized as one of the fundamental building blocks for the growth and development of any economy, the education sector has received a lot of attention globally. Continuous attempts are underway to improve the quality and accessibility of education. The digital boom, penetration of the internet, accessibility to cheap smart hand-held devices, and advancements in the field of science and information technology have brought about remarkable changes to the education system. Technology has played a very important role in bringing about the desired changes in the education system and ensuring deep learning in a very personalized way. However very little has been achieved and there is a long way to go before a substantial difference in the quality of education can be felt and seen, the field of science and technology though is doing its best to make the difference.

8.3 The Role of Technology in Transforming the Education Sector

With the digital boom, the education industry also marked its virtual presence. The digital world offered numerous opportunities to people willing to learn overcome barriers of distance, geographies, and language. As a result, the online mode of education became immensely popular over the years and the 2020 pandemic only further fueled its growth and popularity as a reliable and effective mode of education. Continuous and consistent efforts in the field of EdTech have led to noteworthy innovations one after the other.

8.3.1 Massive Open Online Courses (MOOC)

What began as an experiment at the University of Illinois in the year 1960, evolved into the launch of open universities, today online learning is a full-fledged industry. The global market size of the online learning industry was estimated to be 187.87 billion dollars in 2019 and it has witnessed a 400% growth over the past six years [14]. The availability of MOOCs, Free online courses where anyone irrespective of their discipline and educational background can enroll have transformed the way people look at education. Interdisciplinary and any time anywhere education is the new way of learning that has been only made possible with technology.

8.3.2 Virtual Reality (VR) in Education

Virtual reality is another technology that has a huge role to play in transforming the entire teaching–learning process. With VR, students can gain a memorable and immersive experience even in a classroom setup. VR can promote a better understanding of the subject leading to deeper learning and internalization. The use of VR in educating young minds inspires, sparks creative thinking and imagination, and provides realistic experiences leading to a more inclusive classroom. From primary education, museum, and monument visits to learning to fly using simulators [15], VR as a technology has been in use in the education industry and its use only continues to grow as it becomes more accessible and affordable [16].

8.3.3 Augmented Reality (AR) for Immersive Learning

AR has taken the VR experience beyond the virtual world, by integrating virtual objects into the real world. With AR concepts can be simplified to make learning easy and provide a more immersive learning environment. The use of AR has been incorporated in many ways by the EdTech players to provide a more appealing realistic experience to learners. From interacting with dinosaurs to experiencing 3D volcanoes and tornados, AR is revolutionizing the learning space by pushing away rote learning with realistic experiential learning at its best [17].

8.3.4 Artificial Intelligence (AI) in Education

AI as a technology aims at producing machines that can learn and acquire human-like thinking, reasoning, and decision-making capabilities. AI offers huge benefits, the most important being minimization of human effort and reduction in cost and hence its applicability is ubiquitous. AI has already made its way into every business and every industry worldwide, and the education industry is no exception. Though the perception of AI in education was of robots replacing educators, the actual use and application of AI are far more diverse.

Artificial intelligence in Education AIED is a vast term that encompasses everything from AI-powered personalized instructions and dialogue systems to AI-enabled exploratory learning and intelligent game-based assessments. AI assessment and analysis of students writing for creating personalized development and learning plans. To Chatbots for 24/7 Student-support and AI-based student educator matching, enabling students to take complete control of their learning. Besides AIED also includes one-to-one students' interaction with computers, use of handheld devices by students outside the classroom, the whole school approach, and much more [18]. AIED can also shine a light on learning and educational practices and suggest practical approaches to drive change and improvement with data crunching and analysis. In a nutshell, AIED has the potential to incorporate all the latest technologies for improving and transforming the education sector in ways beyond our comprehension. In addition to helping students, AI has the potential to make things smooth for educators and the management of educational institutes as well. Learning is a complex process and requires a lot of preparation, patience, and dedication from the educators who are also involved in responsibilities outside the classroom that demand a lot of their time and effort. There are several ways in which AI can streamline and infuse fluidity and transparency to the system of education while taking away the burden of monotonous mundane tasks that eat away a lot of time and effort of the educators, administrators, and management of the educational institutes. AI adoption in Education has a wide range of benefits for everyone involved in the process, it is surprising to see how AI can transform the entire teaching–learning space.

8.4 Leveraging AI for Transforming the EdTech Space

AI has proven and established itself as the key driver for growth and as an undisputed source of competitive advantage for businesses and industries worldwide. It is the best example of disruptive technology; it touches our lives every day in ways we are not even aware of. Predictive text, voice assistance, interactive catboats are just a few examples of how AI influences and interacts with us every day. The applicability of AI is so vast that every business, every industry, and every sector is in the race of adopting AI-augmented systems and software. The education sector though has been slow in catching up with the AI trend, the Pandemic, however, has fueled this pace and accelerated it many folds. By late, the education industry has realized the importance of AI and its ability to transform the Education landscape resulting in AI adoption taking the center stage (Fig. 8.2).

There is a multitude of benefits that AI can offer to the education sector. AI capabilities have the potential to transform the entire gamete of tasks and activities involved in Education thereby benefiting all the stakeholders of the teaching–learning process.

8.4.1 *Benefits of AI for Students*

Students are at the center of the education system, the focal point. With AI the learning journey of students can be simplified to a great extent enhancing the outcome. From suggesting the right courses, simplifying complex concepts, providing feedback based on assessments, and extending continuous support and assistance, AI can significantly streamline the learning journey for students. Some of the key benefits can be identified as [19].

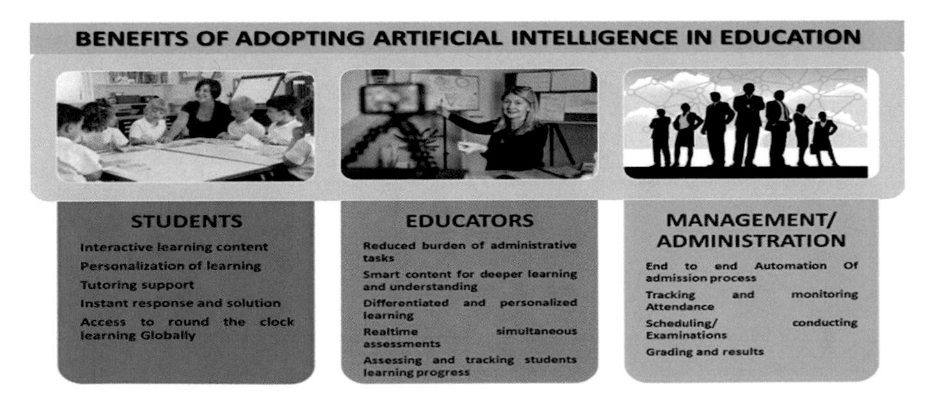

Fig. 8.2 Benefits of adopting artificial intelligence in education

Interactive learning content: One of the major applications of AI is in creating Audio Visual learning content with the inclusion of AR and VR, which makes learning a lot more interesting and fun for the learners.

Personalized learning: Besides, the creation of more appealing audiovisual interactive content, AI content can also be personalized to cater to the different learning styles that students are comfortable with. By analyzing students' past performance and learning history, AI-augmented solutions can identify their strengths and weaknesses and personalize the learning content to complement their speed of learning and level of knowledge ensuring better learning outcomes.

Tutoring support: AI catboats and tutors can provide extended support to students and address their needs beyond the school or university working hours and outside classrooms. Even in the absence of an educator, a one-on-one learning experience can be extended to the students with the help of AI solutions.

Instant response and solution: AI-augmented solutions can ensure continuous learning with no waiting time. Students often have common repeated doubts and questions that they expect resolutions to on an immediate basis which is not possible in a regular classroom setup, however with conversational intelligence and automation support, AI can help resolve student queries within seconds.

Access to round-the-clock learning Globally: With AI learners of any age, discipline, Country can have access to high-quality learning content irrespective of the geography and time zone. Students can learn at their own pace anytime from anywhere. Today education is a 24/7 activity, with students, professionals, elderly accessing courses of their choice and learning at their own pace from the convenience of the space of their choice thanks to AI.

8.4.2 Benefits for Educators

In the traditional education system, educators have often found themselves burdened with excessive workload coupled with additional administrative tasks. Addressing a poor student–teacher ratio has been one of the major challenges of the education sector which ultimately results in the deterioration of the quality of education and the learning outcome. The adoption of AI in education can immensely benefit educators and help them achieve better results and outcomes [20].

Reducing the burden of administrative tasks: Educators are involved in various tasks besides teaching. Maintaining attendance, grading exam papers, sharing notes, communicating with parents, and much more demands a lot of their time. AI-augmented learning systems can free up a lot of this burden by simply automating all of these and many more of such tasks that do not need to be carried out personally by the educators, enabling them to focus more on the important bit that is delivering productive learning to the students.

Smart content for deeper learning and understanding: Numerous concepts and ideas are difficult to convey and teach using the traditional lecture method of delivery, however, AI-enabled smart content can solve this problem. Various forms of digital content, digital books, video and audio versions of textbooks, video lectures, and video conferencing enable educators to deliver learning in a better more understandable manner that students find easy to grasp and retain.

Differentiated and personalized learning: Classrooms are a complex mix of students with different interests, acumen, and learning speed, and adjusting learning based on these specific aspects for large class size is a challenge that educators have faced for years. AI enabled digital platforms to come with the intelligent instruction design to offer learning, assessments, and feedback to students in addition to identifying knowledge gaps, redirecting them to new topics by assessing their readiness allowing every student to learn and grow at their own pace, considerably bringing down the effort of the educators. With the help of AI, educators can have a more in-depth understanding of their student's progress and assist them individually to perform better.

Realtime simultaneous assessments: With the help of AI, educators can administer practical problem-solving assessments, simultaneously to multiple students and also collate assessment results in no time. Every student can be given different assessment questions about the same subject and of similar difficulty levels ensuring fair assessment while eliminating any scope of malpractice.

Assessing and tracking students learning progress: One of the most important capabilities of AI systems is analyzing and interpreting data to predict future outcomes and this feature when applied to student assessments. AI algorithms can deep dive into assessment results for each student and identify their strengths, weaknesses, subject areas that need more focus and attention and provide meaningful insights to educators about the developmental needs for every student allowing them to guide each one of them in the right direction.

8.4.3 Benefits for Management and Administrators of Education Institutes

Education is an ongoing process involving millions of learners that in the current scenario are spread across the globe. Universities and educational institutes today operate round the clock and run multiple different courses with different curriculums across various disciplines in multiple modes such as online, distance, offline, involving millions of students, thousands of educators, and administrative staff. Managing, maintaining, monitoring, and updating the voluminous data related to all these stakeholders is a momentous task for every organization in the education sector. This is where AI comes in as a great source of relief for the management of gigantic educational institutes [21].

Automation: Automation of the entire admission process from collecting and collating student information from digital forms, offering courses, tracking fee payments, and rolling out admission letters, the entire admission, and school leaving process can be automated with AI.

Tracking and monitoring Attendance and absenteeism: Attendance tracking and reporting absentees is a huge administrative task for any educational institution. The performance of students highly depends on their regularity, and any deviation needs to be called out. It is the responsibility of educational institutes to keep parents and guardians informed about any irregularities of their wards, at least at the school and graduate levels. This herculean task can be put on autopilot mode with AI systems shooting regular attendance reports and flagging any concerns regarding performance and attendance.

Scheduling/conducting Examinations: Conducting examinations is an important and integral part of the education process. However, it is not a simple task and requires a lot of preparation and planning. From checking fee clearance to releasing admit cards, scheduling examinations to invigilation duties, a lot of time and effort gets consumed in the entire process. Though, with the help of AI systems, this astounding activity can be managed seamlessly with little monitoring and intervention from the examination cell.

Grading and results: Exams are followed by grading and result declaration that should be communicated to students and their guardians. Grading software with AI capabilities can accomplish this task in no time generating grades and report cards for every student. Not only this, AI grading solutions can compare students' past and current performance and also provide relative comparisons with other learners. All this and much more can be accomplished with minimum human involvement in a very small amount of time.

Due to the innumerable benefits AI can offer, it is rapidly making inroads into the EdTech sector bringing forth a paradigm shift to the traditional approach of teaching and learning, subsequently altering the future of education. However, the adoption of AI systems and solutions into the education sector should be carried out with some caution. While AI undoubtedly offers profuse benefits, its application can raise some challenges as well, and therefore it is imperative to identify and understand the challenges and limitations of the adoption of AI in education and address these issues before they go out of hand.

8.5 Assessing Tech Readiness to Embrace AI Using the SAMR Model

While there is no doubt about the limitless contribution that AI can make to the education sector, a hurry to incorporate and include AI may cost more than what it is worth. Understanding the readiness of the students, educators and the system is important before jumping on to the bandwagon of AI integration. While the primary reason to seek out education technology especially AI is for the substantial positive impact it can have on the performance of the learners, improving their test outcomes and enabling educators to assess them efficiently, the bigger issue is finding a roadmap to integrate technology considering the many tools and systems available in the market

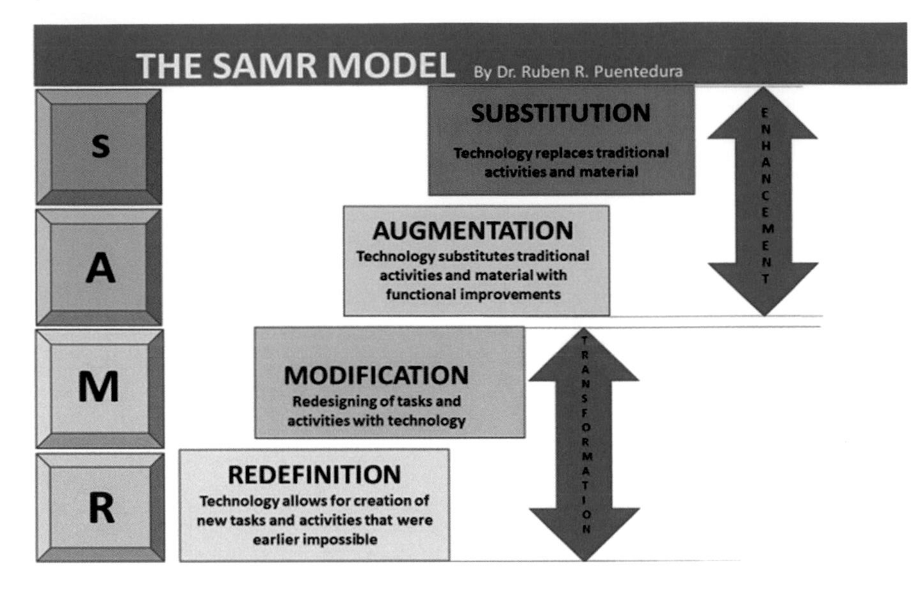

Fig. 8.3 SAMR model by Dr. Ruben Puentedura [22]

and a lack of training and development of the current teaching workforce obstructing the effective use of technology productively. The SMAR model by Ruben Puentedura developed in 2010 offers a conceptual framework for assessing preparedness for tech integration.

SAMR Model lays out four tiers of online education with increasing complexity and transformative power: substitution, augmentation, modification, and redefinition can assist educators in evaluating the current tech-savviness of the students and teachers and thereby understand how technology can support learning productively. The SMAR model is like a toolbox that should be used to analyze which is the best tool or Application for achieving the intent behind considering the adoption in the first place (Fig. 8.3).

Substitution is about replacing traditional materials and activities like paper worksheets or in-class lectures with digital versions. no considerable change is made to the content, only the way it is delivered changes.

Augmentation involves including interactive digital enhancements and features like comments, multimedia, and hyperlinks while keeping the content unchanged, however, enabling the students to take advantage of the digital features enhancing the lesson delivery. For example, students use digital portfolios to create multimedia presentations, allowing more space for creativity and expression, gamify quizzes with digital tools like Kahoot and Socrative. Creating virtual bulletin boards with the help of apps such as Padlet where pictures, links pictures, and questions can be posted by students.

The modification would involve the use of a learning management system such as Moodle, Schoology Google Classroom, or Canvas for managing logistical facets of running a classroom, such as messaging students tracking grades, creating a calendar, and posting projects and assignments.

Redefinition fundamentally transforms learning, enabling actions that were earlier impossible to perform in a traditional classroom. For example, with Virtual Pen Pals students can connect with other students and experts in a field anywhere across the globe. Virtual field trips visit locations such as the Louver Museum, Amazon, or the Pyramids. With technology, Students can create a write their blogs and wikis to which others write and respond with the help of Platforms such as Quad blogging that connect different distant classrooms.

With a clear understanding of how technology can create deeper engagement, improve lesson delivery, and help deliver more authentic realistic learning, making the right choice of which tools and learning management systems to choose becomes relatively easy allowing the investor to derive maximum benefit for everyone using the technology.

8.6 The Challenges and Limitations of AI in Education

Like any other technology, AI and its application in the education sector do have some limitations and concerns around it [23]. While there is a definite clarity around the benefits AI can lend, it is equally important to stop and think as to what can go wrong and where to draw the line (Fig. 8.4).

Expensive AI Software: While the intent of adopting AI in the education system is to bring down the cost of attaining quality education, acquiring AI software for

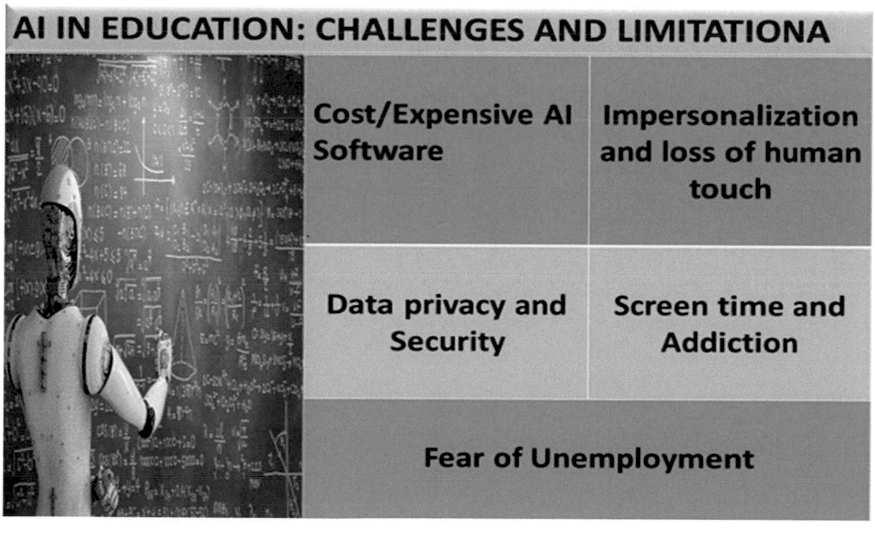

Fig. 8.4 Challenges and limitations of AI in Education

the same purpose might itself cost a fortune. Coupled with the cost of Installation repair, upgrade, and maintenance, AI adoption is surely an expensive affair, that as of today only well-funded education institutes can afford. Though one of the major benefits of including AI is assumed to be enabling the reach of quality education to everyone including the poor and underprivileged, in the current scenario this seems like a distant dream.

Impersonalization and loss of human touch: Learning is a social process and at the heart of learning lies the student–teacher interaction that facilitates the transfer of knowledge with love care and empathy. With AI tutors replacing or reducing the interaction among educators and their pupils, the risk of over-reliance on technology, weak social skills, and compromising thc holistic development of students cannot be written off.

Screen time and Addiction: As AI software requires students to engage and learn on digital devices, their excessive exposure to screen time, the internet, and possible addiction to technology are concerns that are real and need to be addressed efficiently. A large part of the education system deals with very young students, and they cannot be left unattended without any supervision or guidance with access to the internet and technology for long hours.

Data privacy and Security: In today's world as data is power, protecting important and confidential data would remain to be a challenge even after adopting AI solutions as there is a constant attempt by hackers to breach through robust system firewalls and protocols to gain access to data. When it comes to student data, it can be manipulated and misused in multiple different ways which is a real and serious concern.

Fear of Unemployment: As AI can automate a host of important tasks related to management teaching and administration, the application of AI in the education sector may lead to sizeable downsizing and unemployment.

The limitations or the concerns of including AI-augmented systems in education should be addressed seriously as the implications can be quite appalling. AI should be brought in to assist the educators to improve the process of learning and to make it more fun, and not with the intent of replacing them altogether. As the end-users of technology, in this case, are predominantly young students, human assistance and monitoring are important to avoid any misuse of technology. Besides, the leadership of the educational institutions should have utmost clarity before jumping on to the bandwagon of AI adoption, about the real purpose of leveraging this remarkable technology to be able to reap its benefits and still be able to mitigate any risks.

With time and use, both the benefits and the risks of AI adoption into the education sector have started to surface. As a result, many IT and AI experts are relentlessly working on making improvements to the existing AI algorithms, to make them more effective and to address issues around data privacy and security while creating novel and unique AI applications for the education sector in particular. AI education solutions available today, come with multiple functionalities suitable for different budgets and requirements. It is quite fascinating to learn the latest developments in the field of AI-enabled learning systems and the amazing things they can do.

8.7 Top AI Solutions Their Key Features, and Benefits

AI is a big business and the scope of AI application in Education is huge. As a result, many Software companies are now focusing their effort on capturing revenue from this space. While companies of all sizes big and small are entering the market space of education software, some of the renowned players providing best-in-class AI solutions for the education industry can be identified as below [24].

The world of Education has only recently opened its doors to AI and realized that application and implications are endless. It is just the start of how AI has begun to transform the traditional education system, there is a long way to go for the benefits to reach the grassroots level. The major focus of AI developers working in creating AI Learning solutions would be to design software that addresses a wide range of disciplines, learning needs of different types of learners and learners with special needs while keeping the costs low to ensure that the benefits of this superior technology just do not remain confined to a privileged few (Table 8.1).

Table 8.1 Five key developers providing AI education solutions

Company	Functionality	Key features/benefits
Bitwise Academy	Next-generation AI-powered eLearning platform Provides adaptive and personalized eLearning Addresses learning needs of students at schools and colleges level	AI learning solutions for a wide range of subjects Applied CS, STEM, music, art, games Programming language MIT Scratch, Code.org AppLab Studio, BlueJ for Java
CampusLogic	AI-augmented advising platform Aims at providing financial success to students pursuing higher education Employs automation, advising, and analytics	Online scholarship management Simplified financial aid verification Instant financial aid insights
SightLine	AI-enabled Predictive data modeling Solutions for Small to medium-sized institutions with limited funds and resources Bridges the technology-education gap	Forecasts student success, Expand institutional financial performance Improve Graduation rates, Improved Alumni engagement Eliminates the need for steep learning curves or software
DreamBox Learning	Pioneers in Intelligent Adaptive Technology Especially focuses mathematics education at the Addresses learning needs of pre-kindergarten through 8th-grade students	Offer millions of teaching routes Personalized learning Can dynamically adjust and individualizes training in real-time

(continued)

Table 8.1 (continued)

Company	Functionality	Key features/benefits
Exam Soft	Superior assessment solutions Offers, data collection, aggregation, and analysis Derives insights through advanced assessment analytics	Insights allow faculties to impact learning outcomes Improves students' learning and retention

8.8 Conclusion

Like every other field, AI has begun to make inroads into the world of education as well. The popularity of AI in the education sphere is on a constant rise due to the realization of the immense benefits that it can bring forth for all the stakeholders of the education system. From creating high-quality audiovisual and interactive learning content, personalizing learning to automating mundane administrative workload, AI adoption in the field of education can provide flexibility, and agility to the otherwise rigid system, substantially improving the output. However, most importantly, the ability of AI-augmented learning solutions to enable deep immersive learning for students and address the learning needs of the differently-abled is at the crux of what this technology can help achieve to make a true difference to society. AI comes with the capability to lend a new promising future to the education system and society at large. AI is an astounding technology, which can nurture learning and assist the future generation in a way that enables them to help explore the creativity of the human mind and unleash its complete potential.

References

1. Artificial Intelligence (AI) – What It Is and Why It Matters | SAS India. https://www.sas.com/en_in/insights/analytics/what-is-artificial-intelligence.html. Last accessed 28 Sep 2021
2. 15 Social Challenges AI Could Help Solve. https://www.forbes.com/sites/forbestechcouncil/2019/09/03/15-social-challenges-ai-could-help-solve/?sh=205f76cb3533. Last accessed 28 Sep 2021
3. AI in Education Market Size, Share and Global Market Forecast to 2023 | MarketsandMarkets. https://www.marketsandmarkets.com/Market-Reports/ai-in-education-market-200371366.html. Last accessed 20 Jan 2022
4. COVID-19: Schools for more than 168 million children globally have been completely closed for almost a full year, says UNICEF. https://www.unicef.org/press-releases/schools-more-168-million-children-globally-have-been-completely-closed. Last accessed 14 Jan 2022
5. AI's Potential in Education. https://www.forbes.com/sites/forbestechcouncil/2020/03/03/ais-potential-in-education/?sh=5089b4915201. Last accessed 28 Sep 2021
6. UNESCO: 264 Million Children Don't Go to School | News | DW | 24.10.2017. https://www.dw.com/en/unesco-264-million-children-dont-go-to-school/a-41084932. Last accessed 16 Jan 2022
7. Third of World's Poorest Girls Denied Access to School - BBC News. https://www.bbc.com/news/education-51176678. Last accessed 14 Jan 2022

8. 4 Types of Learning Styles: How Do Students Learn Best? | BAU. https://bau.edu/News/types-of-learning-styles/. Last accessed 28 Sep 2021
9. Using Artificial Intelligence in Education: Pros and Cons. https://theknowledgereview.com/using-artificial-intelligence-in-education-pros-and-cons/. Last accessed 28 Sep 2021
10. Half of Children with Disabilities Still Excluded from the School System | HI. https://hi.org/en/news/half-of-children-with-disabilities-still-excluded-from-the-school-system. Last accessed 16 Jan 2022
11. TechnoFunc - Challenges in Education System. https://www.technofunc.com/index.php/domain-knowledge/education-domain/item/challenges-in-education-system. Last accessed 28 Sep 2021
12. Rote Learning vs. Meaningful Learning | Oxford Learning. https://www.oxfordlearning.com/difference-rote-learning-meaningful-learning/. Last accessed 28 Sep 2021
13. Skilling the Youth Ecosystem: Can It Be an Antidote to Society's Overstretched Job Scenario? | TERI. https://www.teriin.org/article/skilling-youth-ecosystem-can-it-be-antidote-societys-overstretched-job-scenario. Last accessed 16 Jan 2022
14. A Brief History of Online Education | Adamas University. https://adamasuniversity.ac.in/a-brief-history-of-online-education/. Last accessed 28 Sep 2021
15. 10 Best Examples of VR and AR in Education. https://www.forbes.com/sites/bernardmarr/2021/07/23/10-best-examples-of-vr-and-ar-in-education/?sh=2b9bd56c1f48. Last accessed 28 Sep 2021
16. 10 Ways Virtual Reality Is Already Being Used in Education | InformED. https://www.opencolleges.edu.au/informed/edtech-integration/10-ways-virtual-reality-already-used-education/. Last accessed 28 Sep 2021
17. Augmented Reality In Education - eLearning Industry." https://elearningindustry.com/augmented-reality-in-education-staggering-insight-into-future. Last accessed 28 Sep 2021
18. Cumming, G.: Artificial Intelligence in Education: An Exploration, vol. 14, no. 4 (1998)
19. 7 Benefits of AI in Education -- The Journal. https://thejournal.com/articles/2021/06/23/7-benefits-of-ai-in-education.aspx. Last accessed 28 Sep 2021
20. (14) 5 Ways To Improve Learning Outcomes With Artificial Intelligence And Machine Learning... | LinkedIn. https://www.linkedin.com/pulse/5-ways-improve-learning-outcome-artificial-machine-abhinav-singh/. Last accessed 28 Sep 2021
21. Wang, Y.: Artificial intelligence in educational leadership: a symbiotic role of human-artificial intelligence decision-making. J. Educ. Adm. **59**(3), 256–270 (2021). https://doi.org/10.1108/JEA-10-2020-0216
22. SAMR: A Powerful Model for Understanding Good Tech Integration | Edutopia. https://www.edutopia.org/article/powerful-model-understanding-good-tech-integration. Last accessed 20 Jan 2022
23. 15 Pros and 6 Cons of Artificial Intelligence in the Classroom – LiveTiles. https://livetilesglobal.com/pros-cons-artificial-intelligence-classroom/. Last accessed 28 Sep 2021
24. Top 10 Artificial Intelligence Companies | by Education Technology Insights | Medium. https://medium.com/@educationtechinsights/top-10-artificial-intelligence-companies-11586ed7afcb. Last accessed 28 Sep 2021

Part III
Emotion and Mental Health

Chapter 9
Using Deep Learning to Recognize Emotions Through Speech Analysis

Arion Mitra, Ankita Biswas, Ananya Ghosh, Ahona Ghosh, Souptik Kumar Majumdar, and Jayati Ghosh Dastidar

Abstract Emotion recognition is the identification of emotions usually through verbal communication and facial expressions such as happy, angry, sad, etc. Not only on the basis of a wide spectrum of moods, but different emotions can also be recognized in order to track mental health of as many people as possible for societal well being. Inside positive it detects specific emotions like happiness, satisfaction, or excitement -depending on how it's configured. The main principles involved in the implementation of our sentiment recognition system that identifies various emotions: anger, happiness, depression, neutral, etc. are audio content and identification of the emotion associated with it. The application developed takes audio input, applies Mel-Frequency Cepstral Coefficients (MFCC) algorithm on it, compares them with those of the content of the existing audio file database depicting various human sentiments, and presents output in the text the emotion expressed by the user. The input from testing was gathered and meaningful spectral coefficients were extracted and stored in a database for comparison with future audio samples. The application extracts the coefficients of the external audio sample and matches it with those present in the database. MFCC algorithm is used to extract the spectral coefficients which are good and can be used for feature matching purposes discarding any static and background noise if present. We have done comparative analysis on our models for their performance evaluation, using four classification metrics and also presented the confusion matrix for better understanding.

A. Mitra · A. Biswas · A. Ghosh
Department of Computer Science and Engineering, University of Calcutta, Calcutta, India

A. Ghosh (✉)
Department of Computer Science and Engineering, Maulana Abul Kalam Azad University of Technology, West Bengal, India
e-mail: ahonaghosh95@gmail.com

S. K. Majumdar
Deloitte USI, Bengaluru, India

J. G. Dastidar
Department of Computer Science, St. Xavier's College, Kolkata, India
e-mail: j.ghoshdastidar@sxccal.edu

A. Biswas et al. (eds.), *Artificial Intelligence for Societal Issues*, Intelligent Systems Reference Library 231, https://doi.org/10.1007/978-3-031-12419-8_9

Keywords Speech analysis · Emotion recognition · Mel-frequency cepstral coefficients · Long short term memory · Convolution neural network

9.1 Introduction

In the age of artificial intelligence and machine learning, communication between machines and humans is becoming increasingly important. Starting from Semantic Search Engines on the web to Natural Language Processing or Personal Intelligent Digital Assistants in our mobile devices, fetching information about anything and everything has become quite easy. However, human communication is not only about fetching information. Communication is also about understanding one another. Therefore, the next step in achieving better communication between machines and humans is for the machines to understand in what context the communication is taking place, which is evident from [1]. From [2, 3], we can conclude, this has been already achieved to quite some extent through Digital Assistants in our devices. However, there is still one basic attribute missing in human-machine communication: Sentiment. The ability of a machine to understand one's sentiment while communicating is still quite limited or doesn't exist. Mishra et al. [1] have shown that most of the existing systems try to understand sentiment by linguistics. This method might be inaccurate as one may hide to try their true emotions by speaking dishonest words or their statements can be wrapped with sarcasm, where the tone or the context might be important to understand the actual emotion [4]. Similarly detecting emotion through video capturing may not always be the best option. It is a usual trend among people to become self-conscious when in front of a camera [5, 6]. This hides the candid emotion that the person might be actually feeling. Also, capturing and analyzing emotions through video in real-time requires much higher computing power thus resulting in hardware limitations [7]. An alternate way to recognize sentiment is to understand how one is speaking by analyzing the voice of the user. This overcomes the problem of people hiding their true sentiments with false words. Our paper on emotion recognition focuses on this aspect of emotion recognition. Incorporating this feature in digital assistants may further improve human-machine communication in daily lives.

Our chapter endeavours to explore the field of speech analysis to identify the various features of speech to understand the multitude of human emotions. A deep learning model was developed which was trained with a large number of recorded voices in datasets and classify the various emotions and accordingly detect the emotion in real-time from users. The purpose of this paper is to create an application that will make the recognition of human sentiments possible through speech only. That is to understand "how" the user is speaking rather than identifying "what" the user is speaking.

Emotions play an essential rolc to identify the mental state of human beings. There have been several opinions about how emotions or sentiments of a person affect his/her behavior or mental health, i.e., what are their direct impacts on self-confidence, creativity or social-commitment, indirectly mental health [8]. Although the concept of emotions in behavioral science has been stated less explanatory and confusing [9], compared to the concept of activation; contradictory arguments have also been raised by the statement, sentiments contribute a lot toward motivating, organizing, and sustaining behaviors of a person [10, 11]. Now coming to the mental health issues, although symptoms of depression mostly relate to threat processing, on the other hand, the main regulators of threat processing, for example, prosocial and affiliative communications with self and others should also be analyzed and integrated into psychiatric formulations and interventions, which may improve the ways of understanding issues of mental health and their prevention and cure eventually [12]. Thus, identifying the emotions from the speech of users can also be valuable in the medical field to detect any mental issues such as depression, anxiety, etc. So, if it is possible to map which emotions tend to cause depression then we can prevent depression from becoming severe, in advance. Even capturing speech samples and identifying any speech anomalies by an automated system can also help in providing early diagnosis of any disease. This paper's contributions are listed as follows:

- It was successful in identifying human emotions/sentiments from the input speech in real-time
- It pointed out the limitations of using these algorithms in a noisy environment as well as limitations of dialect i.e., sentiments depend upon culture and location too.
- It pointed out the lack of sufficient audio data in order to train the deep learning models.

The next section discusses the related works, followed by the proposed methodology in Sect. 9.3. Section 9.4 contains experimental results, Sect. 9.5 discusses the performance evaluation, Sect. 9.6 contains the conclusion, and finally concluded by Sect. 9.7 discussing the future scopes of our paper.

9.2 Related Works

Identifying or recognizing emotions expressed by human beings is referred to as emotion recognition. We can use artificial intelligence or tools to understand and recognize emotions. When it comes to the resolution of mental health problems, as per the scholarly articles [13–16], emotions can create three major dominating functions, classified as:

1. Emotions, serving the functions of detecting threats and as a result, analyzing defending and protection-based strategies. For example, disgust or anger.
2. Emotions, serving the functions of seeking/acquiring resources, energizing, where survival is the driving force. For example, excitement or vitality.
3. Emotions, serving the functions of contentment, calming, settling, being satisfied. For example, safety, happiness.

These classes can be seen as rooted in patterns of (neuro) physiological activation that are always co-regulating and bending, plus, affiliative relations can be related to all these three. This can be explained with three examples for each class, being in the presence of our caring loved ones makes us feel safe, happy, and calm even during depression; A threat to our loved ones raises sorrow, anxiety, fear, or anger to those who are threatening; Spending some moments with family or friends, i.e. the loved ones can be enjoyable, energizing and exciting; [17–20] So, It is evident that a social relationship, signalling help/positive support or threat, have powerful psycho-physiological regulating impacts. The main reason behind this lies in our root of evolution of being socially interactive; So, a conclusion can be drawn here, that the driving force behind our intelligence has always been the sociality of humans [21]. This is why identifying the emotions has always been an interesting area of research.

We now have speech recognition as something that is commercially successful and has found its use in our everyday lives, the biggest example being Google's digital assistant, Google Assistant, and Samsung's digital assistant, Bixby. Collaborating emotion recognition with such successful speech recognition applications can make human-machine communication even more natural and easy. Several existing machine learning classification algorithms have been reviewed by Ingale et al. [22], to classify different moods or emotions. To identify the emotions of a speaker, several parameters of speech are taken into consideration [23] and these speech features are extracted to detect the presence of a particular emotion. SVM, KNN, HMM, GMM all these popular ML algorithms are reviewed in different literarure [22, 24]. Apart from these methods, vector quantization (VQ) and deep neural nets are also reviewed in existing study [25]. Similarly, in the survey paper of Basharirad [26], many more ML algorithms on different datasets are explored using different datasets but these models' performances are tallied by the emotion recognition rate metric. A dataset has been prepared from 30 experimental subjects by Davletcharova et al., where an interesting inference is drawn that rather than a group of people, if data from an individual subject is collected in different situations, then recognition of emotion should be more accurate. A sequence of phases to process speech data and how the retrieved results can be applied to real-life problems are discussed by VH et al. [27], along with their analysis (Table 9.1).

Logan et al. [28] have gone one step further to analyze speech data for music classification and Nandi et al. [29] analysed speech parameters to identify different musical instruments, instead of emotion recognition. Coming back to the sentiment analysis by speech, the speech data used by Maghilnan et al., is somewhat different from the others, as the entire analysis is done on conversations between two individuals, where at a particular time instant, only one person speaks and his/her emotion has been classified into negative, neutral and positive sentiments. The scholarly article [30], has performed sentiment analysis on audio-visual data, where the audio part is extracted using Automatic Speech Recognition system and after conversion of this audio into text, a text-based sentiment extraction system is used for emotion classification into different categories. But, this approach fails, in case the speaker uses different tones or high variation in voice pitching to express their thoughts, as texts have the limitation of not being able to use speech parameters (like speaking tone, pitch, volume) for sentiment analysis.

9.3 Proposed Methodology

The proposed approach is divided into two parts. Feature extraction using Mel-Frequency Cepstral Coefficients and classification of the inputs based on the emotions. The first part i.e. the feature extraction consists of arranging the dataset properly by dividing the audios based on the 5 emotions that we are trying to predict with our model. From these 5 sources, we then read and store the audio files. These audio files are then passed through the MFCC algorithm. For each audio file, the algorithm generates a corresponding vector of coefficients(Mel-coefficients). We store the required number of coefficients (13 in our case) and discard the rest. So we have now obtained a data matrix where the rows identify the audio files whereas the columns identify the corresponding coefficients. This matrix is then used to train our deep learning models, LSTM and CNN individually. The best models produced are saved for future predictions. To predict an emotion from a new recorded audio file, we use the saved model generated which performs the prediction. The general structure of this model is displayed as a flowchart in the Fig. 9.1.

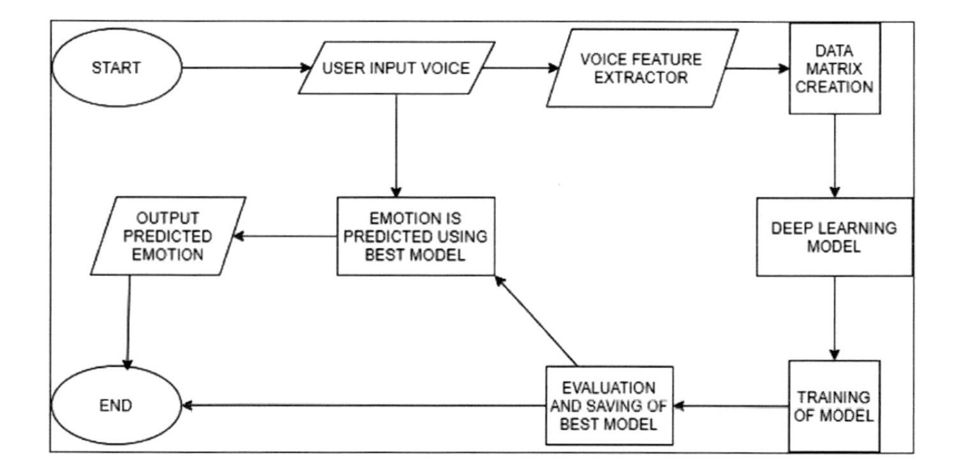

Fig. 9.1 The flow of control and data in the proposed application

Table 9.1 Comparative study of past works

Refs.	Objective	Methods used	Result	Benefits	Loopholes
[25]	Recognising emotion from different speech databases through deep learning	SVM, ANN, Hybrid Rule-Based K-mean clustering, Convolutional DNN, etc.	Most preferred feature: MFCC, Most suitable dataset EMO-DB.	Precise comparison among deep learning models	There is a need to improve the accuracy.
[26]	Comparative performance and limitation based study of the speech emotion recognition approaches	Bayesian Logistic Regression, Decision Tree, Random forest, k-nearest neighbors, Linear discriminant	Hidden Markov Model considering the LFPC speech feature gave the best result	Various types of speech features are taken into consideration	Even after considering the best model, the accuracy will ultimately depend upon the size and type of the dataset
[31]	Comparative study to observe speech feature variants for different emotions state of speech	Naive Bayes, Lazy IB1, Radial Basis Function Network, Logistic, Ada Boost M1, Random Tree	After Area under the ROC curve analysis, It was found that taking data from a single subject was more efficient than group	Observation was Recognizable speech features' association with a particular emotion	Data obtained from various subjects for sentiment analysis results in lower accuracy
[28]	Examining the application of MFCC for music modeling	Discrete Cosine Transform in order to decorrelate the Mel-cepstral vectors	MFCC gave significantly better results than linear-based cepstral features statistically	Verified that MFCC for music modeling is not harmful if not better	Not confirmed if MFCC is specifically appropriate for music modeling or not
[29]	To classify sounds of musical instruments by extracting MFCC features	Support Vector Machine by using Minimum Distance Classification scheme	Extracted MFCC features from the audio, removing noise, and fed to an SVM classifier for instrument recognition	Butterworth filter of order 1 used to remove noise, did not show ringing effect	MFCC is not robust against noises
[32]	For Analysis of various techniques to discriminate the speaker and analyze sentiment finding efficient algorithms	Naive Bayes, Linear SVM, VADER (Valence Aware Dictionary for Sentiment Reasoning)	Audio chat between two people was converted into text by speech recognition tools. The user and their emotions were identified	For Twitter and Movie Review Dataset, VADER gave the best accuracy, followed by linear SVM	Doesn't work well with larger dataset. Can't handle scenarios involving more than two people or if they talk at the same time
[30]	Analyzing sentiment from a given speech by converting it into text	Naive Bayes, SVM	Relevant audio features are extracted from videos, and the sentiment is detected using a classifier	Wide spectrum of sentiment classes	Text-based approaches may fail if the user tries to obscure emotions.
[1]	Automatic extraction of readers' gaze and used as features + textual features for detection of sarcasm and sentiment polarity	CNN	From labelled datasets of gaze, CNN fitted on i) Text features alone ii) Both auto-learned text and gaze. Tallied based on handcrafted textual and gaze features	CNN with both gaze and text data yielded better precision and f1 score compared to others	No optimization technique was applied, such as hyperparameter tuning. The recall score came better for some other model

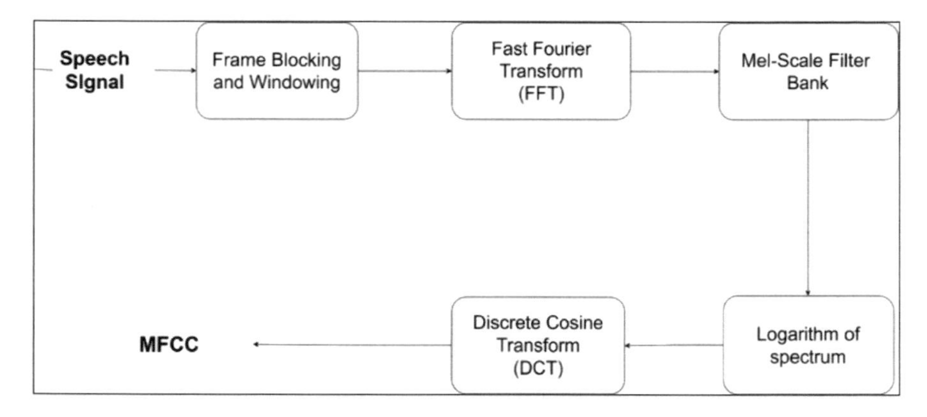

Fig. 9.2 Block diagram of MFCC

9.3.1 Mel-Frequency Cepstral Coefficients

Mel-Frequency Cepstral Coefficients (MFCC) works with the basic principle as to how the human ear perceives sound [33]. MFCC features recognize the critical bandwidths which the human ear can recognize by using frequency filters that are spaced linearly in lower frequencies and spaced in a logarithmic way at higher frequencies. This is done to preserve the phonetically important properties of the speech signal. Since these signals contain tones of different frequencies, each tone with an actual frequency along with the pitch is computed on the Mel scale. The Mel scale has linear frequency spacing below 1000 Hz Hz and logarithmic spacing above that value. Pitch of 1 KHz and 40 dB is defined as 1000 Mels and is used as a reference point for the calculation. Figure 9.2 illustrates the working of the MFCC algorithm and its various components.

The steps which are carried out for signal processing using the MFCC algorithm can be summarized as follows:

1. The input audio signal is at first converted from a 2-dimensional array to a 1D array.
2. This array is broken into frames containing N samples each.
3. In order to prevent any sort of loss of information, the frames are separated by M samples where M is strictly less than N.
4. Thus, initially, the first frame will be having N.M samples and then onwards all of them will begin from M samples after previous samples. Therefore, these frames overlap with N-M in order to prevent loss of transformation. All the values for N are taken as 256 and all the values of M are taken as 100.
5. Each frame now consists of N samples which are now transformed into the frequency domain by using FFT(Fast Fourier Transform). FFT is basically a fast algorithm that is used to implement the conversion to the frequency domain from the time domain.

6. After this, the DFT(Discrete Fourier Transform) is applied to find out the magnitude spectrum, and further along, it is again transformed into the Mel frequency.
7. The last step is to take the logarithm of the spectrum in order to get the cepstral coefficients by performing DCT(Discrete Cosine transform). Only 2–13 coefficients obtained from DCT are kept, the rest are discarded.

9.3.2 Prediction Models Using Neural Networks

Neural Networks or Artificial Neural Networks (ANNs) form the basis of deep learning algorithms. These are loosely structured as human brains which mimic the way neurons send signals to one another. ANNs comprise an input layer (depending upon the input size of the training set), several hidden layers, and an output layer (depends on the number of classes since we are dealing with multi-class classification).

Recurrent Neural Networks with Long Short-Term Memory. LSTM or Long short-term memory follows the architecture of RNN, i.e., artificial recurrent neural networks [34]. Normally, one input gate, one output gate, one forget gate, along with a cell, altogether compose a common unit of LSTM as shown in Fig. 9.3. For some random time period, the cell remembers values stored in it, and the regulation of information flowing in and out of the cell is done by the three gates. For our emotion classification based on speech data, LSTM networks are very suitable, as in these cases chances of lagging between some important events for some unknown amounts of time are high. Overall other sequence learning methods, hidden Markov models, and RNNs, due to its relative sensitivity for the distance between two consecutive

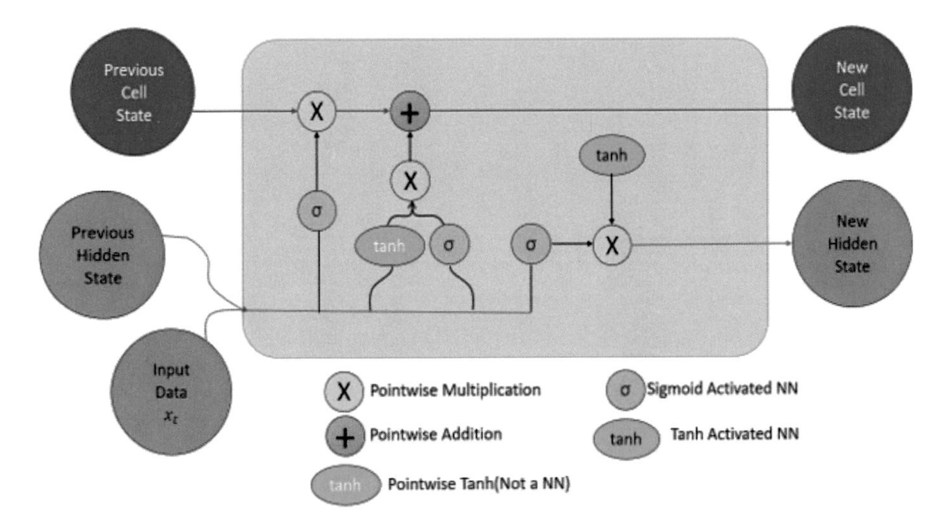

Fig. 9.3 Architecture of LSTM model

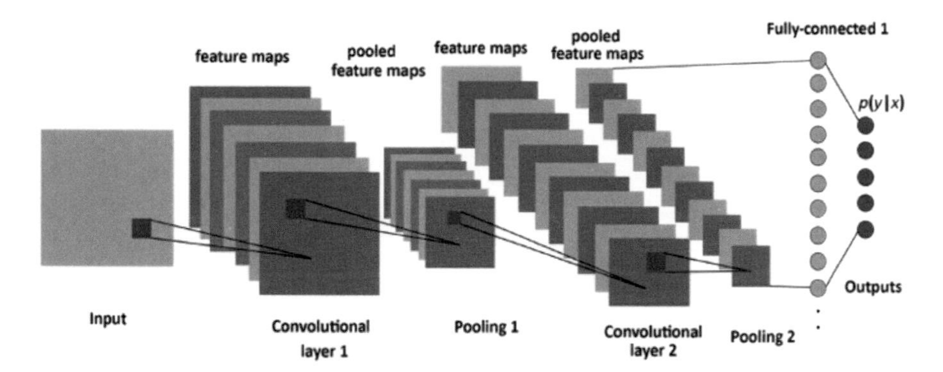

Fig. 9.4 Architecture of CNN model

gaps, insensitivity towards vanishing and exploding gradients, LSTM proves to be more beneficial in numerous applications. In LSTM models due to the presence of the memory cells, it can understand the context of the input data along with capturing long-range dependencies in the continuous stream of data such as in our case where it is imperative to understand the stream of MFCC encountered and identify the context between them which is used to understand the emotion of the input audio stream.
Convolutional Neural Networks. In a class of deep neural networks called CNN [35] or ConvNet or convolutional neural networks, Various perceptron of multiple layers are designed in such a way that the lowest amount of pre-processing will be required, instead they learn the filters themselves, so that prior knowledge or manual effort to design features is not required. The biological procedures in which resemblance can be found between the organization of the visual cortex of animals and pattern connectivity among the neurons inspired the birth of ConvNet. Each Cortical neuron, separately, comes up with a response to stimuli only in the receptive field, i.e. a particular area of the visual field, which is restricted as shown in Fig. 9.4. To cover the entire visual field, partial overlapping of the receptive fields of separate individual neurons are observed.

CNN is predominantly used to determine the features from an image. However, just like an image can be represented as an array of pixel values, here, we are using the array of cepstral coefficients for each audio stream. For sequential data just like this, we have used one-dimensional convolutions to identify the patterns from the data.

9.3.3 Performance Metrics

We have evaluated the performances of our models in terms of the classification metrics recall(sensitivity), precision, f1 score, accuracy, and specificity. We have also generated confusion matrices for a better understanding of our results [36, 37].

To visualize the performances of our models in terms of specificity and sensitivity, the ROC (Receiver Operating Characteristic) curve has also been plotted for each model [36, 38]. The performance metrics used are as follows

1. Accuracy: Percentage of correct predictions.
2. Precision/Positive Predictive Value: Positive Predictive Value = True Positive / (True Positive + False Positive)
3. Specificity/True Negative Rate: True Negative Rate = True Negative / (True Negative + False Positive)
4. Sensitivity / Recall/Hit Rate/True Positive Rate: True Positive Rate = True Positive / (True Positive + False Negative)
5. F1-Score: F1-Score = 2(Precision * Recall) / (Precision + Recall).

9.4 Experimental Result

The experimental outcome will be discussed in this section and the performance of our proposed approach will be evaluated.

9.4.1 Dataset Preparation

For creating the dataset, we obtained audio data from

- Ryerson Audio-Visual Database of emotion speech and song. (RAVDESS)
- Berlin Database of Emotional Speech
- Surrey Audio-Visual Expressed Emotion (SAVEE).

The audio dataset contains multiple audio files in .wav format containing various human emotions of various test subjects (both male and female). The following steps are followed for preparing the raw data for training:

- Dataset is divided according to various emotions:
  - Happy
  - Sad
  - Neutral
  - Angry.
- Every file is read and for each of the files, a separate matrix of coefficients is generated which is used for training the neural network.
- Each file contains a specific emotion i.e., the sound files contain categorical data.

It is extremely necessary to pre-process the data before being put into the machine learning model for training and testing. Machine Learning is based on mathematics

Table 9.2 Audio file (wav) properties

Property name	Value
Sampling Rate	16000 Hz
Channels	1 (Mono)
Chunk (Frames/Buffer)	1024
Length of audio taken for prediction	3 s
Format	Wav (Uncompressed)

and therefore relies on mathematical equations. However, data used for training a model can contain categorical classifications. In our case, we define an emotion by a 2-dimensional matrix. It is mathematically impossible to equate a matrix with a category (string) without proper encoding. Therefore, each emotion in our model is encoded to a simple number first by Label Encoder Class and then further encoded into a binary code by a One Hot Encoder Class. The above data design is used for testing, training, and predicting purposes. The full dataset is divided into 2 parts randomly:

- Training set: The part of the dataset which is used for the training of the model.
- Test set: The part of the dataset which is used for the testing of the model.

The ratio of the size of the training set to the size of the testing set is kept a modest 4:1. The audio file properties are summarized in Table 9.2.

9.4.2 *MFCC Extraction*

Windowing of signal, using the Discrete Fourier Transform, considering the magnitude's log, and finally warping of frequencies on a Mel scale, and using the inverse Discrete Cosine Transform are all part of the MFCC feature extraction technique. Depending upon an assumed mean signal length, same number of frames are extracted for each audio file and then Mel-frequency coefficients are stored in a matrix. Below is a full discussion of the steps to perform the MFCC feature extraction.

1: Start reading the audio files present in the folders
2: For each audio file present, store the values of frame size, signal, and signal length
3: A mean signal length of 32000 frames is assumed for our model. Each wav file is read and the length of the signal is calculated. If the number of frames is greater (>) than the mean signal length, then only the first 32000 frames are taken. Otherwise, if the signal length is lesser (<) than 32000 frames, the signal is padded with blank frames
4: The Mel-frequency coefficients are then stored in a 2D matrix

5: The dataset obtained is divided into 2 parts: test and train where the test contains 20% of the total data.

9.4.3 Training of Neural Network Model

The dataset has been divided into training and testing. x and y became the independent and dependent variables respectively. x contains the Mel-frequency coefficients whereas the emotion labels are stored in y. Below is the step wise procedure followed to the neural network model.

1: Four variables: xtrain, ytrain, xtest, ytest are taken as input from the previous module. xtrain and xtest are 2-D matrices containing the Mel-frequency coefficients. ytrain and ytest are 1-D matrices containing binary encoded labels for emotions corresponding to each of the audio files
2: The four variables are fed to the model
3: Model is trained on the same data approximately 50 times, weights of the Linkages being adjusted on each round (epoch)
4: Model is evaluated after training on the xtest and ytest data set
5: Accuracy is calculated for the model
6: Model is saved using Javascript Object Notation (JSON). The weights are saved using .h5 file

9.4.4 Prediction Using Model

Analysis of emotions, captured from real time audio are carried out by MFCC extraction of wav files, followed by predicting them through different neural network models.The proposed algorithm based prediction procedure has been described step by step below.

1: Record user's voice
2: Save recording as .wav file with properties as mentioned in Table 9.2
3: Saved wav file is read and processed the same way the training set is processed
4: Mel frequency coefficients of the saved file is stored in a variable: xpred
5: Saved model is loaded from disk
6: xpred is fed into the model for prediction
7: Predicted emotion is stored in a variable: ypred which is a 1-D array
8: Predicted emotion is initially a probability distribution. Each probability is linked with a specific emotion. The mode of the probability distribution is considered and the corresponding emotion is taken as predicted emotion.

9.5 Discussion

We have exploited the flexibility to adjust predefined keras models to lean the future of AI algorithms. We observed the internal working of our models empirically by changing the parameters, adjusting the epochs, etc. We also tried to train our models by using subsets of the dataset. This empirical approach has not only provided insight into the functioning of the models but also helped in boosting the accuracy. The performance metrics played a huge role to provide the improvements. The original dataset consisted of four emotions, namely, Neutral, Angry, Happy, Sad.

First, our model was trained using only two emotions, Happy and Sad, which are two opposite emotions with distinct differences in the speech, and hence the model prediction was assumed to work well. The performance accuracy is mentioned in Sect. 9.5.1. The model was then trained on the entire dataset containing four emotions: Neutral, Angry, Happy, Sad, the results of which are discussed in Sect. 9.5.2.

9.5.1 Performance Comparison of CNN and LSTM on Two Emotions

The accuracy of the two deep learning models, LSTM and CNN respectively, are 0.898435 and 0.859375, while predicting 2 emotions i.e. happy and sad. Looking at the values we can understand that the higher accuracy of LSTM denotes that it performs better than the CNN for the prediction of the emotions.Observing the confusion matrices and plotting them as a heatmap in Fig. 9.5, we can infer the performance of our models. The diagonal elements denote the total number of correct predictions made by the model for a particular class. For example, the number of correct predictions for happy (class 0) is 58 whereas the number of correct predictions for sad (class 1) is 57. As the color bar represents, the higher the number, the lighter is the color of the cell. That means diagonals with lighter colors represent higher accuracy of correct predictions, and in Fig. 9.5a, we have lighter colors compared to Fig. 9.5b, which means the LSTM performs better compared to the CNN model. Similar color in the diagonal of Fig. 9.5 indicates a similar performance of LSTM for both the classes, however, CNN performs better in identifying sad emotions than happy.

From precision values, it can be understood, for the Happy class, LSTM and CNN predicted correctly in 91% and 93% cases respectively, out of all positive predictions; for the Sad class, LSTM and CNN predicted correctly in 89% and 81% cases respectively, out of all positive predictions. From recall values, it can be understood, for the Happy class, LSTM and CNN predicted correctly in 89% and 78% cases respectively, out of all positive classes; for the Sad class, LSTM and CNN predicted correctly in 90% and 94% cases respectively, out of all positive classes. As f1 score takes both precision and recall into account hence it is evident that with the increasing f1 score, the model's performance also gets better. For both happy and sad classes, LSTM

Table 9.3 Performance metrics comparison of LSTM and CNN for two emotions

	Precision		Recall		F1 score		Specificity	
	LSTM	CNN	LSTM	CNN	LSTM	CNN	LSTM	CNN
Happy	0.91	0.93	0.89	0.78	0.90	0.85	0.90	0.94
Sad	0.89	0.81	0.90	0.94	0.90	0.87	0.89	0.78

performed better than CNN, with a score of 0.90. From specificity values, it can be understood, for the Happy class, LSTM and CNN predicted correctly in 90% and 94% cases respectively, out of all negative classes; for the Sad class, LSTM and CNN predicted correctly in 89% and 78% cases respectively, out of all negative classes. Overall for two emotions, LSTM outperformed CNN in terms of accuracy, by almost 4% (Table 9.3).

Figure 9.6a shows the ROC curve for LSTM where the 2 emotions along with the baseline prediction (represented by dotted lines), plotted using different coloured curves between the no. of incorrect predictions (along X-axis) and correct predictions (along Y-axis) out of all positive classes. Figure 9.6b shows the ROC curve for CNN where the two emotions along with the baseline prediction (represented by dotted

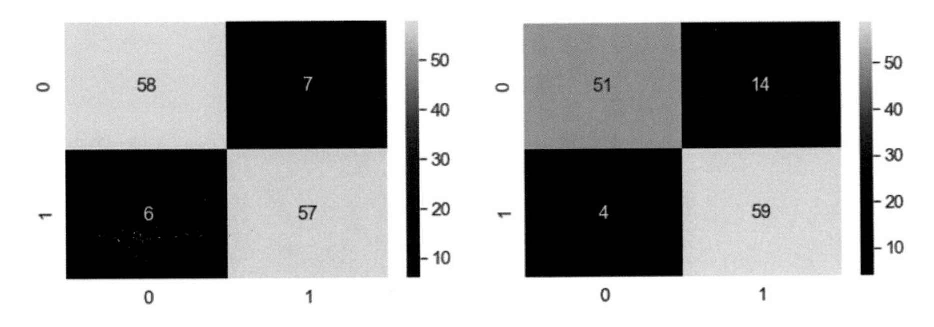

Fig. 9.5 Confusion matrix of LSTM and CNN for two emotions

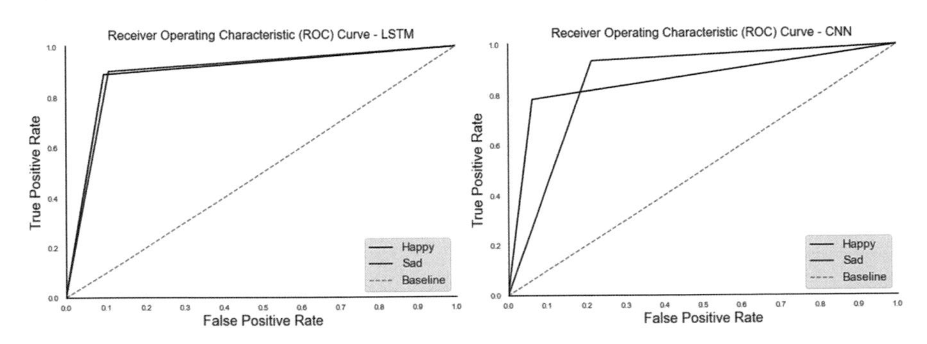

Fig. 9.6 ROC curve of LSTM and CNN for two emotions

lines), plotted using different coloured curves between the no. of incorrect predictions (along X-axis) and correct predictions (along Y-axis) out of all positive classes.

9.5.2 *Performance Comparison of CNN and LSTM on Four Emotions*

The accuracy of the two deep learning models, LSTM and CNN respectively, are 0.837209 and 0.853821, while predicting 4 emotions i.e., neutral, angry, happy, and sad. Looking at the values we can understand that the higher accuracy of CNN denotes that it performs better than the LSTM for the prediction of the emotions. Figure 9.7 provides us the heatmap representation of the confusion matrix generated for the LSTM and CNN models for predicting the 4 emotion classes and help us in understanding the performance of the models in depth. As mentioned before, lighter colors in the diagonal indicate better performance of the models. We observe that for both CNN and LSTM we have similar colors in the diagonal. For the neutral emotion, both the models perform best among all other classes, with CNN performing marginally better. Class 3 or the sad emotion is predicted worst by both models. Angry and happy have similar values with CNN performing marginally better than LSTM. Analyzing the different accuracy of the various classes, it has been discovered that, the best performing class, neutral, has the most number of training data (487 audio files) whereas sad has the lowest number of training data (314 audio files).

From precision values, it can be understood, for the Neutral class, LSTM and CNN predicted correctly in 88% and 83% cases respectively, out of all positive predictions; for the Angry class, LSTM and CNN predicted correctly in 95% and 87% cases respectively, out of all positive predictions; for the Happy class, LSTM and CNN predicted correctly in 82% and 94% cases respectively, out of all positive predictions; for the Sad class, LSTM and CNN predicted correctly in 89% and 88% cases respectively, out of all positive predictions. From recall values, it can be understood, for the Neutral class, LSTM and CNN predicted correctly in 87% and 91% cases respectively, out of all positive classes; for the Sad class, LSTM and CNN predicted correctly in 84% and 89% cases respectively, out of all positive classes.; for the Happy class, LSTM and CNN predicted correctly in 90% and 84% cases respectively, out of all positive classes; for the Sad class, LSTM and CNN predicted correctly in 72% and 78% cases respectively, out of all positive classes (Table 9.4).

As f1 score takes both precision and recall into account hence it is evident that with the increasing f1 score, the model's performance also gets better. For both happy and sad, CNN performed better than LSTM, 2% and 3% better respectively. For Neutral, CNN and LSTM yielded the same 87% of the score, whereas in the case of the Angry class, LSTM performed 1% better than CNN. From specificity values, it can be understood, for the Neutral class, LSTM and CNN predicted correctly in 94% and 91% cases respectively, out of all negative classes; for the Sad class, LSTM and CNN predicted correctly in 99% and 96% cases respectively, out of all negative

Table 9.4 Performance metrics comparison of LSTM and CNN for four emotions

	Precision		Recall		F1 score		Specificity	
	LSTM	CNN	LSTM	CNN	LSTM	CNN	LSTM	CNN
Neutral	0.88	0.83	0.87	0.91	0.87	0.87	0.94	0.91
Angry	0.95	0.87	0.84	0.89	0.89	0.88	0.99	0.96
Happy	0.82	0.94	0.90	0.84	0.86	0.88	0.94	0.98
Sad	0.89	0.88	0.72	0.78	0.80	0.83	0.98	0.98

Table 9.5 Actual value versus predicted value with LSTM

Post ID	Neutral	$Neutralpred$	Angry	$Angrypred$	Happy	$Happypred$	Sad	$Sadpred$
27	1.0	0.999998450	0.0	0.000000438	0.0	0.000000438	0.0	0.000000580
213	0.0	0.000007709	0.0	0.000054127	1.0	0.999933839	0.0	0.000004331
162	0.0	0.000003773	0.0	0.000003996	0.0	0.000006698	1.0	0.99998576
271	1.0	0.999997735	0.0	0.000000609	0.0	0.000000648	0.0	0.000001044
246	0.0	0.000001520	1.0	0.000002700	0.0	0.000002880	0.0	0.000002847
12	0.0	0.000002590	0.0	0.999839664	1.0	0.000133042	0.0	0.000024704

classes.; for the Happy class, LSTM and CNN predicted correctly in 94% and 98% cases respectively, out of all negative classes; for the Sad class, LSTM and CNN predicted correctly in 98% and 98% cases respectively, out of all negative classes (Table 9.5).

Figure 9.8 shows the ROC curve for LSTM and CNN, where the 4 different emotions along with the baseline prediction (represented by dotted lines), plotted using different coloured curves between the no. of incorrect predictions (along X-axis) and correct predictions (along Y-axis) out of all positive classes (Table 9.6).

Tables 7 and 8 have outlined some of the predictions performed by the models LSTM and CNN respectively. We have randomly chosen a few audio files which are unknown to the model and we have predicted the result. Although we have observed

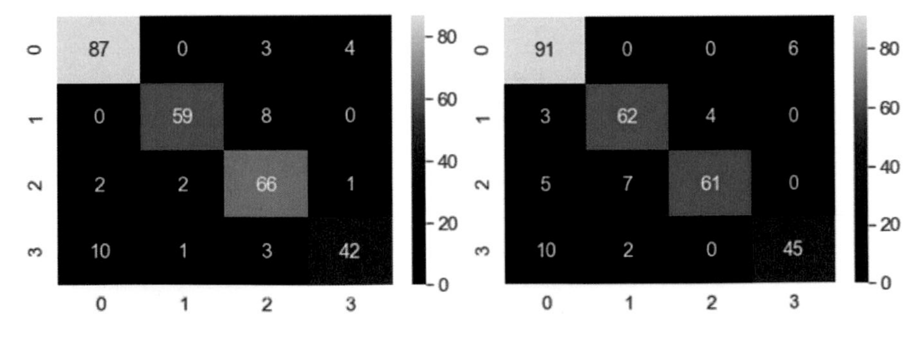

Fig. 9.7 Confusion matrix of LSTM and CNN for four emotions

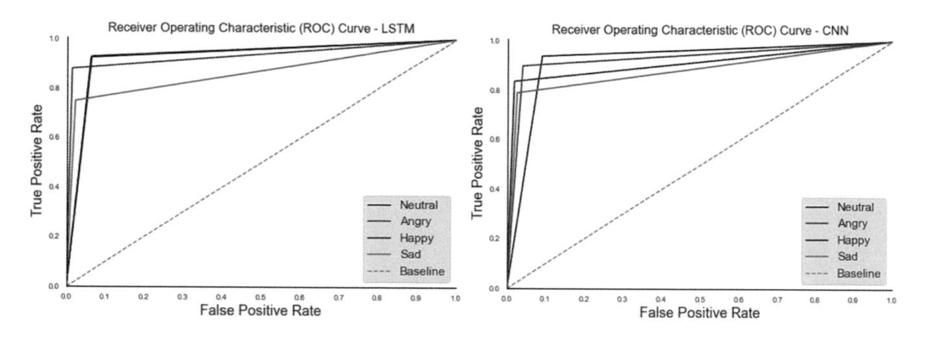

Fig. 9.8 ROC curves of LSTM and CNN for five emotions

Table 9.6 Actual value versus predicted value with LSTM

Post ID	Neutral	$Neutral_{pred}$	Angry	$Angry_{pred}$	Happy	$Happy_{pred}$	Sad	Sad_{pred}
27	1.0	0.998117447	0.0	0.000007290	0.0	0.000005551	0.0	0.001869680
213	0.0	0.000105195	0.0	0.074740238	1.0	0.924470067	0.0	0.000684450
162	0.0	0.000554621	0.0	0.000000178	0.0	0.000083255	1.0	0.999361932
271	1.0	0.410528719	0.0	0.000031837	0.0	0.000565737	0.0	0.588873684
246	0.0	0.000010467	1.0	0.999142647	0.0	0.000837936	0.0	0.000008987
12	0.0	0.000000017	0.0	0.000068238	1.0	0.999926329	0.0	0.000005376

a fairly acceptable accuracy on the testing data, however, while trying to predict emotions from audio files which were live recorded and were fed to the model, we observed much lower accuracy than that was achieved. We have tried to identify the reasons behind this and one of the major drawbacks of our model is the lack of enough data to train the model. Most deep learning models require a huge amount of training data to work with, which is usually considered a drawback of these models. So, the lack of enough data has resulted in a very narrow spectrum of emotions and intonations which are identified properly by our models and has resulted in a lack of a generalized model which can be used by any user with acceptable accuracy. Moreover, a lack of consistency in prediction with real-time data has been observed. This is pertaining to the fact that the model is trained using emotions portrayed by particular speakers in a particular way and in a particular intonation which is not a generalized method. Due to this reason when the pronunciation or the way of speaking varies, the model struggles with the data and makes incorrect predictions.

9.6 Conclusion

The algorithm developed in this chapter can be used to improve the interaction between humans and computers by enabling the computers to recognize the emotions in the voice of the user thus making the communication much more natural. In the 21st Century, the field of speech i.e., voice recognition has already crossed various milestones in the field of technology. It has evolved significantly and is being used daily in our digital life. More complex and specialized applications of voice recognition have come into use. The biggest examples are Google Assistant and Amazon's Alexa. Until recently, voice recognition has mostly been used for speech-to-text conversion, recognition of specific words, etc. We have used some already established principles in our paper to classify audio inputs and recognize the specific emotion depicted by the audio input i.e., human voice. Our aim is to improve upon human-computer interaction as well as diversify this paper into the medical fields whereby capturing speech samples any anomalies or problems can be detected by automated systems and provide an initial diagnosis.

To generalize the models to provide much more consistent predictions, we need to increase the size of the dataset. Moreover, we need to have equal proportions of male and female voice samples for better predictions. Also, the samples must include a variety of intonations and accents of similar sentences. This is to ensure better prediction results for samples that are quite different than those used for training the models. We also aim to include more categories of emotions and compute each of their recognition rates individually. We can further look at other efficient deep learning models for better prediction in terms of accuracy. Multimodal analysis can also prove to be valuable, to understand whether the speech represents the speaker's own voice of opinions or prompts views of another person, or is sarcastic. To incorporate it, we can analyze audio-visual datasets, the actions or facial expressions of the speaker can be the key to differentiate between such situations and also reduce erroneous prediction.

References

1. Mishra, A., Dey, K., Bhattacharyya, P.: Learning cognitive features from gaze data for sentiment and sarcasm classification using convolutional neural network. In: Proceedings of the 55th Annual Meeting of the Association for Computational Linguistics (Volume 1: Long Papers), pp. 377–387 (2017)
2. Rodden, T., Cheverst, K., Davies, K., Dix, A.: Exploiting context in HCI design for mobile systems. In: Workshop on Human Computer Interaction with Mobile Devices, vol. 12 (1998)
3. Squire, K.: From content to context: videogames as designed experience. Educ. Res. **35**(8), 19–29 (2006)
4. Leggitt, J.S., Gibbs, R.W.: Emotional reactions to verbal irony. Discourse Process. **29**(1), 1–24 (2000)
5. How Vanity Affects Video Communication | Highfive. Access Time: 3:20 am Saturday, 15 May 2021 (IST)
6. Somerville, L.H., Jones, R.M., Ruberry, E.J., Dyke, J.P., Glover, G., Casey, B.J.: The medial prefrontal cortex and the emergence of self-conscious emotion in adolescence. Psychol. Sci. **24**(8), 1554–1562 (2013)
7. Salih, H., Kulkarni, L.: Study of video based facial expression and emotions recognition methods. In: 2017 International Conference on I-SMAC (IoT in Social, Mobile, Analytics and Cloud)(I-SMAC), pp. 692–696. IEEE (2017)

8. Izard, C.E.: The psychology of emotions. Springer Science & Business Media (1991)
9. Duffy, E.: Activation and behavior (1962)
10. Izard, C.E.: The face of emotion (1971)
11. Izard, C.E., Tomkins, S.S.: Affect and behavior: anxiety as a negative affect. Anxiety Behav. **1**, 81–125 (1966)
12. Gilbert, P.: Affiliative and prosocial motives and emotions in mental health. Dialogues Clin. Neurosci. **17**(4), 381 (2015)
13. Depue, R.A., Morrone-Strupinsky, J.V.: A neurobehavioral model of affiliative bonding: implications for conceptualizing a human trait of affiliation. Behav. Brain Sci. **28**(3), 313–349 (2005)
14. Le Doux, J.: The Emotional Brain. London: Weidenfeld and Nicholson. Deutsch: Das Netz der Gefühle, München: Deutscher Taschenbuch-Verlag (2001)
15. Panksepp, J.: Affective neuroscience of the emotional Brain. Mind: evolutionary perspectives and implications for understanding depression. Dialogues Clin. Neurosci. **12**(4), 533 (2010)
16. Gilbert, P.: The compassionate mind. Robinson (2009)
17. Gilbert, P.: The evolution and social dynamics of compassion. Soc. Pers. Psychol. Compass **9**(6), 239–254 (2015)
18. Gilbert, P.: Human nature and suffering. Routledge (2016)
19. Keltner, D., Kogan, A., Piff, P.K., Saturn, S.R.: The sociocultural appraisals, values, and emotions (SAVE) framework of prosociality: Core processes from gene to meme. Annu. Rev. Psychol. **65**, 425–460 (2014)
20. Gilbert, P.: The origins and nature of compassion focused therapy. Br. J. Clin. Psychol. **53**(1), 6–41 (2014)
21. Dunbar, R.I.: The social role of touch in humans and primates: behavioural function and neurobiological mechanisms. Neurosci. Biobehav. Rev. **34**(2), 260–268 (2010)
22. Ingale, A.B., Chaudhari, D.S.: Speech emotion recognition. Int. J. Soft Comput. Eng. (IJSCE) **2**(1), 235–238 (2012)
23. Shen, P., Changjun, Z., Chen, X.: Automatic speech emotion recognition using support vector machine. In: Proceedings of 2011 International Conference on Electronic & Mechanical Engineering and Information Technology, vol. 2, pp. 621–625. IEEE (2011)
24. Shaikh Nilofer, R.A., Gadhe, R.P., Deshmukh, R.R., Waghmare, V.B., Shrishrimal, P.P.: Automatic emotion recognition from speech signals: a review. Int. J. Sci. Eng. Res. **6**(4) (2015)
25. Gunawan, T.S., Alghifari, M.F., Morshidi, M.A., Kartiwi, M.: A review on emotion recognition algorithms using speech analysis. Indones. J. Electr. Eng. Inform. (IJEEI) (IJEEI) **6**(1), 12–20 (2018)
26. Basharirad, B., Moradhaseli, M. (2017) Speech emotion recognition methods: a literature review. In: AIP Conference Proceedings, vol. 1891, No. 1, p. 020105. AIP Publishing LLC
27. VH, A., Marimuthu, R.: A study on speech recognition technology. J. Comput. Technol. 2278–3814 (2014)
28. Logan, B.: Mel frequency cepstral coefficients for music modeling. In: Ismir, vol. 270, pp. 1–11 (2000)
29. Nandi, S., Banerjee, M., Sinha, P., Dastidar, J.G.: SVM based classification of sounds from musical instruments using MFCC features. Int. J. Adv. Res. Comput. **8**(5) (2017)
30. Murarka, A., Shivarkar, K., Gupta, V., Sankpal, L.: Sentiment analysis of speech. Int. J. Adv. Res. Comput. Commun. Eng. **6**(11), 240–243 (2017)
31. Davletcharova, A., Sugathan, S., Abraham, B., James, A.P.: Detection and analysis of emotion from speech signals. Procedia Comput. Sci. **58**, 91–96 (2015)
32. Maghilnan, S., Kumar, M.R.: Sentiment analysis on speaker specific speech data. In: 2017 International Conference on Intelligent Computing and Control (I2C2), pp. 1–5. IEEE (2017)
33. Mermelstein, P.: Distance measures for speech recognition, psychological and instrumental. Pattern Recognit. Artif. Intell. **116**, 374–388 (1976)
34. Hochreiter, S.: JA1 4 rgen Schmidhuber (1997)."Long Short-Term Memory". Neural Comput. **9**(8)
35. Atlas, L., Homma, T., Marks, R.: An artificial neural network for spatio-temporal bipolar patterns: application to phoneme classification. In: Neural Information Processing Systems, pp. 31–40 (1987)

36. Fawcett, T.: An introduction to ROC analysis. Pattern Recognit. Lett. **27**(8), 861–874 (2006)
37. Sokolova, M., Lapalme, G.: A systematic analysis of performance measures for classification tasks. Inf. Process. Manag. **45**(4), 427–437 (2009)
38. Bradley, A.P.: The use of the area under the ROC curve in the evaluation of machine learning algorithms. Pattern Recognit. **30**(7), 1145–1159 (1997)
39. Mishra, A., Dey, K., Bhattacharyya, P.: Learning cognitive features from gaze data for sentiment and sarcasm classification using convolutional neural network. In: Proceedings of the 55th Annual Meeting of the Association for Computational Linguistics (Volume 1: Long Papers), pp. 377–387 (2017)
40. http://mirlab.org/jang/books/audiosignalprocessing/speechFeatureMfcc.asp?title=12-2%20MFCC. Access Time: 10:30 pm Tuesday, 2 April 2019 (IST)
41. https://medium.com/mlreview/understanding-lstm-and-its-diagrams-37e2f46f1714 . Access Time: 10:30 pm Tuesday, 2 April 2019 (IST)

Chapter 10
Face Emotion Detection for Autism Children Using Convolutional Neural Network Algorithms

K. M. Umamaheswari and M. T. Vignesh

Abstract Over the past few decades, emotion detection in a real-time environment has been a progressive area of research. This study aims to identify physically challenged individuals and the cognitive gestures of Autism children using Convolutional Neural Network (CNN) based on facial landmarks by creating the emotion detection algorithm. The algorithm is used in real-time by using virtual markers that perform efficiently in the uneven spotlight and head rotation, multiple backgrounds and distinct facial tones. Emotions of faces such as happiness, sorrow, rage, Surprise, Disgust, Fear using virtual markers are collected. Initially, the faces are detected using a cascade classifier. After face image detection, the image is given to the pre-processing stage to remove the noise present in the image. This process is used to increase classification accuracy. Finally, the images are given to the convolution neural network (CNN) classifier to classify the emotions such as happiness, sorrow, rage, Surprise, Disgust and Fear. The performance of the proposed approach is analyzed based on accuracy, precision, recall and F-measure. The proposed Emotion detection using CNN has achieved a cumulative recognition rate of 99.81%.

Keywords Emotion · Autism child · Convolution neural network · Cascade classifier

10.1 Introduction

Computer animated agents and robots acquire new measurement human PC connection which makes it indispensable as what PCs can mean for our social life in everyday exercises. Face-to-face communication is an ongoing interaction working at a time scale in the request for milliseconds. The degree of vulnerability is extensive, making it fundamental for humans and machines to depend on tangible, rich perceptual natives instead of slow emblematic derivation measures. Look recognition is an essential capacity for excellent relational relations and a significant subject of

K. M. Umamaheswari (✉) · M. T. Vignesh
SRM Institute of Science and Technology, Kattankulathur, India
e-mail: umamahek@srmist.edu.in

A. Biswas et al. (eds.), *Artificial Intelligence for Societal Issues*, Intelligent Systems Reference Library 231, https://doi.org/10.1007/978-3-031-12419-8_10

study in human turn of events, mental prosperity, and social change. Indeed, emotion recognition assumes a critical part in the experience of sympathy, the expectation of master social conduct, and the capacity model of emotional insight. Furthermore, the writing shows that impedances in emotional articulation recognition are related to a few unfortunate results, such as challenges in recognizing, separating, and portraying sentiments. For instance, a few investigations have shown a relationship of shortages in emotional look handling with mental problems in grown- ups and kids, alexithymia, and troubles in social functioning.

Emotional recognition is one of the most important skills for promoting social interactions and empathy. It has been suggested that people with autism need a controlled ability to perceive emotions for a wide range of problems "with the hypothesis of the mind". Numerous studies have shown that children and adults with autism have a barrier to perceiving emotions from appearance (1,2), although some studies are reluctant to reproduce these findings.

Several attempts have demonstrated emotional recognition skills in children and adults with autism. Many use recommendations based on unrelated media such as PC or television. Swettenham (1996) cites three explanations for fostering interest in PC use with children with autism: (1) it includes no social components; (2) it is reliable and unsurprising; and (3) it permits the youngster to control the work at their speed. One could contend that for the abilities to sum up to this present reality, the instructing ought to (1) include social elements, (2) be capricious (as are social connections) and (3) the speed is haggled with different members (very much like, all things considered). Consequently, we planned our showing program explicitly to gather intercession that would require the youngsters to rehearse the abilities with different kids.

In particular, a few examinations have proposed that near-adult levels of recognition are accomplished before adolescence. Notwithstanding, it is worth focusing on that formative analysts have utilized a few techniques to gauge how autism youngsters perceive looks of changing emotions, like the separation worldview, the coordinating with strategy and free marking. Subsequently, the capacity to perceive most emotional articulations, to some degree part of the way, appears to be subject to task requests. Moreover, while recognition of, and responses to, looks has been significantly analyzed among youngsters and grown-ups in regular and clinical settings.

The proposed approach's main objective is to effectively identify an autism child facial expression using the CNN classifier. Here, the data are collected from the dataset, and the cascade classifier accurately detects the face images. Then, the noise present in the image is removed. Finally, the image is given to CNN classifier to classify different types of emotion present in the image. The rest of the paper is organized as follows; the various exiting works related to our topic are analyzed in Sect. 10.2, and the multi-model system is explained in Sect. 10.3. The proposed approach is clearly described in Sect. 10.4, and experimental results are interpreted in Sect. 10.5. Finally, the conclusion is presented in Sect. 10.5.

10.2 Literature Survey

One way of expressing their thoughts is through emotions. Recognition of facial expression is one of the most robust, natural and immediate means of transmitting their emotions and intentions to human beings. In some instances, people may be prevented from expressing their feelings, namely illness peoples, or because of shortcomings; hence, greater appreciation of human emotions can result in successful communication.

Natural language processing is used by Intelligent Personal Assistants (IPAs) to communicate with strangers; however, it raises the degree of successful contact and human- level when supplemented with feelings about intelligence. The technology is actively used in many fields because of the rapid development of machine learning and artificial intelligence (AI) used in spam detection, which uses spam classifier to reorganize email based on some basic requirements to transfer unwanted messages and unsolicited archives for emails [1]. Along with being substantially utilized in machine learning for market research to help the broad quantity of data generated daily to detect the possibility of fraud via the customer insurance against fraud [2]. For instance, the enhanced Fraud Miner uses 'Lingo' clustering process that recognizes frequent patterns [3]. Moreover, machine learning-based medical innovations, like patient data, management data, concentrate on affluent clinical data areas [4, 5].

In addition, machine learning has played a vital role in pattern classification and recognition problems, particularly facial expression recognition, in the last few decades [6–8]. Several areas of computing and beyond have been revolutionized, like human interaction with a computer (HCI) [9, 10]. Interaction of human computers has been in our routine life.

Emotion data is the primary aspect of the interaction between humans because it is cost- effective, has standard detection, and has minimum computational time compared to other advantages [10]. Evaluation of facial emotion is attractive and challenging to solve the problem. It has a significant impact on applications like the medical field and man-to- machine interactions. Many works lie on facial detection to retrieve features essential for emotion recognition [16]. The main framework for 68 face images is to determine three different face emotions in reality: positive, negative, and blank using the camera.

The proposed work focused on determining the emotions from 79 features and 26 geometrical features results in approximately 70.65% of accuracy. It calculates 32 geometric face features that contain eccentricity, slope, polygonal and linear [18], lying on 20 face image data for automatic emotion detection. However, a few relatable works have been done to detect facial expressions more accurately in real-time systems that cover many application areas like video gaming, machine vision and behavioural analysis. So, Human-machine interactions are processed.

The deep neural network is used for many fields such as human action recognition, classification, and face recognition and segmentation process. In [20], extreme learning machine approach used for clinical human gaint classification, Boltzmann machine approach was used for joint angle trajectories for humanoid locomotion

[21], for identifying different activities of human deep ensemble learning approach was used [22]. Human activity recognition using wearable sensor is explained in [23–27]. Similarly, deep learning algorithm based human action recognition is explained in [28–31].

The proposed work focused on determining the emotions from 79 features and 26 geometrical features result in approximately 70.65% accuracy, which calculates 32 geometric face features containing eccentricity, slope, polygonal, and linear [18] lies 20 face image data for automatic emotion detection. However, a few relatable works have been done to detect facial expressions more accurately in real-time systems that cover many application areas like behavioural analysis, machine vision, and video gaming. So, Human-machine interactions are processed using human gestures [10, 12–14]. Finally, the still images and emotions are relatable by measuring the distance between the lips and eyes and using many other literature strategies [10, 11, 15].

10.3 Background of the Research

An existing classifier mathematical expression and multi-model system are clearly explained in this section. After this background explanation, the proposed methodology is presented.

10.3.1 Existing Classifier

In this paper, a convolution neural network is utilized for emotion recognition. We compare the proposed CNN classifier with the existing ANN and SVM classifier for comparative analysis.

Artificial Neural Network. The ANN is a computational model that seeks to calculate the parallelism of the human brain. ANN is a network of highly interconnected processing components (neurons). The biological nervous system inspires this method. Like nature, interconnected connections often determine network performance. The element of processing is called a layer in a network of subgroups. It contains of three layers: input, hidden, and output. Let us consider the input neurons $[A_1, A_2, \ldots, A_a]$, Hidden neurons $[B_1, B_2, \ldots, B_b]$ and output neurons $[C_1, C_2, \ldots, C_c]$.Three layers are interconnected based on the weight values. The weight between hidden layers and input is represented, W_{ij}^h and the weight value is represented between the hidden and the output layers W_{jk}^o. The hidden layer output is calculated using the below equation;

$$B_j = \sum_{i=1}^{a} A_i * W_{ij}^h \tag{10.1}$$

After the hidden layer output, the activation function is applied to attained the result. The obtained output value is given to the activation function. The outcome is calculated as follows;

$$F(x) = \frac{1}{1 + e^{-x}} \tag{10.2}$$

$$F(B_j) = \frac{1}{1 + e^{-B_j}} \tag{10.3}$$

Then, obtained hidden layer output is multiplied with weight to obtain the output. The output calculation is given in below equation;

$$C_k = \sum_{j=1}^{b} W_{ij}^{o} F(B_j) \tag{10.4}$$

$$F(C_k) = \frac{1}{1 + e^{-C_k}} \tag{10.5}$$

After the output calculation, an error is calculated. The error is calculated between the target value and obtained output.

$$E = \frac{1}{2} \sum_{k=1}^{c} (T_k - C_k)^2 \tag{10.6}$$

where T_k represents the target value, and C_k represents the output value. The error value will minimize by training the network using the back propagation algorithm.

Support Vector Machine (SVM). The SVM was first proposed by Wapnik and which is mainly used for classification [2]. SVMs are related supervisory learning methods utilized for regression and classification [2]. They belong to the standard linear classification family. A unique property of SVM is that SVM simultaneously reduces the experience classification error and increases the geometric margin. Therefore, SVMs are called upper margin ratings. Consider the training sample (a_i, b_i), $(i = 1, 2, \ldots, m)$. Based on the training sample, the optimal hyperplane is calculated. The hyperplane function is given in Eq. (10.7).

$$f(u) = \omega.\varphi(u) + a \tag{10.7}$$

where; a $\rightarrow$ Threshold value
$\omega \rightarrow$ Weight factor

10.3.2 Multi-model System

Multimodal information is used to build facial expressions [11]. There are drawbacks to facial recognition, such as light strength, face orientation, and shifts in the context. EEG is still suffering from restrictions, such as noise, objects, wired/wireless sensor location, speech noise and interference, speed of gesture-light, and background modifications. Therefore, the combination of both signals introduces a new layer of complexity, which fulfills the necessities of two Modalities. Generally, the difficulty of computing has increased exponentially on account of multimodal networks. The multimodal approach best identifies emotions using the rate's single modal method [19]. However, combined with other modalities, facial expressions from voice signals, gesture emotions cannot be effectively observed in dumb, deaf and biosignals; owing to the above problems, patients are paralyzed while designing intelligent Human Machine Interface (HMI) assistive systems. The presence of HMI is best suitable for physically disabled persons and the patient with additional needs while looking for aid support. To date, many approaches execute offline and are not applicable in many applications [10, 11]. Currently, it is primarily based on creating an intelligent HMI multimodal system. Using facials, it can identify feelings, Machine learning and Deep neural EEG-based expressions and EEG Methods of Networks.

There are seven human facial expressions: joy, sadness, Fear, Surprise, Anger, hatred, and neutrality. During the expression of the face, facial muscles are activated. Signals are sometimes subtle but complex, and the expressions often contain a lot of information about our mood. With facial recognition, we can efficiently and cost-effectively measure the impact of content and services on visitors/users. For example, vendors can get clients' ratings using these facial expressions. Facial expression recognition consists of some steps, which are explained below;

- Initially, Tracing faces in the scene (this is the same as the face detection).
- Features are extracted from the detected facial area (e.g., Shape feature, texture features of face).
- Based on the features, the emotions are classified using convolution neural network. Analyzing the movement of facial features and the appearance of facial features and

Many projects have already been done in these fields. Our goal is to create an automated facial recognition system and improve the accuracy of this system compared to other methods available.

10.4 Proposed Emotion Detection Model

The main aim of this work is to identify the emotions of Autism children from facial expressions effectively. Autism Spectrum Disorder is a complex neurological, behavioral condition. People with autism in this condition cannot make social interactions.

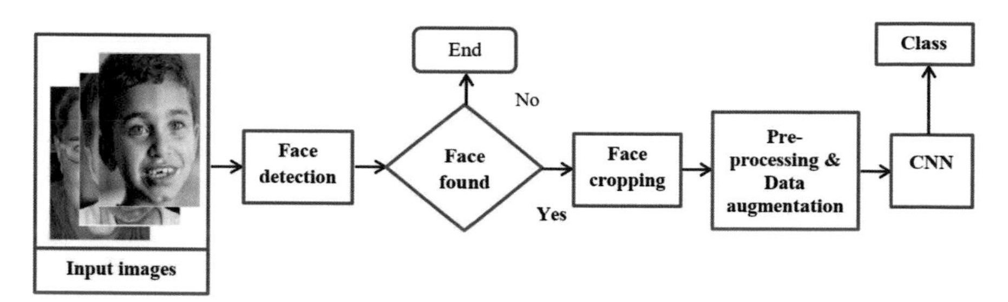

Fig. 10.1 Flow diagram of the proposed approach

People with autism tend to do the same thing over and over again. This study works on the emotional diagnosis of children with autism from facial expressions. This study works on happiness, Anger, Disgust and neutrality. The objective is achieved by using a CNN architecture. The flow of the proposed approach is given in Fig. 10.1.

A_i : i = 1, 2, ..., N be a database image where i the number of images from the database A_c is one of the images in the database A_i. In this proposed approach, initially, the images are collected autism childrenâŁ™s images from the Kaggle website, which consists of 224 × 224- pixel grayscale images of the face. Then, the faces of the collected images are detected by Cascade Classifier [17]. If The face image is correctly detected, the image is given to the cropping process. After the cropping process, the image is given to the pre-processing stage. Here, the images are filtered, and unwanted portions and noise are removed. Data have been augmented by the Image Data Generator function offered by the Keras API. At last, the augmented dataset is fed into CNN to predict the class.

10.4.1 Face Detection

Identifying faces in an image is essential in many computer vision applications, such as recognition and facial expression analysis. Due to variation in the face and environmental conditions, Automatic face detection is a problematic binary classification. We propose a layered facial detection method based on Oriented Gradient (HOG) histograms to gradually exclude facial features using various features and classifiers [17].

10.4.2 Face Cropping

After the detection process, the face portions are cropped. In this process, background and non-face areas are cropped from the face image. The image's cropped face portion is considered input in the next step to enhance facial properties.

10.4.3 Pre-processing and Data Augmentation

Pre-processing is an essential process for classification. It will remove the noise present in the cropped image and resize it. On the other hand, data augmentation is a viable strategy, particularly for image data of making new information from the accessible data for this situation. New data is created by turning, moving, or flipping the first image. The thought is that if we turn, shift, scale, or flip the first image, this will, in any case, be a similar subject, yet the image isn't equivalent to previously. In pre-processing, two steps are available: histogram equalization and feature point extraction.

Histogram Equalization. Histogram equalization is a method of adjusting the variability of a digital image. This step involves applied to change each pixel and obtaining a new image from an independent function on each pixel.

Feature Point Tracking. Feature point tracking system integrates input emotional expressions according to feature point tracking. Figure 3.3 shows how to extract more than 20 features from a home model from a video stream and then monitor these feature points with particle filters to create a variable 3D exposure recognition model. Feature points are evenly distributed without coordination. The CNN matching algorithm is designed to step by step so that the target has a transparent attitude and can quickly monitor the competition when there is a change in size. Extracted features points are given in Fig. 10.2.

10.4.4 Convolution Neural Network-Based Emotion Detection

The input images are fed into the CNN architecture to classify an input image emotion. The main advantage of our proposed work is that it can organize any classification method. CNN [18] is an in-depth learning neuroscience network. The observation

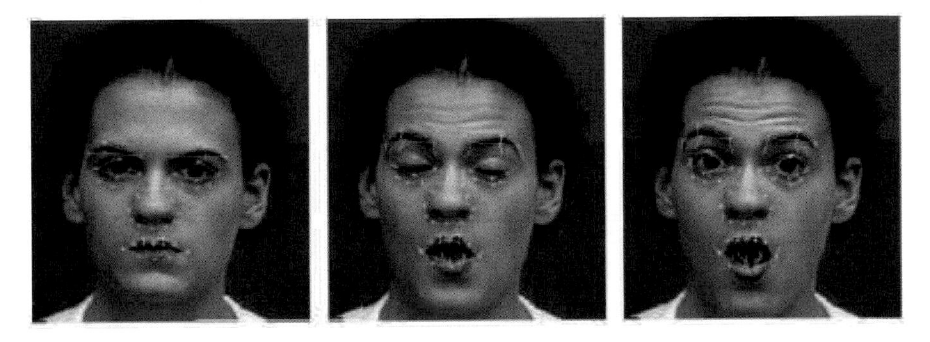

Fig. 10.2 Feature extraction

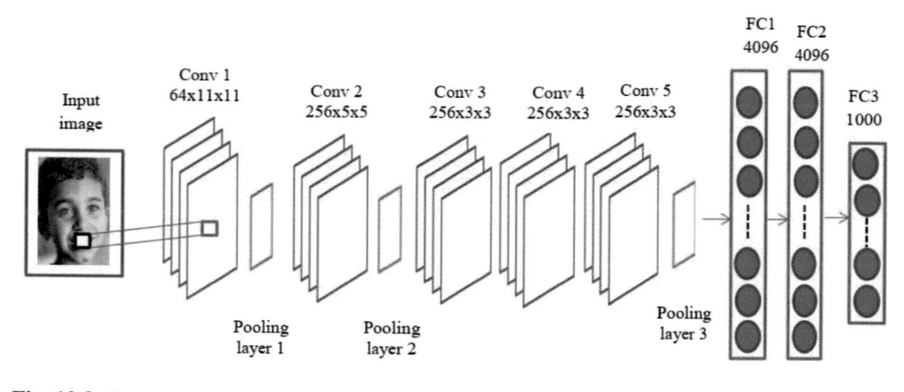

Fig. 10.3 Proposed block diagram in CNN

of the biological process followed this. It reflects the functions of various layers in the human brain. CNN has proven to be very efficient in image processing applications such as face recognition, emotion detection, and method recognition. The output from the rolling layer is provided as input to the highest pooling layer that removes unwanted pixels. The layers involved in CNN is listed below;

- Convolutional (Conv)
- Rectifying linear unit (ReLU)
- Pooling
- Fully connected.

The structure of CNN is given in Fig. 10.3. The proposed structure consists of five convolution layers with three fully connected layers and a softmax layer at the end. Here, the input image is resized to 224 × 224. Initially, the input is filtered through the first displacement layer by 4 pixels, measuring 64 cores 11 × 11 × 3. The filtered output of the first compression layer is taken input by the second curve layer, which again filters with 256 cores. The size of each center of the second displacement layer is 5 × 5 × 64. The output of the fifth coil layer is connected to the fully bonded first layer, which passes through the relay and the voting layer. There are 4096 neurons in the first and second fully connected layers. The second and third fully integrated layers combine the previous and Dropout layers. The input image representing a function goes through five displacements and three fully connected layers, which separate the spatial and temporal properties of the image. Finally, the software layers generate repeat probability scores used for the classification process.

Convolution Layer. The essential layer in the CNN structure is the convolutional layer. The conversion matrix is the filter used to create the feature change map for the input image. Usually, the filter size is taken as 3 × 3 or 5 × 5; the method proposed here is to take the filter size as 5 × 5. The following equation refers to the output of the change layer.

$$B^{b} = \sum_{j \in F_j} B_j^{b-1} \otimes \xi_{ij}^{b} + L_i^{b} \tag{10.8}$$

where; $\otimes \rightarrow$ Convolution Factor
$\xi_{ij}^{b} \rightarrow$ Weight value of ith filter of the bth convolutional layer
$L_{i}^{b} \rightarrow$ Bias of the ith filter of the bth convolutional layer
$B_{i}^{b} \rightarrow$ Activation Map

Pooling Layer. The pooling layer is placed following the convolutional layer in the layered architecture of CNN. It has the same planes as the conventional layer. This layer reduces the space size of the feature map, thereby reducing the computational function of the network. The pooling layer runs separately on each feature map. The most common methods of performing the pooling process are maximum pooling and average pooling. Max Pooling maintains the abstract format of the input image while balancing the average pooling images and making it harder to identify the sharp features of the image. Maximum pooling for image data works best compared to the average pooling layer. Therefore the recommended method uses the maximum pooling layer. Here the image is divided into a matrix of 2 × 2 volumes. An M-size pool feature map is produced during peak pooling operations by passing the F-size feature map over the pooling layer. The below equation represents the pooled map.

$$P_i = Max_{j \in R_j} F_j \tag{10.9}$$

Fully Connected Layer. In this paper, for fully connected layer feed-forward neural network is used. The output of the maximum pooling layer is provided as the input for the maximum attached layer. At the end of the release, a software function is used, and the selected features are presented to the detection level. The input image is spread over five transition layers and three fully integrated layers, which isolate the spatial and temporal components of the image. Finally, the Softmax layers generate rear probability scores. The attained score value is used for further processing.

10.5 Results and Discussion

The results achieved from the proposed approach are analysed in this section. The implementation has been written in python programming language. The performance of proposed methodology is implemented using different metrics namely, accuracy, sensitivity, specificity, precision, recall and F-measure.

10.5.1 Evaluation Metrics

The proposed classifier is used to detect emotions from the autism child. Using the following constraints, the confusion matrix is calculated.

- The quantity of the cases that are effectively named positive class True Positive $(TRUE^{P_o})$.
- The quantity of the cases that are effectively delegated negative class True Negative $(TRUE^{N_e})$.
- The quantity of cases that are in the negative class and mistakenly named positive class which named as False Positive $(FALSE^{P_o})$.
- The quantity of cases in the positive class and mistakenly named negative class named as False Negative $(FALSE^{N_e})$.

10.5.2 Comparative Analysis

The efficiency of recommended approach is compared with other emotion classification approach with different metrics (Table 10.1). Accuracy about Emotion detection is obtained from the facial expression is given in Table 10.2. The individual ranges between 20 to 25, They are students from undergraduate studies (25 men, 30 women).

Then, the reliability was for the collected data, found at 100 epochs in both normalized cases as illustrated in Table 10.3, and not normalized results. The equation used to normalize the data is given in Eq. (10.10).

$$T_i = \frac{((S_i) - Min(S))}{(Max(S) - Min(S))} \tag{10.10}$$

Table 10.1 Proposed performance metrics

Performance metrics	Equation
Accuracy (A)	$\frac{True^{Po}+True^{Ne}}{True^{Po}+True^{Ne}+False^{Po}+False^{Ne}}$
Sensitivity (Se)	$\frac{True^{Po}}{True^{Po}+False^{Ne}}$
Specificity (Sp)	$\frac{True^{Ne}}{True^{Ne}+False^{Po}}$
Precision (Pe)	$\frac{True^{Po}}{True^{Po}+False^{Po}}$
Recall (R)	$\frac{True^{Po}}{False^{Ne}+True^{Po}}$
F-measure (F)	$\frac{2*P*R}{P+R}$

Table 10.2 Performance analysis based on the accuracy

Dataset	Avg. accuracy percentage
30 images are training (Not Normalized)	90.14
Normalized 30 images are training	96.72
Normalized 55 images are training	93.23

Table 10.3 Confusion matrix representation of facial emotion recognition

Measures	0	1	2	3	4	5
Precision	97.25	98.15	99.45	97.45	97.89	99.46
Sensitivity	96.45	96.78	97.16	98.45	98.56	98.89
Specificity	99.72	97.44	96.45	96.89	99.89	97.86
F-Score	98.16	96.15	97.55	97.88	97.89	96.89
Accuracy	97.45	98.41	98.55	99.89	98.78	98.56

The proposed models were executed until 300 epochs checked the curve shape and achieved the biggest precision for detecting emotions. As illustrated in Fig. 10.4, identifying emotions utilizing facial points has a clear correlation up to 400 Epochs. Hence, 99.81% of the system can perceive facial emotions landmarks at 400 Epochs. As shown in Fig. 10.5 shows a direct correlation between the proposed model accuracy and the number of iterations up to 100, and the curvature stops increasing after 100 epochs. Finally, at 100 Epochs, the maximum emotion detection precision is 87.25

The confusion matrix representation of facial representation is given in Table 10.3. Each emotion success factor is labeled as 0 is Anger, 1 is Fear, 2 is Disgust, 3 is Sad, 4 is Smile, and 5 in Surprise. The success of each emotion is represented as 0 Anger, 1 Fear, 2 Disgust, 3 Sad, 4 Smile and 5 Surprise. The proposed work exposes the acceptable accuracy and the amount of true positive and negative feelings that are correctly remembered. For other researchers, this data remains the benchmark. 10 virtual distances are taken from facial points. The table corresponding graph is given in Fig. 10.6.

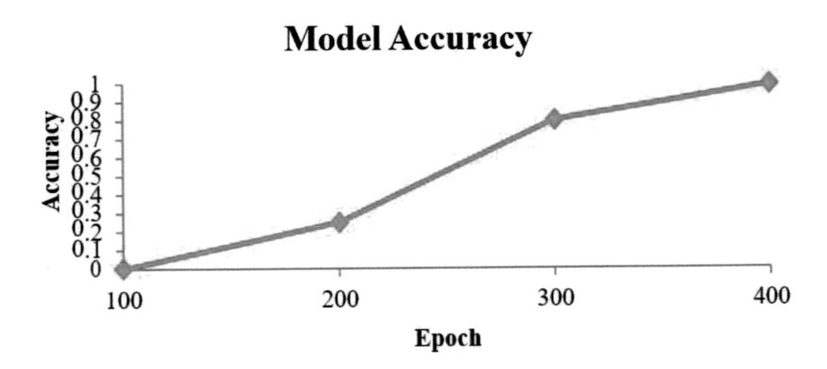

Fig. 10.4 Accuracy versus Epoch

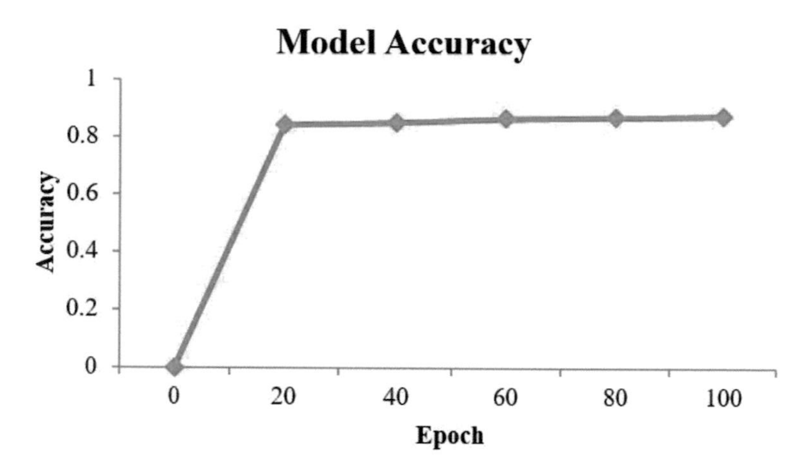

Fig. 10.5 Accuracy versus epochs (up to 100 epochs)

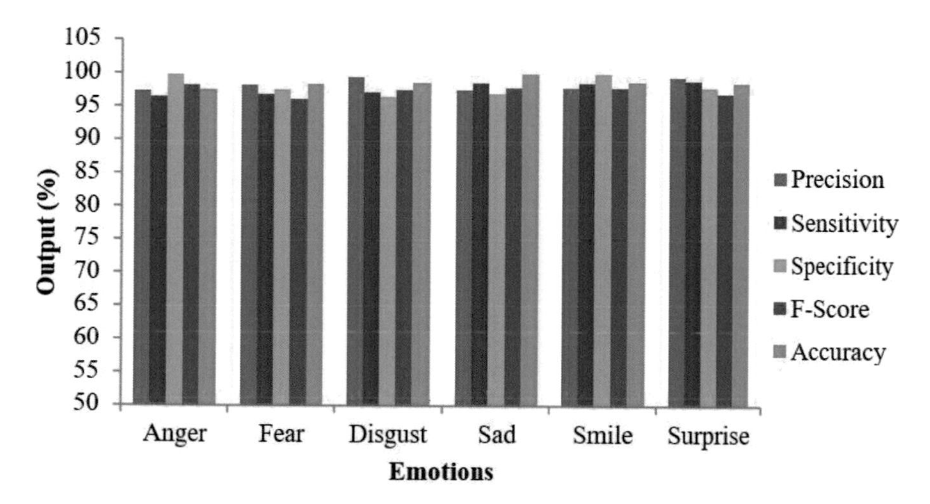

Fig. 10.6 Performance analysis based on different metrics

10.5.3 Comparative Analysis with Other Classifiers

To prove the efficiency of our suggested CNN-based emotion recognition approach, we compare our work with exiting two classifiers such as SVM and ANN. Table 10.4 compares our work performance with SVM-based face emotion recognition and ANN-based face emotion recognition. As per Table 10.4, the recommended approach attained the better accuracy of 99.81%, 92.45% for SVM-based face emotion recognition and 90.32% for facial emotion recognition. Similarly, we obtained the maximum sensitivity of 97.71%, specificity of 98.04%, F-score of 97.42% and

Table 10.4 Comparative analysis results

Measures	CNN	SVM	ANN
Precision	99.81	92.45	90.32
Sensitivity	97.71	93.1	91.43
Specificity	98.04	94.4	92.5
F-Score	97.42	92.5	90.8
Accuracy	98.6	93.23	91.02

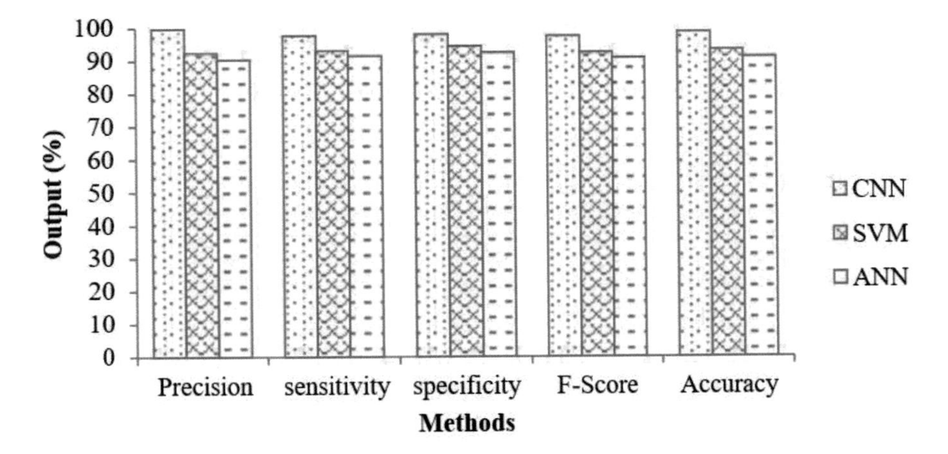

Fig. 10.7 Graphical representation of comparative analysis

accuracy of 98.6%. The table corresponding graph is given in Fig. 10.7. We clearly understand from the results that our proposed approach achieved better results than other methods.

10.6 Conclusion

Past several decades an efficient emotional recognition has been developed. The intention of this work is to classify the emotional expressions of children with disabilities (deaf, dumb and bedridden) and autism based on facial expressions. For achieving this concept, convolution neural network has been developed. At the beginning, the facial data are collected from different autism people and the face has been detected using cascade classifier. The detected face has been given to the CNN classifier to classify an emotion as, happiness, sorrow, rage, surprise, disgust and fear. The effectiveness of suggested approach has been analysed based on accuracy, precision, recall and F-measure. The proposed approach attained the cumulative recognition rate of 99.81%. This approach works perfectly on random lightning and various background, subject head rotation, and different skin tones. This method is aimed at helping the physically challenged and helps children with autism to recognize the feelings of others. In future we will plan to do speech signal based emotion recognition using deep learning algorithm.

References

1. Dada, E.G., Bassi, J.S., Chiroma, H., Abdulhamid, S.M., Adetunmbi, A.O., Ajibuwa, O.E.: Machine learning for email spam filtering: review, approaches and open research problems. Heliyon **5**(6), e01802 (2019)
2. Xie, M.: Development of artificial intelligence and effects on financial system. J. Phys. Conf. **1187**, 032084 (2019)
3. Hegazy, O., Soliman, O.S., Salam, M.A.: A machine learning model for stock market prediction. Int. J. Comput. Sci. Telecommun. **4**(12), 16–23 (2014)
4. Beckmann, J.S., Lew, D.: Reconciling evidence-based medicine and precision medicinein the era of big data: challenges and opportunities. Genome Med. **8**(1), 134–139 (2016)
5. Weber, G.M., Mandl, K.D., Kohane, I.S.: Finding the missing link for big biomedical data. Jama **311**(24), 2479–2480 (2014)
6. Loconsole, C., Chiaradia, D., Bevilacqua, V., Frisoli, A.: Real-time emotion recogni- tion: an improved hybrid approach for classification performance. Intell. Comput. Theory 320–331 (2014)
7. Huang, X., Kortelainen, J., Zhao, G., Li, X., Moilanen, A., Seppanen €, T., Pietik € ainen, M.: Multi-modal emotion analysis from facial expressions and electro encephalo gram. Comput. Vis. Image Understand **147**, 114–124 (2016). https://doi.org/10.1016/j.cviu.2015.09.015
8. Raheel, A., Majid, M., Anwar, S.M.: Facial expression recognition based on electroencephalography. In: 2nd International Conference on Computing, Mathematics and Engineering Technologies (iCoMET), Sukkur. Pakistan 1–5 (2019)
9. Vassilis, S., Herrmann J.: Where do machine learning and human-co mputer interaction meet? (1997)
10. Keltiner, D., Ekrman, P., Lewis, M., Haviland Jones, J.M. (eds.) Facial Expression of Emotion, Hand Book of Emotions, pp. 236–49. Gilford Press, New York (2000)
11. Ekman, P.: Darwin and Facial Expression: A Century of Research in Review, p. 1973. Academic Press Ishk, United State Of America (2006)
12. Ekman, P., Friesen, W.V.: Constants across cultures in the face and emotion. J. Pers. Soc. Psychol. **17**(2), 124 (1971)
13. Ekman, P.: Darwin and facial expression: a century of research in review, p. 1973. Academic Press Ishk, United State of America (2006)
14. Ekman, P., Friesen, W.V., Ancoli, S.: Facial signs of emotional experience. J. Pers. Soc. Psychol. **39**, 1123–1134 (1980)
15. Ekman, P., Friesen, W.V., Ancoli, S.: Facial signs of emotional experience. J. Pers. Soc. Psychol. **39**, 1123–34 (1980)
16. Nguyen, B.T., Trinh, M.H., Phan, T.V., Nguyen, H.D.: An efficient real-time emotion detection using camera and facial landmarks. In: 2017 Seventh International Conference on Information Science and Technology (ICIST) (2017). https://doi.org/10.1109/icist.2017.7926765
17. Loconsole, C., Miranda, C.R., Augusto, G., Frisoli, A., Orvalho, V.: Real-time emotion recognition novel method for geometrical facial features extraction. In: Proceedings of the International Conference on Computer Vision Theory and Applications (VISAPP), pp. 378–385 (2014)
18. Palestra, G., Pettinicchio, A., Coco, M.D., Carcagn, P., Leo, M., Distante, C.: Improved performance in facial expression recognition using 32 geometric features. In: Proceedings of the 18th International Conference on Image Analysis and Processing. ICIAP, pp. 518–528 (2015)

19. Zhang, J., Yin, Z., Cheng, P., Nichele, S.: Emotion recognition using multi-modal data and machine learning techniques: a tutorial and review. Inf, Fusion (2020)
20. Patil, P., Kumar, K.S., Gaud, N., Semwal, V.B.: Clinical human gait classification: extreme learning machine approach. In: 2019 1st International Conference on Advances in Science, Engineering and Robotics Technology (ICASERT), pp. 1-6. IEEE (2019)
21. Raj, M., Semwal, V.B., Nandi, G.C.: Bidirectional association of joint angle trajectories for humanoid locomotion: the restricted Boltzmann machine approach. Neural Comput. Appl. **30**(6), 1747–1755 (2018)
22. Jain, R., Semwal, V.B., Kaushik, P.: Stride segmentation of inertial sensor data using statistical methods for different walking activities. Robotica 1–14 (2021)
23. Bijalwan, V., Semwal, V.B., Mandal, T.K.: Fusion of multi-sensor-based biomechanical gait analysis using vision and wearable sensor. IEEE Sens. J. **21**(13), 14213–14220 (2021)
24. Bijalwan, V., Semwal, V.B., Gupta, V.: Wearable sensor-based pattern mining for human activity recognition: deep learning approach. Ind. Robot.: Int. J. Robot. Res. Appl. (2021)
25. Dua, N., Singh, S.N., Semwal, V.B.: Multi-input CNN-GRU based human activity recognition using wearable sensors. Computing **103**(7), 1461–1478 (2021)
26. Jain, R., Semwal, V.B., Kaushik, P.: Stride segmentation of inertial sensor data using statistical methods for different walking activities. Robotica 1–14 (2021)
27. Semwal, V.B., Gaud, N., Lalwani, P., Bijalwan, V., Alok, A.K.: Pattern identification of different human joints for different human walking styles using inertial measurement unit (IMU) sensor. Artif. Intell. Rev. 1–21 (2021)
28. Challa, S.K., Kumar, A., Semwal, V.B.: A multibranch CNN-BiLSTM model for human activity recognition using wearable sensor data. Vis. Comput. 1–15 (2021)
29. Bijalwan, V., Semwal, V.B., Singh, G., Mandal, T.K.: HDL-PSR: modelling spatio-temporal features using hybrid deep learning approach for post-stroke rehabilitation. Neural Process. Lett. 1–20 (2022)
30. Semwal, V.B., Gupta, A., Lalwani, P.: An optimized hybrid deep learning model using ensemble learning approach for human walking activities recognition. J. Supercomput. **77**(11), 12256–12279 (2021)
31. Dua, N., Singh, S.N., Semwal, V.B., Challa, S.K.: Inception inspired CNN-GRU hybrid network for human activity recognition. Multimed. Tools Appl. 1–35 (2022)

Chapter 11
Prevention of Global Mental Health Crisis with Transformer Neural Networks

A. Rajagopal, V. Nirmala, J. Andrew, Muthuraj V. Arun, and A. Piush

Abstract The COVID-19 pandemic is causing monumental effects on mental well-being worldwide. Literature calls for action to avert an impending global mental health crisis. This chapter presents the challenge, and presents a new approach to avert the crisis in mental health. The objective is to design a solution by applying advances in Deep Learning literature. The advances in Transformer neural network enable pandemic scale screening, mental health diagnostics and counseling. A synthesis of Multimodal Deep Learning architecture to analyze thinking patterns is presented. The use of BERT for diagnosis is well established in the literature. By approaching the modeling of Cognitive Triad and Cognitive Behaviour Therapy with an Encoder-Decoder Transformer architecture, this chapter initiates future research by interested communities to avert the global mental health crisis.

Keywords Mental health · Multimodal deep learning · Natural language processing · Cognitive triad · Cognitive therapy

---

A. Rajagopal
Indian Institute of Technology, Chennai, India

V. Nirmala (✉)
Queen Marys College, Chennai, India
e-mail: gvan.nirmala@gmail.com

J. Andrew
Karunya Institute of Technology and Sciences, Coimbatore, India
e-mail: andrewj@karunya.edu

M. V. Arun
National Institute of Technology, Trichy, India

A. Piush
UNICEF, Chennai, India

A. Biswas et al. (eds.), *Artificial Intelligence for Societal Issues*, Intelligent Systems Reference Library 231, https://doi.org/10.1007/978-3-031-12419-8_11

11.1 Introduction

This chapter motivates anyone interested in applying AI for social challenges to explore a new challenge in mental health, which transitioned from an invisible problem to a global crisis. The priority on mental health for societies across the globe has increased significantly since the COVID-19 pandemic. Mental health topic was added into UN SDG goals in 2015, and got highlighted in UNICEF's 2021 annual report. Experts are calling for action to avoid an impending global crisis in mental health. Millions of families need help in improving mental wellness.

Is it possible to prevent this global mental health crisis? As per literature, implementation of early intervention is a practical challenge for 21st century mental healthcare. A call for action to avert the crisis was placed by experts and notable editorials. To address this call to action, this chapter introduces a promising new approach to avert the crisis. The objective is to design a breakthrough solution to avert the global mental health crisis in all its dimensions.

The research question is how to design a mental health solution that can support millions of families in a timely intervention without compromising privacy? This chapter presents the design of Deep Learning solution for mental health. Modelling of Cognitive Triad and Cognitive Therapy using Encoder-Decoder Transformer neural network architecture is explored.

This chapter is organized by defining the problem statement in Sect. 11.2, then the rest of chapter focuses on design approaches to avert the crisis. As mental health challenge has risen from an invisible issue to a global challenge, a dedicated section articulates the challenge. To address the challenge, a design approach is introduced in Sect. 11.3. The design explores how AI can be employed to model thinking patterns of an individual by modelling of Cognitive Triad , and then help transform one's mindset to a healthier thinking pattern. This promising new approach combines Deep Learning advances in Transformer architecture to model both Cognitive Triad and Cognitive Behavior Therapy. The three aspects of the solution namely screening, diagnosis, therapy are expanded in Sects. 11.4, 11.5 and 11.6 respectively. In short, the chapter lays a rich landscape for future research by presenting a synthesis of a novel Deep learning solution for mental health. This novel design is backed up by findings from literature. The chapter ends with a treatment of potential for Multi modal Transformers in mental health. The chapter is designed to motivate multidisciplinary research communities to avert the pandemic scale crisis in mental wellness.

11.2 Background

11.2.1 Motivation

The COVID-19 pandemic is having a profound effect on mental health as per a panel of experts [1]. The UN COVID-19 policy document [2] in May 2020 called for redressing the historic underinvestment in mental health. UN calls for action on mental health to reduce immense suffering among hundreds of millions of people, to mitigate long-term social and economic costs to society [2]. As per WHO facts [3], almost 800000 people die by suicide every year. The global economy loses about US$ 1 trillion per year in productivity due to depression and anxiety [3].

Experts are calling for action to prevent this impending global crisis in mental health [4]. Given the urgency and call to action to address this global challenge, there is a need for research and development. This chapter is authored to aid the research community and interested technologists to make progress towards addressing this global challenge.

Rising to this challenge will require research across disciplines as per Lancet [4]. Given the scale of COVID-19 pandemic worldwide, the urgent need is to improve mental healthcare for a billion families with timely intervention. The scope of the challenge is immense, hence the need for research.

11.2.2 From an Invisible Problem to a Global Crisis

Between 2015 to 2021, experts have increased the priority on mental health.

2015: Realizing the burden of disease, mental health was included in UN SDGs (Sustainable Development Goals) since 2015. SDG #3 includes an indicator of suicide. As per WHO fact sheets [3], around 1 in 5 of the world's adolescents have a mental disorder, and 5% of adults suffer from depression. As per statistics from John Hopkins medicine, mental health disorders account for the top causes of disability in the US, and 26% of Americans suffer from a mental health disorder.

2020: As per UN [1], though COVID-19 is a physical health crisis, it has seeds of a major mental health crisis, if action is not taken. In its 2020 policy brief, the United Nations urges urgent priority on mental health crisis as a central theme of every country's response to the COVID-19 pandemic [2] .

2021: By Sep 2021, a sharp decline in mental health triggered by the pandemic was reported by National Health Service [5] . The NHS UK data showed shows the alarming pressure young people are facing with their mental health. By Oct 2021, the UNICEF report of State of the World Children identified mental health as the key issue of concern among children and young [6].

11.2.3 Can COVID-19 Pandemic Seed a Global Mental Health Crisis?

Mental health and mental wellness concern is emerging as a significant and urgent need for a vast majority of the world population as per the World Health Organization (WHO) [7]. COVID-19 has a significant consequence on mental health across a vast population as per the JAMA Internal Medicine article [1]. As per the editorial in the Lancet Infectious Diseases, the COVID-19 pandemic is causing monumental effects on mental wellbeing worldwide [8]. Fear about the perceived threat of COVID-19 infection, prolonged uncertainty and fear of being infected by an asymptotic coronavirus carrier, loss of a family income or job, physical isolation from friends due to lockdown, quarantine of family members, long closures of schools, fear of infecting loved ones by nurses caring for COVID-19 patients, uncertain future, risks due to asymptomatic virus are affecting mental health [1, 4, 7, 8]. More than half of young people have trouble sleeping in the UK as per the NHS report published on Sep 2021 [5].

11.2.4 Call for Action by Editorials and Experts

Editorials in reputed journals such as Nature Medicine [9], Lancet Global Health [10] and The Lancet Infectious Diseases [8] have highlighted the need for action as there is mounting evidence of a widespread impact of the pandemic on mental health. An urgently deployable intervention is called by a panel of experts in Lancet Psychiatry [4]. The need for prevention is brought forward in JAMA Internal Medicine [1].

The editorial of Nature Medicine [9] called for researchers across multiple domains to address the impact of COVID-19 on mental health. The editorial of Lancet Global Health [10] says 1 billion people suffer from a mental disorder. The Lancet editorial [10] also restates that sooner or later, health systems will be challenged to face a widespread demand for mental health due to COVID-19. By May 2020, United Nations published a policy brief [2] strongly urging the need for action on mental health. By Oct 2020, WHO calls the world to action on mental health. In short, multiple experts and leaders have forewarned about the impact of the pandemic on mental health. By Oct 2021, UNICEF report spotlights mental health as the key issue affecting children across countries in their annual flagship report [6].

11.2.5 Dimensions of the Global Crisis in Mental Health

The challenge in front of us is monumental. The challenge of addressing this crisis must be understood in the background of dimensions of the challenge, namely

1. *Timely intervention*: Practical implementation of early intervention is a challenge for 21st century mental health as per JAMA Psychiatry [11].
2. *Scalable*: The sheer size of the problem is significant. Around 1 billion people are impacted as per BMJ editorial in 2021 [12].
3. *Privacy*: Privacy requirements cannot be ignored as mental health therapy often involves handling sensitive personal information.
4. *Practical*: Given populations across countries are impacted, the solution needs to be usable by people speaking different languages.

Is it possible to design an AI solution to address all the dimensions of the challenge? In Oct 2020, the Lancet editorial [8] said it is unclear how the world will deal with this forthcoming mental health crisis, given the shortage in the capacity to respond. WHO found disruption of mental health services across many countries, especially for the vulnerable [7]. With many millions of families impacted by the COVID-19 pandemic, the shortage of capacity, and the criticality of providing timely health intervention, combined with the privacy requirements, the challenge of averting the forthcoming mental health crisis is a monumental one.

11.3 Design of Deep Learning Solution for Mental Health

11.3.1 Key Ideas in Deep Learning for Mental Health

The role of applying advanced Deep Learning modeling [13] to create a solution for mental health is the focus of this chapter. The creation of a breakthrough model for Cognitive Triad and Cognitive Therapy by applying Transformers is presented. Both Transformers based Natural Language Understanding (NLU) and Natural Language Generation (NLG), combined with Multimodal Transformers hold the promise to improve mental wellness for millions of families.

Key ideas discussed in this chapter are

1. Strategy to avert global mental health crisis using Deep Learning
2. Design of a solution approach for improving personal mental health
3. Approach to Mental health Screening, Diagnosis and Therapy
4. Architecture for Cognitive Triad using Multimodal Transformers
5. Architecture for Cognitive Therapy using Transformers

As shown in Fig. 11.1 and Table 11.1, the components of the AI solution for mental health are

1. ***Screening at pandemic scale***: Given the pandemic scale impact, it is important to identify who needs help. So an effective screening strategy is required. By analyzing social media interactions [14–16], using fine tuned NLU models, it is possible to spot those in depression or anxiety [17].
2. ***Timely diagnosis of the situation or mental illness***: Like in any healthcare setting, early detection of mental health conditions is essential. To analyze any mental

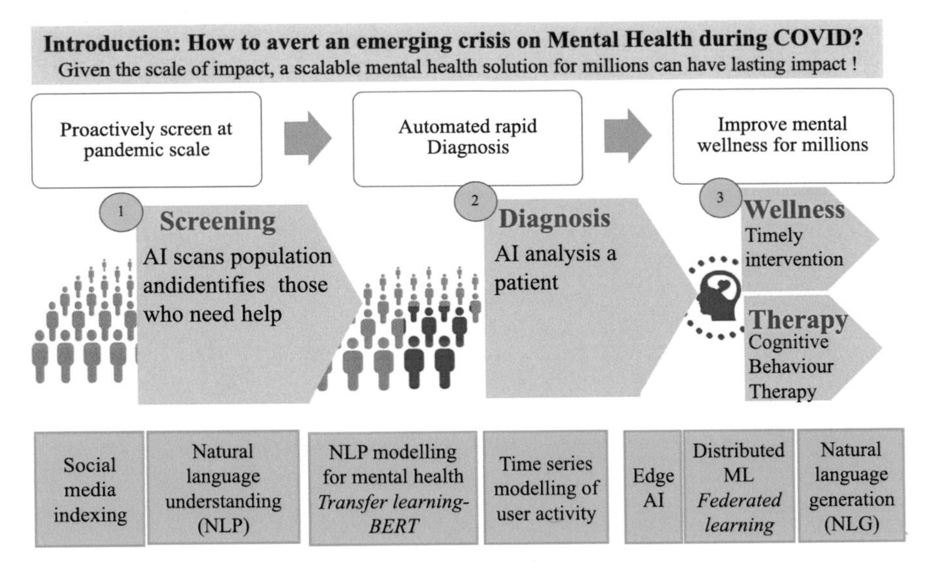

Fig. 11.1 Deep learning framework to avert the mental health crisis

Table 11.1 Overview of mental health use cases

	Use case	Approach
1	**Screening at scale of the pandemic** Identify those who need help in a timely manner among populations	Social media analysis and Cloud based deployments with fine tunned language models such as BERT to detect families who need help. ***Keywords: NLU, Transformers/ Transfer Learning, Social media analysis, Cloud AI, BERT***
2	**Rapid Diagnosis** Analyze mental health condition of an individuals	On-device ML inference on smartphones with DistillBERT allows privacy safe diagnosis of the condition of the individual. Federated Learning allows for privacy safe learning in a community. In addition, progression of the condition over days can be tracked with time series modelling and Multimodal Learning ***Keywords: NLU, Transformers/ Transfer Learning, Federated Learning, Distillation, Privacy safe on-device ML, Smartphone AI inference, DistillBERT, LSTM temporal modelling of progression, Resilience, Multimodal Deep Learning***
3	**Cognitive Therapy** Therapy to enable the individual to improve his/her perspective of the given situation	On-device ML inference on smartphones with small GPT2 conditioned to talk like a therapist. Federated Learning for privacy safe learning in community. BERT based selective activation of therapy to support neuroplasticity. Help visualize a better future using GAN and DALLE ***Keywords: NLG, Transformers/ Transfer Learning, Federated Learning, privacy safe on-device ML, GPT-2, smartphone AI, Reinforcement Learning, BigGAN, OpenAI CLIP, DALL.E***

illness or cause of depression, it is important to interact with the patient with certain questions. The answers to such questions can be analyzed by fine tuned NLU models to detect the class of problem. Transfer Learning techniques [18] allow for customization of Deep Learning models to the custom tasks. Additionally, the progression of the state of mental health over a time duration can be modeled by time series or sequence modeling.

3. ***Therapy to improve personal mental health***: Therapy in mental health is aimed at counseling the patient. Cognitive Therapy [19] or talking therapy works on the principle that often few words spoken can change the way we see the world around us, and change our perspectives. The words exchanged between two people can change the way we perceive and hence relieve suffering. To address the challenge of shortage the staff [7], the patient can exchange words with NLG models that are conditioned to talk like a human therapist or human counselor [20]. By learning from interactions from a group of individuals with on-device Machine Learning, it is possible to transform communities by applying Privacy safe Federated Learning [21, 22]. A better future can be imagined by DALL-E and presented to user.

11.3.2 Landscape

There are substantial research publications in NLU (Natural Language Understanding) such as BERT [23] for the diagnosis of sentiment [16] or sensing of emotions [14]. However, there are only very few publications on NLG (Natural Language Generation) for mental health therapy. A survey paper [24] took stock of a decade of studies with 139 papers showing a lot of effort has happened on diagnosis, but there is significant scope for future research on novel applications of NLG for mental health. A paper in Nature scientific reports [16] also shows efforts on the topic of diagnosis. It shows progress in the application of Deep Learning for classification problems such as classifying emotions/mental health conditions. However, the opportunity to apply Deep Learning in the synthesis using NLG is a relatively unexplored research theme in the context of mental health, and even more specifically in the context of Cognitive therapy. Figure 11.2 depicts landscape of research progress in NLG and NLU for mental health therapy and diagnosis respectively. Deep Learning architectures such as LSTM, Transformers have been applied for classification problems in mental health, and the future potential is in application for Multimodal Deep Learning for holistic comprehension (Triad) across different datasources. Also the role of Generative Deep Learning approaches can help individuals to visualize a better future for themselves. Generative Adverserial Networks (GAN) can help imagine a better future for people affected by violence. Generative pretrained Transformer models (GPT-3) can chat with patients in a caring way. Reinforcement Learning can optimize for better outcomes. Such Therapy applications hold significant potential given the shortage of staff in mental health.

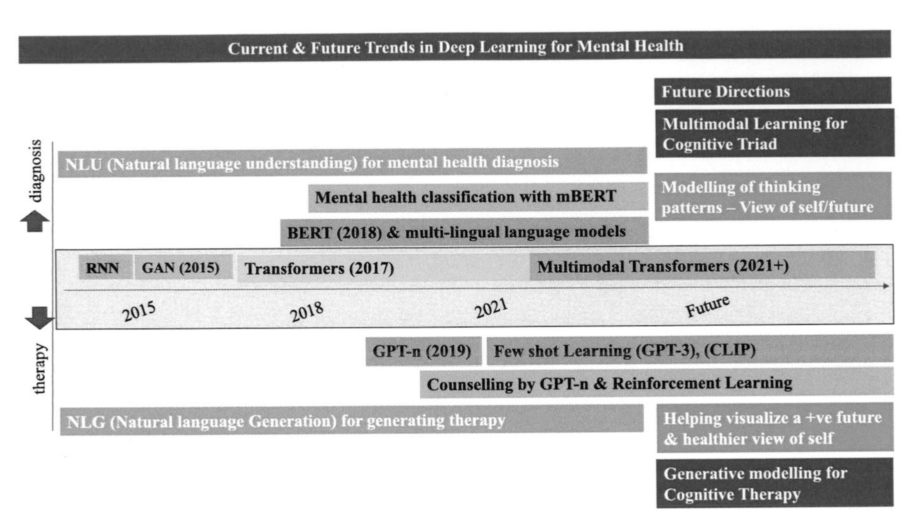

Fig. 11.2 Trends in deep learning and its application for mental health

11.3.3 Design of AI Solution to Avert the Global Mental Heath Crisis

The design approach aims to address the global mental health challenge. Given the dimensions of the challenge such as early intervention for millions of families in a privacy safe way, the design approach is to improve personal mental health for individuals at global scale. Once pandemic scale screening is done, the problem then can be simplified to Cognitive Triad and Therapy for each individual and then scaled for the population through the magic of technology in an affordable and privacy safe way.

The high level approach consists of 3 steps as presented in Fig. 11.3. The solution approach can be conceptualized as follows

1. **Cognitive Triad**:
   A fine-tuned Transformer language model [25] can understand sentences uttered by the individual, and any change in thinking patterns can be modeled with time series modeling, and a Cognitive Model of the individual can be developed to capture his/her view of self & view of the world. Based on these views over time, Cognitive Triad [26] can be performed.
2. **Check for resilience**:
   Before proceeding with any treatment, it is important to recognize the natural capabilities of humans to re-bound back from any traumatic situation over a time horizon. The system needs to wait and analyze for the natural resilience of the human brain [27, 28]. So the AI needs to look for a change in thinking patterns as the individual starts to become resilient. Here a temporal model often comes in handy to explore resilience [29], though the time horizon may differ from

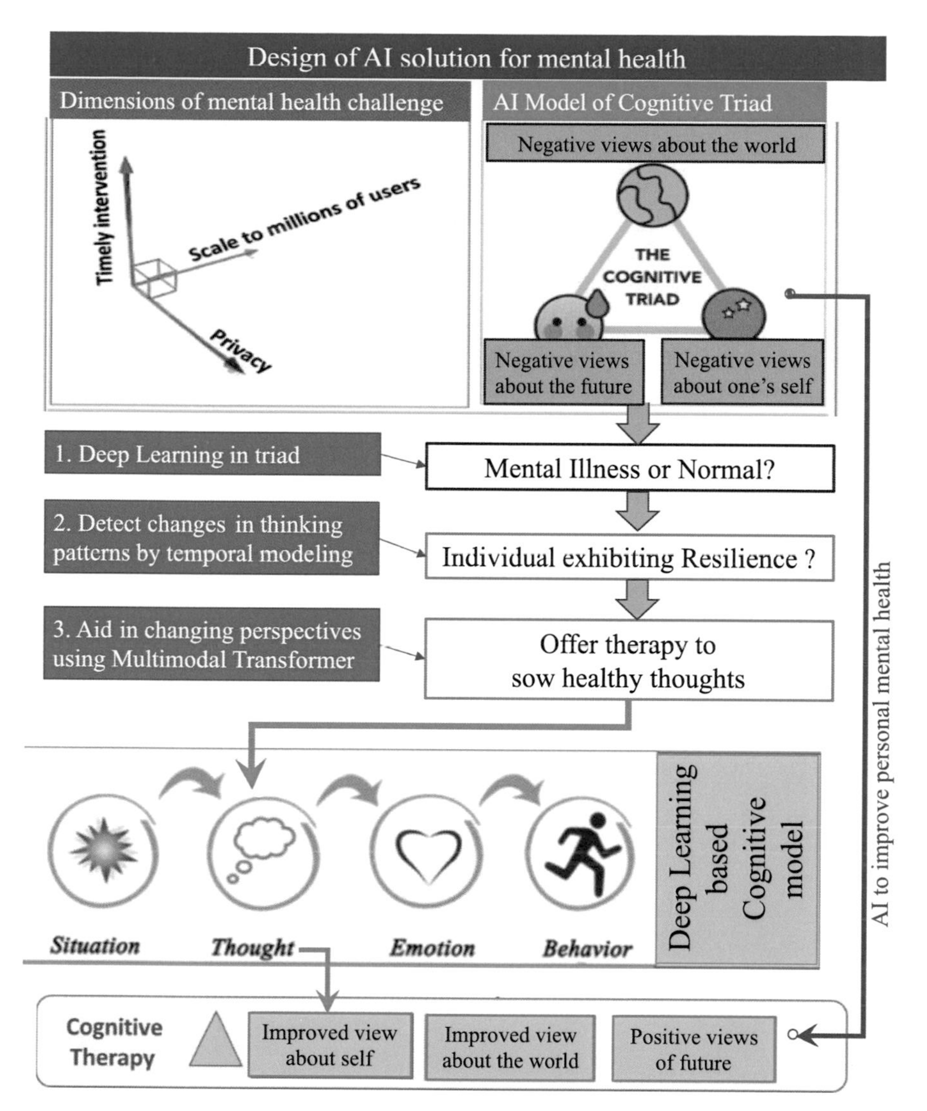

Fig. 11.3 High level design of a AI solution for improving personal mental health

individual to individual. Here, Federated Learning [21] can learn insights from communities of likely duration for a given situation. For example, for a given situation of losing a parent to COVID-19, how children in a particular age group in a particular social-economic background become resilient can be learned by analyzing the behavior of many children in a privacy-safe Federated Learning approach. The important point is to detect natural resilience and to support the rebound.

3. **Cognitive Therapy, only when appropriate**:
Based on resilience, the choice to intervene with therapy can be planned. One popular form of Mental health Therapy is a Cognitive Therapy [30] technique based on a simple exchange of words between a phycological counselor and the patient. With the help of certain words, the counselor helps the individual to re-think the painful situation from a new perspective. This change of perspective often removes the suffering as per the Cognitive Theory [26, 30–32]. Given the severe shortage of staff across many countries, the solution choice delineates to digital counseling by an AI therapist. The AI virtual therapist has to speak like a human therapist. This capability can be implemented by Transformer [33–36] based language models such as GPT [37, 38]. Combined with the power of Transfer Learning, the pre-trained GPT model [34, 39–41] can be customized to speak like a human therapist. A Reinforcement learning approach [42] further can make the words effective. AI can be used to model the "view of self" or the thinking pattern, and based on the resilience, the AI can then influence the perspective and help change the "view of self" by speaking like a human therapist. A wise choice of words in such counselling conversations influences the thoughts, and hence the emotions and then the behavior. This idea of understanding and influencing human thinking patterns with an AI model is shown in Fig. 11.3. This Cognitive modelling is implemented as a Multimodal encoder-decoder and is elaborated later.
4. **Scaling for global population**:
The final step is one of scaling to millions of families. Unlike physical health conditions, since mental health issues are often invisible at least in the early stages, early identification is often a challenge in practical implementation of mental healthcare. By leveraging the omnipresence of smartphones, and given that language used by the individual often gives early clues about the mental health condition, on-device Machine Learning with privacy safe Federated Learning [21, 43] can be employed for proactive screening by smartphone AI based Cognitive Triad.

11.3.4 Design of AI to Improve Thinking Patterns: Views of Self/Future

Modelling of human thinking patterns can open doors to breakthroughs to enable imporve one's mental wellness and one's mindset. Modelling Cognitive Triad [26] by AI simply consists of analyzing thinking patterns such as

1. View of Self
2. View of World
3. View of Future

Since the views are expressed in the form of language in social media interactions or responses to specific questionaries, it is feasible to design a Deep Learning based language model to understand the words spoken by the individual. Further, the changes in views over time can be modeled using time series modeling. Further, the correlation between "the view of self" [19] and "the view of the future" [32] can be made by AI. This can quickly allow for pandemic scale screening of mental health.

Once screened, a more accurate diagnosis can be performed by Multimodal Deep Learning techniques [44, 45] by combining input data from a variety of sources. Once diagnosed and probed for natural resilience of the brain [28], the choice of Cognitive therapy can be made. It is important to intervene in a timely manner while encouraging the opportunity for natural human resilience. And this can be done based on each situation. For example, an individual may be naturally resilient [28] in handling relationships, but may not do well when it comes to financial-related stressors. So the choice of therapy for a given situation can be smartly made based on analysis of the resilience of the individual for the given type of challenge. Here again, models such as BERT can be customized with Transfer Learning to identify the type of situation being faced by the individual. Once Cognitive Therapy [46] is planned, simple use of words by a virtual therapist can help the individual to change his/her thoughts about the given situation. Since self-help to change of perspectives often lead to change in thought patterns, and often change emotion and hence behavior, the mental health of the individual can be affected by the conditional language model [37]. These models can speak the language of the therapist to help the individual to re-look at the situation from a new point of view [30, 31]. These ideas to model thinking patterns using AI and further influence the viewpoints to bring out a better mindset for a healthier cognitive behaviour is illustrated in Fig. 11.3.

11.3.5 Detailed Design

Based on the high level design discussed so far, the detailed design of AI solution for mental health is covered in rest of the chapter as follows.

1. Screening (Sect. 11.4)
2. Diagnosis (Sect. 11.5)
3. Therapy (Sect. 11.6)

11.4 Mental Health Screening at Scale

Given the scale of the pandemic, it is important to first identify those who need help. As mental health condition is often invisible, and many individuals arc often unaware of their state of mental health during the early stage of mental illness and combined with the social stigma in opening up about mental health, combined with a severe shortage of mental health staff, there is a practical challenge in early identification of candidates at scale.

11.4.1 Approaches for Pandemic Scale Screening

In this task, the goal is quickly to identify possible families among millions who require support. Typically analysis can be done on any of the following types of data such as

- clinical data (MRI/ EEG)
- genetic data
- facial visual expression and voice data
- social media data

These 4 data types are explained in the Nature article [48]. Among the above 4 types of data, both scanning social media interactions and facial/voice expression data would be the most cost-effective strategies, given the pandemic scale requirement in mental health screening. Social media interactions can be analyzed using advances in NLP [16, 23]. Facial expression based affective computing can be used to track emotions [45] using CNN models [48]. The approaches to model different types of data is mapped in Table 11.2.

11.4.2 Deep Learning in Mental Health Screening

Identifying those who need help early in disease life cycle is a key challenge when the number of families to be screened is order of tens of millions. By screening social media using NLU (Natural Language Understanding), AI algorithms can identify potential candidates for next level of diagnosis. This often involves detecting if someone is depressed or not [49]. Such problems are typically classification problems.

The abundance of literature in NLU for social media listening [16, 17] is noted in Table 11.2. Until 2018, Deep Neural Networks such as RNN were utilized [17]. Using RNN and LSTM, nature survey paper [48] reported that many authors demonstrated that it is possible to detect depression or suicidal risks from social media posts such as

Table 11.2 Overview of mental health use cases

	Input data	Deep learning approach
a	Mental health analysis from social media interactions	Transformers based NLU for detection of depression [15–17, 23]
b	Mental health analysis from facial expression	Affective computing based emotion detection and tracking based on CNN models [14, 47]
c	Combination of above two types of data	Multimodal Deep Learning

twitter, reddit posts. These authors showed that it was feasible to model this detection as a classification problem in NLP [48].

By 2018, "imagenet moment" arrived in the field of NLP [23]. This brought in the power of pre-trained language models and transfer learning techniques to NLP. So it was possible to finetune a pre-trained language model and quickly tailor it for the custom task at hand. The emergence of Transformer Deep Learning in NLP opened the doors for large models on gigantic datasets using semi-supervised learning approaches. This allowed NLP models to understand a sentence more accurately than before. Combined with the ability of neural networks to understand language, transfer learning allowed for such language models to be customized for specific tasks in NLP. Transfer Learning on BERT like models [23, 25, 50, 51] is proven to be a viable technique for understanding language like humans in any domain [18, 29, 41, 46, 52, 53].This opened the doors for accurate detection of the state of mental health. These ideas from literature can be summarized and can be represented as a architecture diagram in Fig. 11.4.

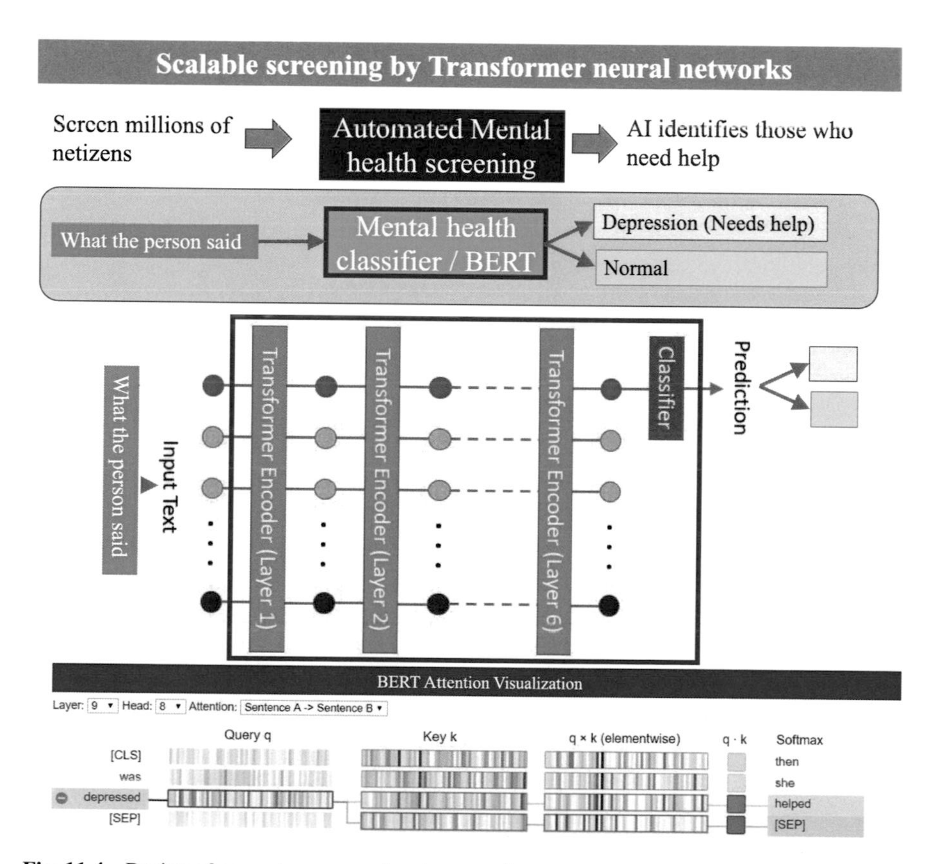

Fig. 11.4 Design of screening approach using transformers

Within the context of social media listening with NLU, the idea of the use of BERT (Bidirectional Encoder Representations from Transformers) is well established recently in 2019. The idea of screening social media during a pandemic using fine tunned BERT can be powerful to identify those who need help. Here language models such as BERT can be customized for specific detection tasks [14, 15, 17, 24, 29, 50] and deployed.

11.5 Mental Health Diagnosis and Resilience Detection

11.5.1 Modelling of Neuroplasticity/Resilience Using Deep Learning

One of the key considerations in diagnosis is to detect if the individual is developing resilience after trauma. The concept of resilience is detailed in [27]. If a person is developing resilience naturally, then chemical drugs/treatment may not be required. So it is important for any diagnostic tool to recognize natural resilience. Checking for resilience is an essential strategy as per the high level architecture approach presented earlier in Fig. 11.3. This is easily done by tracking thinking patterns over a period of time. The neuroplasticity of the brain [28] creates significant dynamics and the necessity to model temporal changes in the thinking patterns. Here such changes over a period of time can be modeled using time series modelling and often fit the sequence modelling paradigm of Deep Learning. So the steps are

1. Step 1: Does she need help?
2. Step 2: Diagnose the type of illness and understand her personal situation
3. Step 3: Is she resilient naturally?
4. Step 4: Offer wise words during counseling to help change her perspective to reduce her suffering/anxiety/suicidal thinking

Typically language understanding models are employed to understand what the user said, and to analyze how these patterns changed over a period of time. For instance, after a loss of a family member, a patient may rebound back into recovery though he may have shown some symptoms of depression. It is important for the AI system to recognize this natural recovery of the person.

Hence, analysis of thinking patterns over a period of time can show real time progress in mental health. This is typically modeled by a combination of a language model and a temporal model as illustrated in Fig. 11.6. As shown in the Fig. 11.5, a fine tuned BERT classifer can identify if the person is depressed or recovering, and a temporal LSTM model can detect the rate of recovery from depression. The ability to diagnose, and analyze temporal patterns of every candidate makes it possible for accurate diagnosis and tracking of mental health conditions over time [15, 16, 20, 29, 46]. Using transfer learning, a BERT [23] based binary classifier identifies if the person is showing the language of a depressed person or exhibiting the signs of a

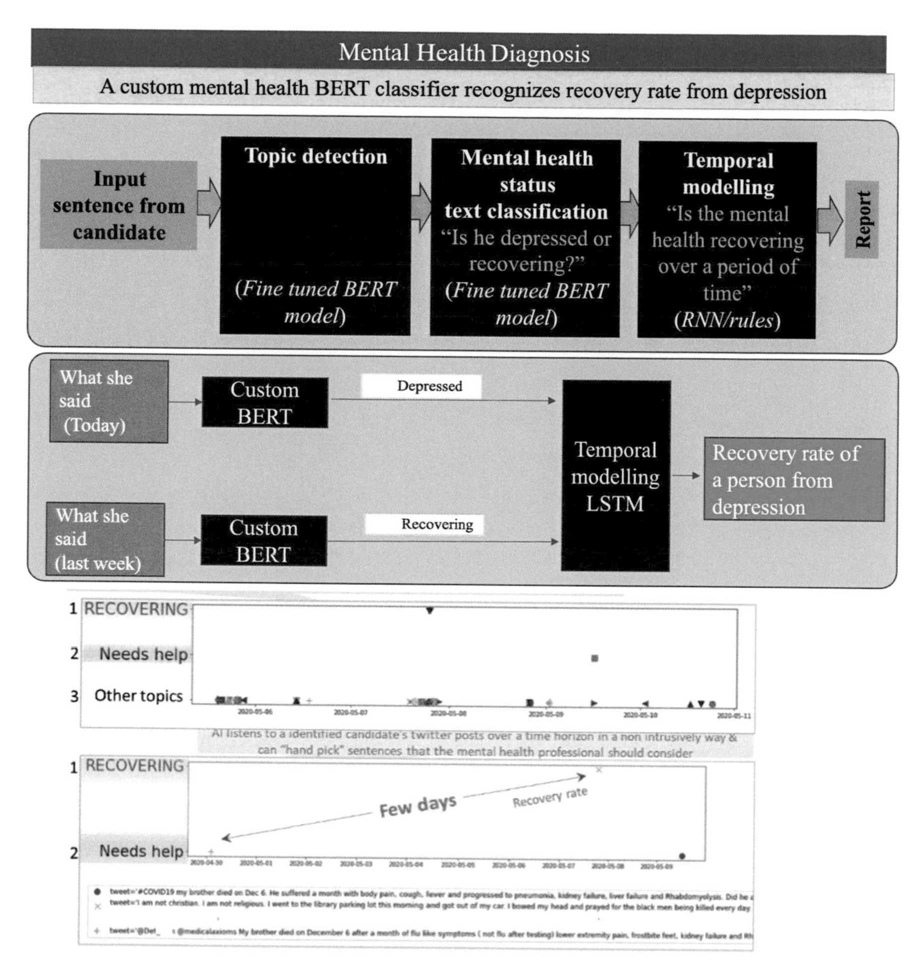

Fig. 11.5 AI architecture for mental health resilience

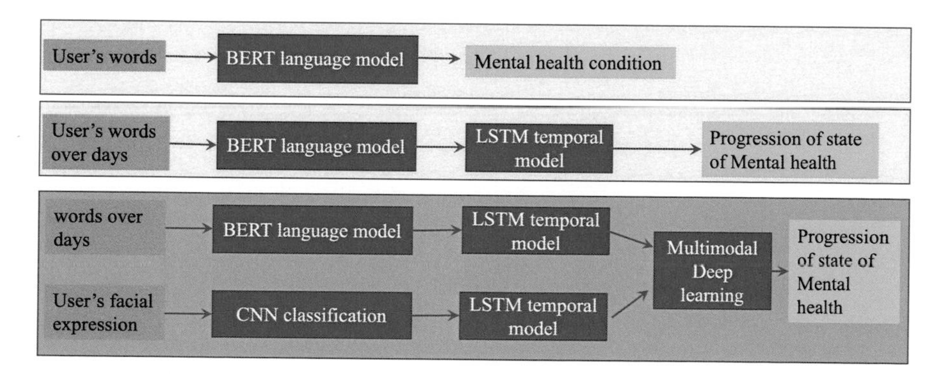

Fig. 11.6 Holistic screening using multimodal transformers

person on a recovery path. The time to recover after a loss of job or family member is also explored. The mental resilience [27] of a candidate can be understood by seeing the trend over a period of time.

Mental illnesses are diagnosed based on answers provided by the individual to specific questionnaires designed for the detection of specific thinking patterns [48]. It can be also diagnosed based on the language used by the individual over a period of time. Additionally, brain scans can unravel neuroplasticity and changes in the structure of brain cells as per nature article [28]. Hence MultiModal Deep Learning for Cognitive Triad is a robust approach here.

11.5.2 Diagnosis with Multimodal Deep Learning

With the increasing trends of digitization, it is possible to combine multiple data sources and derive insights. Such a multimodal approach [44] was implemented in [45] to combine two different data sources. In Multimodal Deep Learning approaches, two or more data sources can be combined to gain insights as depicted in Fig. 11.6.

11.5.3 Modelling of Cognitive Behavior: View of Self and Future

By modelling of the human thinking pattern [19, 32], an individual's "the view of self" and "view of future" can be analyzed by AI as illustrated in Fig. 11.3. Modelling of Cognitive behavior [19] of an individual can be implemented with Transformer architecture based language models [23] as they offer a significant ability to understand the English language, and such pre-trained BERT models [53] can be customized to detect what is the "View of Self". For example, researchers have demonstrated it is possible to detect suicidal thoughts using NLP [48]. Moreover, BERT is applied to build a classification model to type of illness. Researchers have demonstrated that a finetuned NLP classifier model can diagnose any of the 5 types such as depression, bipolar disorder, schizophrenia, and dementia [27].

11.6 Cognitive Therapy

Cognitive Therapy [30–32] can be simply explained as follows: A few words spoken at the right time to a depressed patient can suddenly enable him to re-look at the situation from a fresh perspective, and instantly relieve him of the suffering or harmful thoughts or sucide. This is typically the counseling technique used by phycological counselors. Given the reported shortage of staff by WHO across countries, timely intervention is at risk for millions of families, especially in low income settings [7]. Here, the role of AI to become virtual counselors is explored. The vision of e-therapy is described in [46]. One of the mental health therapies is CBT or "Cognitive Behaviour Therapy" [19]. CBT is a therapy technique that can help people find new ways to behave by changing their thought patterns [26, 30, 31].

11.6.1 Reinforcement Learning and GPT-n for Therapy Conversations

A Deep reinforcement learning approach with GPT-2 was demonstrated by researchers recently in online mental health [42]. Here researchers [42] show how AI can insert words of empathy in conversation with a patient. This is achieved by GPT-2 and Deep Reinforcement Learning. It used the reward function to improve empathy in the conversation. It also used the Transformer language model to re-write a sentence with improved empathy [54]. Advancement in the year 2019 in the field of Deep Learning based NLG is GPT-2 [37]. GPT-n (Generative PreTrained Transformer) [37] is a powerful language model built using Attention mechanism [33] based Transformers [25]. GPT-n [36, 38] is a language model that could perform natural language processing applications such as answering questions, completing the text, reading comprehension, text summarization. GPT-n is capable of generating human-like language. It was demonstrated to create fake news [40] or to create poetry [39] or generate image captions [55]. This was possible due to transfer learning capability [18], of Transformers architecture [37], opening up the "imagenet moment in NLG". Such NLG models can be fine-tuned to talk like an empathetic counselor. Cognitive Behavior Therapy (CBT) [32] is a psycho-social intervention that aims to improve mental health. The cognitive model of depressed individuals self-constructs such a negative view of themselves [19]. Beck's "Cognitive triad" [26, 32] enables the patient to correct their view of the world [30]. By changing unhelpful or inaccurate thinking, Cognitive therapy equips individuals to practice more flexible ways to think to overcome the cognitive distortion [19]. The JAMA article [31] concluded the effect of early intervention using Cognitive therapy. But the number of people who need help is multiple orders of magnitude higher due to the pandemic, hence the JAMA article [1] calls for creative thinking in treatment. The Lancet Psychiatry position paper calls for Digital interventions [4]. This delineates the potential of applying Deep Neural network-based "Digital Cognitive therapy". Such digital therapy or is e-therapy is described in [46]. During such e-therapy, researchers [42] show the use of GPT-2 [37] to re-write the conversations to be more effective, with the effectiveness enabled by reward mechanism of Deep Reinforcement Learning. This mechanism is illustrated as block diagram in the architecture diagram in Fig. 11.7.

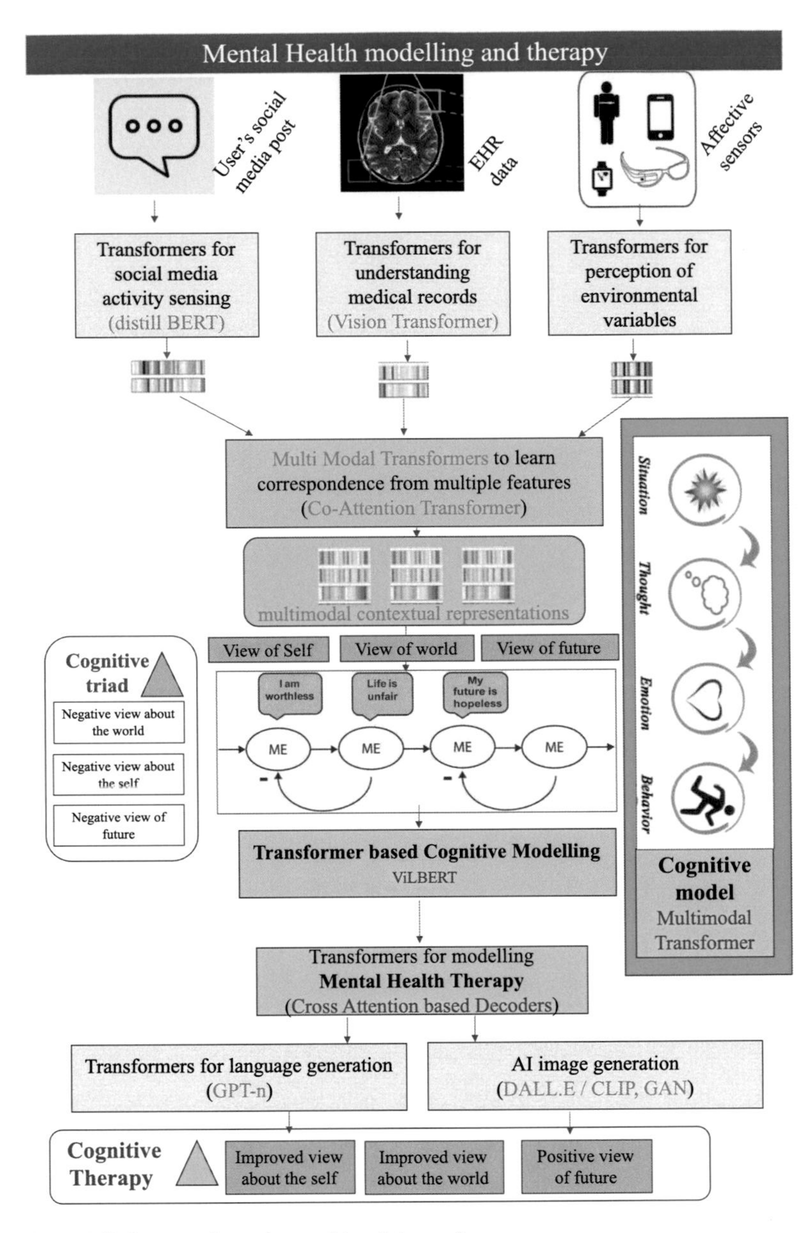

Fig. 11.7 Holistic screening using multimodal transformers

11.6.2 Privacy Safe On-device ML, Distillation Versus Few Shot Learning

The recent advances in Attention [33] based Deep Neural Networks and the concept of Query Key Value based Attention computation in Transformer based neural network architecture has opened the door for modeling real life sequences with relatively lesser amount of data. With Few Shot Learning [38, 56], it is now possible to fine tune a pre-trained GPT3 neural network with even fewer examples. GPT3 models can be primed to "caring" or "empathic" by prompting a GPT3 pre-trained model [56, 57]. Transfer Learning in GPT-2 still requires relatively few examples for fine-tuning to model when compared to earlier RNN [17] based approaches.

While large sized models often require significant compute resources and typically suited for cloud, there is need for on-device ML inference for privacy reasons. Smaller and compact models can be deployed on smartphones with TensorFlow Lite. Distillation techniques [58] enable design of smaller networks such as DistillBERT. DistillGPT2 is lightweight alternative to GPT2. Such lightweight models enables deployment in smartphones and such on-device AI inference support privacy requirements. This makes it easy for deployment on smartphones for scale and affordable access.

11.7 Future Directions: AI Architecture for Mental Health

11.7.1 Triad-Therapy Using Multimodal Encoder-Decoder Modelling

A vision for the architecture for combined modelling of Cognitive Triad and Cognitive Behavior Therapy is presented in Table 11.3. As seen in Fig. 11.7, the architecture can be generalized as Encoder-Decoder Transformer. This architecture combines a combination of Vision Transformers, Multimodal Transformers (ViLBERT) [51, 55], Language models (M-BERT, GPT-n) [18, 22, 23, 39], TensorFlow Federated Learning [21, 22, 43].

Based on the architecture elements covered in literature, Cognitive Triad is about the diagnosis of the state of mental health of an individual based on multiple data sources [48]. The nature of data sources used for diagnosis could range from the language used by the person such as social media posts, radiology scans of the brain, and data from sensors to perceive the environment. The ability to combine learning representations from different types of data sources and learn jointly from the multimodal representation [44, 45] allow for learning all aspects of Cognitive Triad. This idea of multimodal modelling using Multimodal Transformers like ViLBERT [51] is illustrated in Table 11.4. By modeling three features of Cognitive Triad namely views of self, world, future, a Deep Learning model can analyze the state of mental health. The model combines two different views to analyze, namely

Table 11.3 Encoder-Decoder architecture

Architecture	Function	Approach
Encoder	Cognitive Triad	Model understands View of World
Decoder	Cognitive Therapy	Model influences View of Self

Table 11.4 Architecture implementation

		Architecture Implementation
Encoder	Triad	Multimodal Deep Learning (ViLBERT) 1. Language Transformer (BERT) to model View of World by understanding words uttered by user 2. Vision Transformer (ViT) to understand affective state by classifying facial expressions 3. Temporal modelling to understand changes over time
Decoder	Therapy	Generative modelling 1. GPT-n for generating human like text to speak like counselors 2. BigGAN to generate positive visualization of the View of future 3. Open AI DALL-E, CLIP for AI generated images to help patients visualize a better future

1. Modelling of view of self by analyzing the language of user (DistillBERT) like counselors
2. Modelling of view of environment (multimodal Transformer)
3. Modelling of view of future by analyzing the language of user.

The therapy is about enabling the patient to change the perspective of the situation by Generative modeling using NLG and GAN, with the generation attributes such as empathy rewarded using Reinforcement Learning. An instance of such an implementation was demonstrated by [42].

11.7.2 Addressing Needs of Countries with NLP Beyond English Language

Given many local language-speaking populations across countries are impacted by COVID, the approach of AI based mental health intervention has to scale beyond -English speaking population. Here, it is notable to mention that Transformer based language models can learn any language. For example, many multi lingual Transformers models such as mBERT are already pre-trained in many world languages [50], expanding the scope of the population that can be impacted. Hence, the design approach laid out in this chapter can be scaled out to deployment in multiple countries.

11.7.3 Implications of Findings and Scope for Future Work

Based on the literature, the application of Transformer language models for Cognitive Triad is well established by many authors [49]. Researchers have established the idea of employing Transfer Learning [18] on a pre-trained BERT for custom tasks in the diagnosis of mental health conditions. Also many types of data sources have been used in diagnosis. However multimodal Learning with multiple combinations of data sources is yet to be established for mental health, although a handful of papers have attempted some version of multimodality. While a substantial among of literature have demonstrated NLU based diagnosis use case [20, 45–47, 49], there is only a handful of papers that have demonstrated therapy [42]. Based on analysis of literature, we certainly conclude that Transformers-based Deep Learning architectures play a crucial role in mental health, and have the potential for immediate deployment to avert the crisis. The opportunity for future research in modelling "thinking patterns" using Multimodal Deep Learning is immense. Using the patten recognition power of Deep Learning, and the generative power of Generative Deep Learning, a neural network to analyze "thinking pattens on view of world" and to help adopt a "healthier thinking patterns by generating an positive view of the self" using generative modelling techniques in Multimodal Deep Learning is a unexplored theme. A Generative Multimodal Deep Learning based approach can enable in transforming a individual's negative visualization of a situation in the human mind, and can help imagine a positive visualization of Cognitive self and future, thus improving the state of mental health for millions of people.

11.8 Conclusion

This chapter established the role of Deep Learning based approach to address the global mental health challenge. While mental health was included as part of UN SDGs in 2015 due to the disease burden, it changed from an invisible problem to a global crisis by 2020. With mounting evidence of COVID-19 pandemic causing a widespread impact on mental health, the need for action is urgent. This chapter not only surveyed the literature, but also made the analysis from the point of view of averting the pandemic scale mental health crisis. Thus strategies were highlighted from the literature. The main contribution of this chapter is also a synthesis of a modern Deep Learning architecture for mental health based on a generalization of architectural learnings from the various research literature. Also, the chapter made this synthesis based on the latest advances in the field such as Transformers. Thus a unique perspective on AI in mental health was synthesized and discussed, backed up with supporting research articles for each of the elements of the solution approach.

The science of Cognitive Triad and CBT was analyzed and accordingly a solution approach was identified, and then solved using Deep Learning based modelling. The implementation of Triad and Therapy was generalized into an Encoder-Decoder style architecture using Multimodal Transformers for the sequence to sequence translation from the Cognitive "view of the world" to counseling conversations to effect the "view of self". So a translation of Beck's "negative view of world" to a "healthier view of self" was modeled using Encoder-Decoder styled Multimodal Transformer architecture. Various aspects of the generic architecture were simplified and presented through the chapter with supporting research papers for each of the ideas in the solution. The idea of fine-tuning a pre-trained BERT language model for custom tasks in pandemic scale screening & rapid diagnosis was highlighted with supporting research literature. The architecture employed a combination of advanced Deep Learning ideas such as Multimodal Transformers for improving Cognitive Triad, Transfer Learning and Few shot Learning for modelling thinking patterns, Transformer language models for both diagnosis and therapy, temporal modelling for detecting changes in thinking patterns for detecting natural resilience, privacy safe on-device ML. In short, the design of Deep Learning architecture to avert the global mental health crisis was articulated starting with an understanding of the dimensions of the challenge, and an approach to improve mental health at pandemic scale.

This chapter introduced a promising approach by tapping into the advances in Deep Learning to deploy Transformers-based mental health solutions into the hands of a billion people via personal smartphones. This presents a powerful paradigm to avert the crisis, and hence the chapter motivates multidisciplinary research communities to contribute AI based solutions to address this global crisis on mental health.

References

1. Galea, S., Merchant, R.M., Lurie, N.: The mental health consequences of COVID-19 and physical distancing: the need for prevention and early intervention. JAMA Intern. Med. (2020). https://doi.org/10.1001/jamainternmed.2020.1562
2. UNSDG Policy Brief: COVID-19 and the need for action on mental health (2020).https://unsdg.un.org/resources/policy-brief-covid-19-and-need-action-mental-health. Accessed 28 2020
3. WHO Mental Health Facts. https://www.who.int/news-room/facts-in-pictures/detail/mental-health
4. Holmes, E.A., et al.: Multidisciplinary research priorities for the COVID-19 pandemic: a call for action for mental health science. Lancet Psychiatry (2020). https://doi.org/10.1016/s2215-0366(20)30168-1
5. [NHS website]: Mental Health of Children and Young People in England 2021, 30 Sep 2021, NHS, https://digital.nhs.uk/data-and-information/publications/statistical/mental-health-of-children-and-young-people-in-england/2021-follow-up-to-the-2017-survey
6. Keeley, B.: The state of the world's children 2021: on my mind-promoting. Protecting and Caring for Children's Mental Health, UNICEF (2021)
7. WHO team: The impact of COVID-19 on mental, neurological and substance use services, WHO Publication, ISBN: 978-92-4-001245-5 (2020). https://www.who.int/publications/i/item/978924012455
8. [Editorial]: The intersection of COVID-19 and mental health. The Lancet Infectious Diseases (2020). https://doi.org/10.1016/S1473-3099(20)30797-0
9. [Editorial]: Keep mental health in mind. Nat. Med. **26**, 631 (2020). https://doi.org/10.1038/s41591-020-0914-4

10. [Editorial]: Mental health matters. Lancet Glob. Health **8**(11) (2020). https://doi.org/10.1016/S2214-109X(20)30432-0
11. McGorry, P.D., Ratheesh, A., O'Donoghue, B.: Early intervention–an implementation challenge for 21st century mental health care. JAMA Psychiatry **75**(6), 545–546 (2018). https://doi.org/10.1001/jamapsychiatry.2018.0621
12. Ford, T., John, A., Gunnell, D.: Mental health of children and young people during pandemic. BMJ **372** (2021). https://doi.org/10.1136/bmj.n614
13. Durstewitz, D., Koppe, G., Meyer-Lindenberg, A.: Deep neural networks in psychiatry. Nat. Mol. Psychiatry **24**, 1583–1598 (2019). https://www.nature.com/articles/s41380-019-0365-9, https://doi.org/10.1038/s41380-019-0365-9
14. Denecke, K., Vaaheesan, S., Arulnathan, A.: A Mental Health Chatbot for Regulating Emotions (SERMO)—concept and usability test. IEEE Trans. Emerg. Top. Comput. **1** (2020). https://doi.org/10.1109/tetc.2020.2974478
15. Doan, S. et al.: Extracting health-related causality from twitter messages using natural language processing. BMC Med. Inform. Decis. Mak. **19** (2019). https://doi.org/10.1186/s12911-019-0785-0
16. Gkotsis, G., Oellrich, A., Velupillai, S. et al.: Characterisation of mental health conditions in social media using Informed Deep Learning. (Nature) Sci. Rep. **7**, 45141 (2017). https://doi.org/10.1038/srep45141
17. Chancellor, S., De Choudhury, M.: Methods in predictive techniques for mental health status on social media: a critical review. NPJ Digit. Med. **3** (2020). https://www.nature.com/articles/s41746-020-0233-7, https://doi.org/10.1038/s41746-020-0233-7
18. Ranti, D. et al.: The Utility of General Domain Transfer Learning for Medical Language Tasks (2020). arXiv:2002.06670
19. Dobson, K.S.: Handbook of Cognitive-Behavioral Therapies. Guilford Press (2019)
20. Oh, K., Lee, D., Ko, B., Choi, H.: A chatbot for psychiatric counseling in mental healthcare service based on emotional dialogue analysis and sentence generation. In: 18th IEEE International Conference on Mobile Data Management (MDM), Daejeon, pp. 371–375 (2017). https://doi.org/10.1109/MDM.2017.64
21. Yang, T. et al.: Applied Federated Learning: Improving Google Keyboard Query Suggestions (2018). arXiv:1812.02903
22. Li, X. et al.: Multi-site fMRI Analysis Using Privacy-preserving Federated Learning and Domain Adaptation: ABIDE Results (2020). arXiv:2001.05647
23. Devlin, J., Chang, M.-W., Lee, K., Toutanova, K.: BERT: Pre-Training of Deep Bidirectional Transformers for Language Understanding (2018). arXiv:1810.04805
24. Sanches, P. et al.: HCI and affective health: taking stock of a decade of studies and charting future research directions. In: Proceedings of the 2019 CHI Conference on Human Factors in Computing Systems (CHI '19). ACM (2019). https://doi.org/10.1145/3290605.3300475
25. Wolf, T. et al.: HuggingFace's Transformers: State-of-the-art Natural Language Processing (2020). arXiv:1910.03771
26. Keser, E., Kahya, Y., Akn, B.: Stress generation hypothesis of depressive symptoms in interpersonal stressful life events: the roles of cognitive triad and coping styles via structural equation modeling. Curr. Psychol. (2017). https://doi.org/10.1007/s12144-017-9744-z
27. Ungar, M., Theron, L., Murphy, K., Jefferies, P.: Researching multisystemic resilience: a sample methodology. Front. Psychol. **11**, 3808 (2021). https://doi.org/10.3389/fpsyg.2020.607994
28. Manji, H., Moore, G., Rajkowska, G. et al.: Neuroplasticity and cellular resilience in mood disorders. Nat. Mol. Psychiatry **5**, 578–593 (2000). https://doi.org/10.1038/sj.mp.4000811
29. Holt-Quick, C. et al.: A chatbot architecture for promoting youth resilience (2020). arXiv:2005.07355
30. Dobson, K.S., Shaw, B.F.: The effects of self-correction on cognitive distortions in depression. Cogn. Ther. Res. **5**, 391–403 (1981)
31. Ehlers, A., et al.: A randomized controlled trial of cognitive therapy, a self-help booklet, and repeated assessments as early interventions for posttraumatic stress disorder. Arch. Gen. Psychiatry **60**, 1024 (2003)

32. Beck, J.S.: Cognitive Behavior Therapy Basics and Beyond. Guilford Press (1995)
33. Vaswani, A. et al.: Attention is all you need. In: Advances in Neural Information Processing Systems (NIPS), pp. 5998-6008 (2017)
34. Li, X., Li, P., Bi, W., Liu, X., Lam, W.: Relevance-Promoting Language Model for Short-Text Conversation (2019). arXiv:1911.11489
35. Vig, J.: A multiscale visualization of attention in the transformer model (2019). arXiv:1906.05714
36. Fedus, W., Zoph, B., Shazeer, N.: Switch Transformers: Scaling to Trillion Parameter Models with Simple and Efficient Sparsity (2021). arXiv:2101.03961
37. Radford, A. et al.: Language models are unsupervised multitask learners. OpenAI Blog **1**(8), 9 (2019)
38. Brown, T.B., Mann, B., Ryder, N., Subbiah, M., Kaplan, J., Dhariwal, P., Neelakantan, A. et al.: Language models are few-shot learners **4** (2020). arXiv:2005.14165
39. Liao, Y., Wang, Y., Liu, Q., Jiang, X.: GPT-Based Generation for Classical Chinese Poetry (2019). arXiv:1907.00151
40. Zellers, R. et al.: Defending Against Neural Fake News (2019). arXiv:1905.12616
41. Peng, X., Li, S., Frazier, S., Riedl, M.: Fine-Tuning a Transformer-Based Language Model to Avoid Generating Non-Normative Text (2020). arXiv:2001.08764
42. Sharma, A., Lin, I. W., Miner, A.S., Atkins, D.C., Althoff, T.: Towards facilitating empathic conversations in online mental health support: a reinforcement learning approach. In: Proceedings of the Web Conference 2021, pp. 194–205 (2021). https://doi.org/10.1145/3442381.3450097
43. Lim, W.Y.B. et al.: Federated learning in mobile edge networks: a comprehensive survey. IEEE Commun. Surv. Tutor. **1** (2020). https://doi.org/10.1109/comst.2020.2986024
44. Hendricks, L.A. et al.: Decoupling the role of data, attention, and losses in multimodal transformers (2021). arXiv:2102.00529
45. Yang, L., Jiang, D., Han, W., Sahli, H.: DCNN and DNN based multi-modal depression recognition. In: Proceedings of 2017 7th International Conference on Affective Computing and Intelligent Interaction, pp. 484–489 (2017). https://doi.org/10.1109/ACII.2017.8273643
46. Gratzer, D., Goldbloom, D.: Therapy and E-therapy—preparing future psychiatrists in the era of apps and chatbots. Acad. Psychiatry **44**, 231–234 (2020). https://doi.org/10.1007/s40596-019-01170-3
47. Zhu, Y., Shang, Y., Shao, Z., Guo, G.: Automated depression diagnosis based on deep networks to encode facial appearance and dynamics. IEEE Trans. Affect. Comput. **9**, 578–584 (2018). https://doi.org/10.1109/TAFFC.2017.2650899
48. Su, C., Xu, Z., Pathak, J. et al.: Deep learning in mental health outcome research: a scoping review. Nat. Transl. Psychiatry **10**, 116 (2020). https://doi.org/10.1038/s41398-020-0780-3
49. DeSouza, D.D., Robin, J., Gumus, M., Yeung, A.: Natural language processing as an emerging tool to detect late-life depression. Front. Psychiatry 1525 (2021). https://doi.org/10.3389/fpsyt.2021.719125
50. Eisenschlos, J. et al.: MultiFiT: Efficient Multi-lingual Language Model Fine-Tuning (2019). arXiv:1909.04761
51. Lu, J.et al.: Vilbert: Pretraining task-agnostic visiolinguistic representations for vision-and-language tasks (2019). arXiv:1908.02265
52. Maiya, A.S.: ktrain: A Low-Code Library for Augmented Machine Learning (2020). arXiv:2004.10703
53. Clark, K., Luong, M.-T., Khandelwal, U., Manning, C.D., Le, Q.V.: BAM! Born-Again Multi-Task Networks for Natural Language Understanding (2019). arXiv:1907.04829
54. Sharma, A., Miner, A.S., Atkins, D.C., Althoff, T.: A Computational Approach to Understanding Empathy Expressed in Text-Based Mental Health Support (2020). arXiv:2009.08441
55. Xia, Q. et al.: XGPT: Cross-Modal Generative Pre-Training for Image Captioning (2020). arXiv:2003.01473
56. Floridi, L., Chiriatti, M.: GPT-3: Its nature, scope, limits, and consequences. Minds Mach **30**(4) (2020). https://doi.org/10.1007/s11023-020-09548-1

57. Korngiebel, D.M., Mooney, S.D.: Considering the possibilities and pitfalls of generative pre-trained transformer 3 (GPT-3) in healthcare delivery. NPJ Digit. Med. (2021). https://doi.org/10.1038/s41746-021-00464-x
58. Melas-Kyriazi, L., Han, G., Liang, C.: Generation-Distillation for Efficient Natural Language Understanding in Low-Data Settings (2020). arXiv:2002.00733
59. Dalglish, S.L., Costello, A., Clark, H., Coll-Seck, A.: Children in all policies 2030: a new initiative to implement the recommendations of the WHO-UNICEF-lancet commission. Lancet **397**(10285), 1605–1607 (2021)
60. [Editorial]: Mental health of children and young people during pandemic. BMJ (2021). https://doi.org/10.1136/bmj.n614

Chapter 12
Diagnosis of Mental Illness Using Deep Learning: A Survey

Sindhu Rajendran, Ritesh Gandhi, S. Smruthi, Surabhi Chaudhari, and Saurav Kumar

Abstract Deep learning is a robust technique that aims to build systems that can improve automatically based on experience using advanced statistical and probabilistic data. It has led to significant advances in the health sector by collaborating health professionals with data scientists that solved challenges previously not possible. Mental illness deeply impacts the day-to-day living of a person and is found to be very common in society these days. However, it is taken for granted in many cases. To overcome this issue, analytical techniques are implemented on mental health data which has shown broad potential in improving patient outcomes and also, to better understand the existence of psychological conditions in a wider community. With the rapidly growing field of bioinformatics, deep learning plays a pivotal role in detection by enabling speedy and scalable analysis of complex data like images of the brain. While this technique can be employed and worked at the practice for the early diagnosis of a mental illness, the same can be extended for real-time monitoring of the individual and improving the efficiency of treatment. This chapter gives an insight into the involvement of Neural Networks in the medical profession along with the comprehension of bioinformatics. Furthermore, the chapter progresses with the practice for diagnosing mental illness like Alzheimer's, anxiety, depression, schizophrenia, dementia, etc., eventually discussing how deep learning can be implemented in the detection of mental illness. Finally, the chapter summarizes the implications for future work.

S. Rajendran (✉) · R. Gandhi · S. Smruthi · S. Chaudhari · S. Kumar
R. V. College of Engineering, Bengaluru 560059, India
e-mail: sindhur@rvce.edu.in

R. Gandhi
e-mail: riteshgandhi.ec20@rvce.edu.in

S. Smruthi
e-mail: smruthis.ec20@rvce.edu.in

S. Chaudhari
e-mail: surabhihc.bt17@rvce.edu.in

S. Kumar
e-mail: sauravkumar.ec18@rvce.edu.in

A. Biswas et al. (eds.), *Artificial Intelligence for Societal Issues*, Intelligent Systems Reference Library 231, https://doi.org/10.1007/978-3-031-12419-8_12

Keywords Deep learning · Mental illness · Neural networks

12.1 Introduction

A mental disorder often termed as a mental illness or psychiatric condition, is defined as "a syndrome characterized by a clinically significant disturbance in an individual's cognition, emotion regulation, or behaviour that reflects a disfunction in the psychological, biological, or developmental processes underlying mental functioning". The outcome is a pattern of behaviour and thought process leading to substantial distress and impairing the personal functioning of the patient. The symptoms for these disorders might be chronic with the individual's state subsequently deteriorating and fading, or may manifest as one single episode. There have been many illnesses in which both symptoms and indications vary greatly amongst them. Generally, a psychiatrist or a mental health expert can diagnose such illness [1]. The causes of various mental disorders are commonly unknown. Theories may include results from a variety of fields. Mental disorders are often characterized by a persons' behaviour, feelings, perception, or thought process. This may be connected with certain brain areas or processes, frequently in a social setting. One aspect of mental health is a mental disorder. While establishing a diagnosis, cultural and religious views, as well as societal standards, should be considered. Autism Spectrum Disorders and intellectual disability are examples of neurodevelopmental diseases that often appear in infancy or early childhood. Infamy and indifference can aggravate the pain and impairment associated with mental illnesses, resulting in a variety of social groups aiming to raise awareness and combat social exclusion.

After mental health got included in the list of Sustainable Development Goals, the concerns towards mental health have increased as it is a leading factor of instability. Depression has proved to be a significant cause of disability. Suicide has become the second greatest cause of mortality for those aged between 15 and 29. Premature death up to two decades earlier as a result of avoidable physical problems is due to serious mental illness. Despite developments in certain nations, severe human rights violations, discrimination, and stigma are frequently targeting individuals subjecting them to mental illness. Various mental illnesses can be treated at a reasonable cost at the early stages of onset, but the problem arises in the accessibility of such a treatment by the affected. Exceedingly low treatment prices continue to be available for an effective treatment process. An increase in the contribution is needed on all fronts, mainly in instilling awareness amongst people and in the process, reducing the stigma associated with mental health, improving access to effective treatment and medication. Encouraging research in the identification of new methods to treat mental issues is the right step towards sustenance. Patients who are diagnosed with mental diseases early on may be able to avoid acquiring critical psychological issues. Because mental diseases cannot be predicted physically, they can only be recognized by symptoms or manifestation in behaviour or attitude. Although some people exhibit apparent signs of depression, others appear to be normal. People with depression modify the way they move, sleep, and interact with their surroundings.

Effective services are being provided by medical centres and hospitals, where mental health professionals (aka psychologists, psychiatrists, social workers) are performing assessments using various parameters, questionnaires, and therapies. Treatment is carried out under a variety of mental health specialists. The two main treatment methods are psychotherapy and psychiatric medication. Other therapies include dietary modifications, peer support, social interventions, and most importantly self-care. In a few situations, there may be involuntary hospitalization. Prevention strategies have been found to lower depression. Depression is one of the leading causes of mental illness and disability worldwide. Patients with mental illness who are not treated in time also worsen their mental health and sometimes even commit suicide. More than 350 million people suffer from depression, according to the World Health Organization. The number has increased by more than 18 between 2005 and 2015. Mental illness costs a lot of money to the world economy. Mental health spending was $2.5 trillion in 2010 and are expected to reach $6 trillion by the 2030s. On the other hand, mental illness puts a burden on society as a whole. The mental health treatments available are highly inadequate to accommodate the growing number of people with mental health problems. Lack of support for people with mental illnesses, as well as fear of the stigma associated with mental illness, contribute to insufficient access to care [2].

Mental health issues are becoming highly prevalent nowadays, with the estimated number of patients to be about 450 million in the world. Major game players in the deterioration of mental health include ADHD, depression, Alzheimer's, schizophrenia, ASD among others. No specific age group is restrained by developing such illness as it affects adults, children, and adolescents equally. Patient's experiences, behaviour recorded by relatives or friends, and mental status assessments can all be used to make a diagnosis. The majority of studies on mental diseases are based on social media detection. Using natural language processing technologies, scientists have attempted to define the correct link between depression and language. The vast majority of persons suffering from mental illnesses prefer to be alone. As a result, patients with mental illnesses will seek out a location that can accommodate their mind's overflow. For mental patients, social media, particularly Twitter, is a great venue to share their tales. Twitter is thought to be better at expressing the emotions of its users. Statements on Twitter will provide useful information as a tool for determining policy, and text mining on tweet data will be possible [3]. Globally, mental health problems are on the rise. Over the last decade, mental health problems and substance use disorders have increased by 13%, primarily due to demographic changes (until 2017). Mental health problems are leading to be the cause of disability once in every five years. Around 21% of adolescents and children worldwide suffer from mental health issues, with suicide becoming the second greatest cause of mortality among those aged 15–29. One in five people involved in post-conflict situations suffers from mental illness. Many facets of life, including education or work performance, relationships with family and friends, and the ability to engage in social activities, are impaired by mental health conditions. Anxiety and depression, two of the world's most common health problems, cost the global economy $1 trillion each year. Despite the graveness of the situation, the global median of government health expenditure on mental health is less than 2%. This chapter gives an insight into the involvement of Neural Networks in the medical profession along with the comprehension of bioinformatics. Furthermore, the chapter progresses with the practice for diagnosing mental illness like Alzheimer's, anxiety, depression, schizophrenia, dementia, etc., eventually discussing how deep learning can be implemented in the detection of mental illness. Finally, the chapter summarizes the implications for future work.

12.2 Concept of ML and DL

Machine Learning (ML) plays a major role in the analysis of big data. ML was first conceptually introduced by Allan Turing and the term itself was conceived by Arthur Samuel in the 1950s. It is now in widespread use in various fields, with the medical field not being an exception, and has been a part of it since 1990, due to expeditious developments in the concept. Long before the emergence of the importance of big data, large-scale data was studied and scalable algorithms were pre-designed by ML researchers, hence giving merit in handling big data. Some of the available techniques also enable their machine to learn adequately from the given data of various variables as compared to other cases. Thus, ML being an 'an essential part of big data analytics,' is being highly used both in early diagnosis and resolution of diseases and health issues accompanied by real-time patient monitoring, centric caregiving advancement in the treatment process. After the remarkable success of AlphaGo, people started highlighting ML as a magical wand. Recent research in the diagnosis of mental health suggests that ML algorithms could be a better solution for smaller-sized samples and could give a proper treatment with expertise's training [4, 5].

Sometimes the main algorithm of Artificial Intelligence, well known as Deep Learning or Neural Networks algorithm is taken into consideration as ML can create hindrances in research conducted by domain experts. Although, it is obvious that ML techniques are not an elixir that would instinctively yield a solution without the involvement of a large quantity of dataset and instructions by humans for training, for analysis of clinical data, ML algorithms with specific advantages are widely being used along with deep learning techniques. Such implementations would lead to more efficient research work by facilitating smooth communication between clinical and ML researchers.

The development of predictive models, computational algorithms that can be used to extract hidden clues (patterns) from various datasets is the main aim of Machine Learning. The data analysis of the healthcare sector done using ML algorithms and models has witnessed an increasing number in recent years. However, approaches of a significant amount of feature engineering are required for an upgrade in efficiency in most scenarios which consume both time and material. Deep learning, a product of AI and ML, aims at the development of a network of numerous layers that identifies hidden patterns by mapping the input data (raw) to the result. The complete process

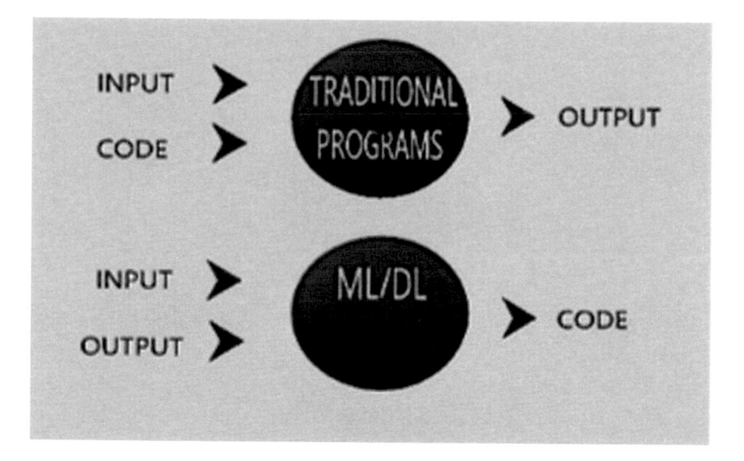

Fig. 12.1 Comparison of deep learning and traditional programming

accompanies the involvement of algorithms exclusive to DL such as recurrent neural network (RNN), convolutional neural network (CNN), fully convolutional neural network (FCN) and autoencoders [4] (Fig. 12.1).

12.3 Deep Learning in Mental Health

Concepts involving Artificial intelligence (AI) and related fields have been especially useful to psychologists to make decisions based on previous data of patients such as behavioural traits, social platforms, medical records. Being one of the most recent versions of AI technology, Deep learning has proven to be better in tasks involving a spectrum of real-time applications including healthcare and computer vision. A considerable amount of completed, as well as ongoing researches, exist on DL algorithms involving mental health studies.

For a better understanding of a particular status of mental health and for providing better care to patients, early detection of mental disorders is of extreme importance. Unlike many chronic disorders, in which diagnosis is carried out through laboratory tests and measurements, mental illnesses are generally diagnosed based on self-report of an individual subjected to specialized questionnaires, designed to detect reactions to social settings and interaction. AI and ML technologies assist clinicians to make necessary decisions and draw conclusions on the mental status of individuals based on the data available.

Deep Learning algorithms have revealed outstanding facets in various applications laden with data. Simply put, the data (input) goes through several layers of the network (non-linear processing units) to discover new meanings to the once raw information. Deep Neural Networks (DNN) are being used for the measurement of brain activities involving an investigation of the status of mental health [6].

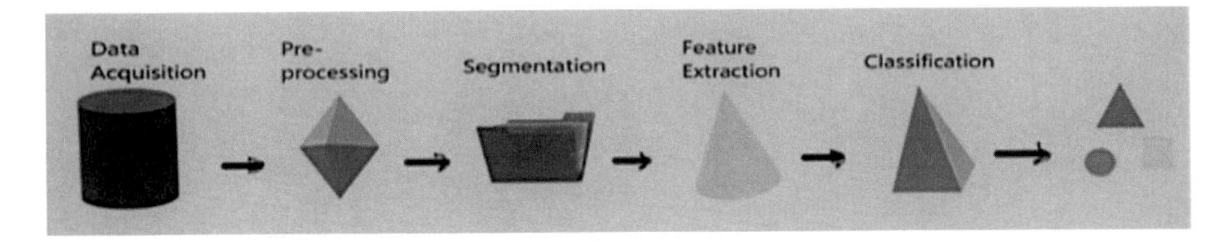

Fig. 12.2 Data validation for deep learning

Brain activities are measured by Deep Neural Networks (DNN) for the investigation of the status of the brain. Deep learning, along with structural T1-weighted images are used to detect and classify patients suffering from schizophrenia as opposed to their healthy counterparts. The Deep learning technique holds good potential for the treatment of mental illness as it has 90% classification accuracy for extraction of features from Deep Belief Networks using raw data and used for clinical brain imaging applications. It may help in the identification of the disease and assess the severity level of it, i.e.; low, medium, or high for early treatment of the disease.

Deep learning algorithms have demonstrated significant capabilities in a variety of disciplines of study in recent years. Deep learning is a method that is "data-driven". Deep learning algorithms' efficiency is determined by the quality and quantity of data supplied to them. Figure 12.2 depicts a generic deep learning technique. Algorithms of deep learning look for patterns in data and then learn them. Deep learning attempts for mathematical approximation of function, which will serve as a model. Deep learning has become a preferred choice for academics in a variety of domains because of its limited constraints and tremendous untapped potential. Lower extremity recognition tasks can be expanded using deep learning and wearable technology applications, such as monitoring patients in clinical or home settings, monitoring babies in maternity clinics, and analysing athlete performance metrics and recovery rates in rehabilitation centres. The demand for a strong and dependable solution in the domain of human activity detection, whether it's personalized healthcare, robust authentication mechanisms, prosthetics, biped robotics, or other applications, is urgent, and deep learning algorithms are ready to meet the challenges. Furthermore, embedding such systems in IoT devices would enable edge computing and small machine learning frameworks [7, 8].

12.3.1 Concept of Bioinformatics in Deep Learning

The current period of "big data" requires the conversion of enormous quantities of data into an effective source of knowledge. The importance of this conversion is increasing enormously in various domains with no exception in the field of bioinformatics. A significant amount of biomedical data has been accumulated which

Table 12.1 Input data and research fields of bioinformatics with deep learning [9, 14]

	Input	Research areas
'Omics'	– Data (DNA, DNase, ChIP, RNA) sequence from the genome. – PSSM-Position specific scoring matrix – Physicochemical properties – 1D structural properties – Contact map – Microarray gene expression – FAC- Atchley factors	**Prediction of protein structure** Properties of 1D structures contact map Quality assessment of structure model **Classification of proteins** Super family Localization of subcells **Regulation of gene expression** Splicing of junction Genetic variants affecting splicing Specificity of sequence **Anomaly detection** Cancer
Biomedical imaging	– MRI- Magnetic Resonance Imaging – Volumetric electron Microscopy image – PET- Positron Emission Tomography – Retinal image – ISH-'in situ' hybridization image – Histopathology image	**Recognition** Anatomical structure cell nuclei Finger joint **Segmentation** Neuron structure Cell structure Vessel map Brain tumour **Anomaly detection** Schizophrenia Cancer Alzheimer's Disease
Biomedical signal analysis	– EMG,ECoG, EMG,EOG,ECG – EEG (raw, wave, frequency) – Features extracted from EEG – Variation of peak – Normalized Decay	**Brain decoding** Behaviour emotion **Anomaly detection** Stages of sleep Seizures Epilepsy Alzheimer's Disease

includes omics, biomedical images, mentioned in Table 12.1, which has resulted in high potential in biological and healthcare research, drawing the interests of industries as well as academia. For example, platforms such as IBM developed Watson for Oncology to assist clinicians along with treatment options after analysing patients' medical information. Google DeepMind, has also been successful in achieving the purpose with AlphaGo in the game of Go, after launching DeepMind Health for the development of effective technologies for health care [9]. For extracting effective knowledge from enormous data while working with bioinformatics, the concept of machine learning is a successful methodology that is most widely being used. Training data is used by various algorithms for uncovering underlying patterns, the building models, and for making predictions that are based on best fit models. Support vector machines, random forests, hidden Markov models, Bayesian networks, Gaussian networks are a few well-known algorithms that have been applied in the field of genomics, proteomics, systems biology, and numerous other illnesses. Data representation known as features is responsible for the proper performance of the existing or conventional machine learning algorithms. Human engineers with extensive domain expertise typically design the required data representation or features, still, identification of an appropriate feature is difficult for a particular task. Deep learning, a fast-emerging domain of machine learning, is based on the analysis of big data involving the powerful technique of parallel and distributed computing, along with advanced algorithms [10]. The academic has witnessed great increment since the early 2000s s due to the overcoming of previous limitations along with major advancement in diverse domains of the Artificial Intelligence (AI) community which has suffered a loss for many years. Image and speech recognition is marked as one of the most important advancements, although dissemination in processing and translation of natural language is done for promising results. Deep Learning can provide much larger benefits to bioinformatics, such as the discovery of splice junctions from DNA sequences, recognition of finger joints from X-ray images, detection of electroencephalography (EEG) signals and so on.

The gait is regarded as one of the most distinctive biometric identifiers of humans. It is a difficult learning process that humans gain through constant interaction with their surroundings. It is one of the more difficult physical activities since it requires the coordination and synchronization of numerous body components. It is changeable and is influenced by a variety of characteristics such as age, gender, weight, and so on. The gait is frequently utilized in neurological illness patients for clinical assessment, early identification of gait anomaly, and so on. It's also utilized in robots to generate walks. It is regarded as suitable for diverse human activity recognition because of its biphasic and bipedal nature [11–13]. Medication, as we all know, contributes to balance and gait issues, especially in vulnerable groups. Medications play a role in at least 30% of elderly people who have balance or gait issues. Polypharmacy (taking more than four prescriptions) is associated with a higher risk of falling, and psychiatric meds are among the worst offenders-and not just the "usual suspects" (tricyclics, benzodiazepines, barbiturates, and antipsychotics).

Selective serotonin reuptake inhibitors (SSRIs), for example, have been linked to falls and are more likely to produce postural instability than tricyclics. We may picture drugs that impact nearly any neurological function interfering with balance and gait, and so contributing to falls if viewed mechanistically. The most common causes of medication-related falls include cognitive impairment, ataxia, Parkinsonism, and hypotension. Antihistaminic and anticholinergic medicines, of course, frequently cause cognitive impairment.

Dementia is intimately linked to the preceding. Balance and gait abnormalities are linked to cognitive deficiencies. In the long run, a sluggish walk indicates cognitive deterioration. Balance and gait disorders are common in the elderly, with 14% of those over 65 years old having them and 50% of those over 85 years old having them. In psychiatry and neurology, alcoholism is very prevalent. At every level of the neural system, it impacts gait. Regardless of medication, schizophrenia is linked to moderate Parkinsonism and ataxia. The gait is frequently slower, the stride length is shorter, and the tandem gait is moderately hindered. Parkinsonism is sometimes noticeable in depressed patients, but it goes away once the depression is gone. Depression patients in general (particularly those over the age of 40 and those with melancholic depression) walk slowly and take tiny steps. As the mood problem improves, the gait returns to normal.

12.4 Mental Health Disorders

12.4.1 Anxiety Disorders

Anxiety is a natural reaction to a stressful situation. It's the brain's mechanism of responding to stress and warning that danger may be lurking around the corner. Anxiety is something that everyone experiences from time to time. Anxiety disorders, on the other hand, are not the same as other mental illnesses. They are a collection of mental diseases that produce extreme anxiety and panic on a regular basis.

Obsessive Compulsive Disorders. OCD is defined by compulsions (to bring oneself to do something over and over again) due to persistent, uncomfortable thoughts, ideas, or feelings (obsessions). Frequently such individuals are seen obsessing over tasks like washing hands, checking on various things, and cleaning. This often degrades their quality of social interaction and their routine. People with OCD have persistent thoughts and strict actions. Abstaining them from their compulsions causes them a lot of stress. Even when they recognize their obsessions are unrealistic, they have difficulty disengaging from obsessive thoughts or quitting compulsive bchaviour.

Post-Traumatic Stress Disorder. PTSD affects persons who have faced unpleasant occurrences like abuse (physical or verbal), death scenarios, or major accidents. People with PTSD have persistent, disturbing, and restless thoughts and perceptions about the traumatic incident even after a long time since it elapsed. Dreams or flashbacks about the experience make them sad, fearful, or angry, as well as detached from others. Victims of this disorder often avoid circumstances that remind them of the incident with extreme stress to inoffensive things like loud noises or unintentional touches.

12.4.2 Mood Disorders

Mood disorder is a common flag covering all types of depression and bipolar disorder. Mood disorders are not restrained to teenagers alone. Children and adults experience them as well. But the symptoms vary with age.

Major Depression. Major depression, a medical condition that instils a negative impact on the behaviour of its victims is undoubtedly becoming common among all. Patients with depression don't seem to have the same enthusiasm toward previously enjoyed activities. They also experience many problems (mental and physical), along with a drain in energy while performing work.

Bipolar Disorder. Bipolar disorder is identified by severe mood swings. One of the signs is mania or an unusually heightened mood. Depressive episodes can also occur as a result of it. Manic depression or bipolar sickness are other names for bipolar disorder. People with bipolar disorder struggle to keep their mood at bay leading to disruptions in routine and relationships with others.

12.4.3 Psychotic Disorders

Psychotic disorders are severe mental illnesses characterized by aberrant thoughts and perceptions. Psychosis causes victims to lose touch with reality often leading them to hallucinations and delusions. Delusions are incorrect beliefs, some examples might be arriving at a conclusion that a person is deliberating scheming against them or the radio has hidden messages in it. False perceptions, (such as hearing, seeing, or experiencing something other than reality) are termed hallucinations.

Schizophrenia. Schizophrenia is a chronic mental disorder in which patients are unable to differentiate between reality and fantasy. The duration, severity, and frequency of such symptoms can vary depending on the condition. But often, patients improve as they get older. However, symptoms of the disorder aggravate when patients refuse to take medications as recommended, or by drinking or using illicit drugs. Stress is a major symptom booster. Therapy can reduce the severity of the symptoms and prevent hallucinations from recurring.

Delusional Disorder. The delusional disease is a severe mental condition that makes an individual fervently believe something unrealistic. But delusions are not always bizarre. Real-life scenarios such as being followed, poisoned, conspired against are also existent. The misinterpretation of perceptions or events is common in these delusions.

12.4.4 Dementia

Dementia is a major ailment that is characterized by loss of linguistic, logical, and memory abilities reducing the quality of one's life. A deterioration in cerebral abilities also termed as cognitive capacities not only affects one's intelligence but also emotions and behaviours towards others.

Alzheimer's Disease. It is a neurological ailment that causes the brain cells' death ultimately reducing their size. This disease has a major role to play in increased. Dementia in individuals. Forgetting recent events or discussions may indicate the inception of Alzheimer's disease. Severe impairment in memory makes the individual incapacitated.

Parkinson's Disorder. Parkinson's disease is a nervous system that affects a person's life slowly. Early indications might include muscle stiffness, trembling, and difficulty in coordination and balance. Later stages disrupt the person's linguistic abilities, thinking process, sleeping patterns, and behaviour.

12.5 Diagnosis Using Deep Learning

The diagnosing procedure necessitates competence and experience that can only be acquired by trained mental health professionals. A person's symptoms could be the result of a variety of factors. As a result, only knowledgeable individuals are capable of recognizing the behavioural patterns required to identify specific diseases. Also, the quality of life of a patient can be improved if mental disorders are diagnosed accurately at the initial stage.

A deep neural network is a multi-layer perceptron with many hidden layers between the input and output layers with fully connected weights. Complex, non-linear relationships can be modelled using deep neural networks. Multiple buried layers contribute to the formation of features from the bottom. One can build deep neural networks using both feedforward and recurrent neural networks and train them using standard backpropagation techniques. The weights of the DNN layer can be updated using the stochastic gradient descent approach. Deep neural network technology is also used to extract robust properties. Deep belief networks (DBNs) are generative models that have multiple layers of latent variables, most of it being binary. There are unidirectional interlayer connections, but no intra-layer connections. Learning is difficult in these networks because deriving a later distribution from the hidden (potential) layer is challenging. To sample the posterior region, Markov chain Monte Carlo (MCMC) algorithms can be utilized, however, they are time draining. Further, is described a deep belief network training technique that is akin to training a constrained Boltzmann machine stack or sequence (with complementary prior distributions to avoid path effects being explained) (RBM). Algorithm 1 shows a deep neural network algorithm (trajectory database: gait data) [7, 15, 16].

In any machine learning project, data, before being fed into the algorithm, is divided into 3 sets:

I. Train set: A set used in the training of algorithms and construction of batches.

II. Dev set: A set used in the finetune process of the algorithm and evaluation of bias and variance.

III. Test set: A set used in the generalization of the error/precision of the final generated algorithm.

According to the size of the data set m, the repartition of the three sets is shown in the given Table 12.2.

Table 12.2 Partition of three sets of algorithm

	Train (%)	Dev (%)	Test (%)
$m = 10^4$	60	20	20
$m = 10^6$	96	2	2

Standard deep learning algorithms necessitate a huge dataset of about lines of samples. Now that the data is ready, we'll look at the training algorithm in the next section. In most cases, before entering the data into the algorithm, the inputs are normalized. Standard data is a widely used method that involves the removal of the variables' mean and division by their standard deviation.

Let X be a database variable, then we can set:

$$X := \frac{X - \mu}{\sigma} \tag{12.1}$$

$$\text{Where } \mu = \frac{1}{m}\sum_{i=1}^{n} x^{(i)} \text{ and } \sigma = \frac{1}{m}\sum_{i=1}^{n} (x^{(i)} - \mu)^2 \tag{12.2}$$

Algorithm 1- DNN Algorithm [16]
procedure DNN – ALGORITHM

$$Initialize\, \omega,\ \beta\ for\ H\ data,\ load\, training\, data\, t\ for\ all\, nodes. \tag{12.3}$$

$$for\ each\, node\, n\, training\ T_k,\ do\ parallel$$

$$(\omega_n, \beta_n) \leftarrow global\, model\ (seed\ value,\ update\, later) \tag{12.4}$$

$$(T_{k\alpha}) \subset F_k\ (F_k\, is\, the\, sample\ for\ all\, iteration) \tag{12.5}$$

$$divide\ (T_{k\alpha})\ in\ (T_{k\alpha}c)\ from\, k_c\ (k_c\, is\, number\ of\ cores) \tag{12.6}$$

$$(T_{k\alpha}c)\ on\, node\, k,\ do\ parallel \tag{12.7}$$

$$\textbf{for}\ each\, node\, i \in T_{k\alpha}c\ \textbf{do} \tag{12.8}$$

$$\omega_{ij} \in \omega_n\ and\ \beta_{ij} \in \beta_n \tag{12.9}$$

$$(\omega_{ij}) = \omega_{ij} - \alpha\frac{\partial L\left(\omega, \beta_{ij}\right)}{\partial \omega_{ij}} \tag{12.10}$$

$$(\beta_{ij}) - \beta_{ij} - \alpha \frac{\partial L(\omega, \beta_{ij})}{\partial \omega_{ij}} \tag{12.11}$$

Electroencephalography (EEG) is used to diagnose mental disorders as it provides brain biomarkers. It is a non-invasive test that records the electrical brain activity generated by the brain's top layers. It is made up of an array of electrodes that are put on the patient's scalp. There are other techniques (Magnetic Resonance Imaging: MRI and Positron Emission Tomography: PET) that capture just a static brain image to provide a complete spatial brain map. EEG provides a high temporal resolution due to its high sample rate efficiency. EEG has several advantages over MRI and PET, some of them being its non-invasive nature, ability to perform an ambulatory diagnosis, and cost effectiveness [17].

Individual differences in behavioural traits and cerebral maturity can be predicted using deep learning techniques. It is a neural network-based extension of classical machine learning methods. It can learn a higher level of artistic notion dynamically by applying consecutive nonlinear modifications to pre-processed data in underlying neural network layers, making it perfect for analysing complicated, dispersed, and nuanced neural activity. It can eliminate the requirement for time absorbing pre-processing and prior feature selection, as well as the model-dependent judgments that go along with it.

Deep learning algorithms (mentioned in Table 12.3) do better than strictly supervised methods when it comes to classifying brain imaging data. To learn a model, the algorithms employ complicated data representations. Using unsupervised learning approaches, deep learning algorithms retrieve key features with minimum human interaction. Unsupervised method classification of patient groups may aid in the discovery of neural arrangement of psychiatric diseases making them less prone to category errors. Presumptive labels are used to build a detector and classify patterns of brain engagement or association related to the labels in supervised approaches. The detector in unsupervised approaches searches population data for trends in the brain that might be linked to a category of clinical subjects; with the subjectivity in label identification discarded.

Neural Network (NN) is a type of machine learning algorithm made up of layers containing artificial neurons or units. Input, hidden, and output layers are the three types of layers found in NNs. The input data is inserted in the input layer. The raw data is then computed in the hidden layer and subsequently transferred to the output layer, where the network returns the result. Because neuronal connections between layers are connected, each of these connections has a different 'weight'. Neurons in the hidden and output layers compute the scalar product of their weights and input, and an activation function (non-linear) receives the result of such a product. Inference in NNs is arrived at using a technique called Forward-Propagation, in which data moves from the input to the output layers. The Back-Propagation is a technique used to train the NNs to alter the parameters of each unit to produce the predicted result (Figs. 12.3 and 12.4).

Algebra of the neural network is given as,

Table 12.3 Various algorithms of deep learning and its application [23–27]

Techniques	Working	Applications
Convolutional Neural Network (CNN)	Artificial neural network with at minimum three types of layers: – Convolutional – Fully connected – Pooling	Disease-specific feature extraction model for prediction and diagnosis. Examples include skin cancer, heart disease.
Fully Convolutional Network (FCN)	In such a network, each neuron has been connected to every neuron of the previous layer.	Textual data (patient's medical records) can be used for the diagnosis of diseases.
Recurrent Neural Network (RNN)	It is made up of at least two stacks of recurrent neural network cells called Long Short-Term Memory units (LSTMs).	Disease diagnosis is done by building models specifically for a disease. Alzheimer's, Parkinson's can be diagnosed under this technique.
Dilated Convolutions	In this technique, all the available weight matrices are of adjustable size (dilation)	This method is widely used in computer vision and image processing. Can be used for disease classification, and segmentation.
Generative Adversarial Network (GAN)	It comprises two neural systems: one that produces the samples and the other, known as the discriminator, that examines them.	Detects diseases using medical images like MRIs, X-rays or CT. Leukaemia and myocardial infraction prediction models.
Auxiliary Classifier GANs	It has two adversaries: generating or generator and discriminative or critic networks. It produces more accurate outcomes than GANs by combining knowledge from both models.	Diagnosis by feature extraction models using symptoms, medical images and records.
Convolutional Auxiliary Classifier GAN	A combination of generator networks suitable for multi-label classification	Generation of medical images, records for diagnosis.
Attention-based Deep neural network (Attentional NN)	It incorporates the context of input data into its output prediction using an attention technique.	Performs complex tasks like treatment planning.
Adversarial Autoencoders (AAE)	It comprises of an encoder and a decoder, where the encoder changes information into important highlights whilst the decoder switches these highlights back to the information.	Disease-specific feature extraction models
GANs with auxiliary output (GAN-AO)	It's a form of discriminative adversarial network that employs a second classifier network to classify patients' disease states from disease images.	In cases of lung cancer, it classifies the malignant and benign tumours, switching to healthy and diseased kidneys in diabetic situations.

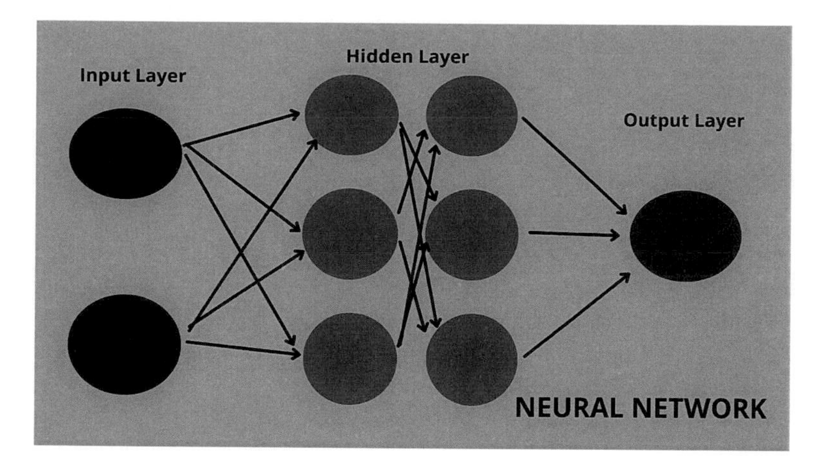

Fig. 12.3 Basic representation of a neural network

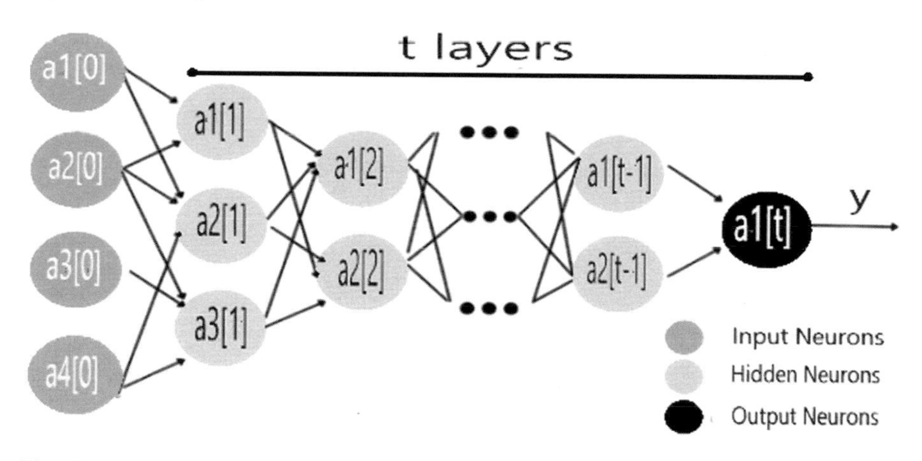

Fig. 12.4 Representation of neural network of t layers

We consider the first node of the second hidden layer denoted as $a_1^{[2]}$ *It's computed using all the neurons of the previous layer as follows:*

$$z_1^{[2]} = \sum_{l=1}^{3} w_{1,l}^{[2]} a_i^{[1]} + b^{[2]} \quad (12.12)$$

$$\rightarrow a_1^{[2]} = \psi^{[2]} \left(z_1^{[2]} \right) \quad (12.13)$$

In general, considering the jth *node of the* ith textit layer we have the following equations

$$z_j^{[i]} = \sum_{t=1}^{n_{i-1}} w_{j,l}^{[i]} a_l^{[i-1]} + b_j^{[i]} \quad (12.14)$$

$$\rightarrow a_1^{[2]} = \psi^{[2]}\left(z_1^{[2]}\right) \quad (12.15)$$

With n_{i-1} being the number of neurons in the $(i-1)th$ layer and W^T is the transpose of matrix W.

Finally, we denote:

- $W^{[i]} = \left[w_1^{[i]}, w_2^{[i]}, \ldots, w_{n_i}^{[i]}\right]$ *where* $\dim\left(w_j^{[i]}\right) = \left[n_{i-1}, 1\right]$
- $b^{[i]} = \left[b_1^{[i]}, b_2^{[i]}, \ldots, b_{n_i}^{[i]}\right]^T$
- $Z^{[i]} = \left[z_1^{[i]}, z_2^{[i]}, \ldots, z_{n_i}^{[i]}\right]^T ; \mathcal{A}^{[i]} = \left[a_1^{[i]}, a_2^{[i]}, \ldots, a_3^{[i]}\right]$
- $\mathcal{A}^{[i]} = \psi^{[i]}\left(Z^{[i]}\right) = \left[\psi^{[i]}\left(z_1^{[i]}\right), \psi^{[i]}\left(z_2^{[i]}\right), \ldots, \psi^{[i]}\left(z_{n_i}^{[i]}\right)\right]^T$

Thus: $\mathcal{A}^{[i]} = \psi^{[i]}\left(Z^{[i]}\right) = \psi^{[i]}\left(W^{[i]^T}\mathcal{A}^{[i-1]} + b^{[i]}\right)$

Where,

$$\dim\left(Z^{[i]}\right) = \dim\left(\mathcal{A}^{[i]}\right) = [n_i, 1] \quad (12.16)$$

$$\dim\left(W^{[i]^T}\right) =^T \dim\left(W^{[i]}\right) = \left[n_i, n_{i-1}\right] \quad (12.17)$$

$$\dim\left(b^{[i]}\right) = [n_i, 1] \quad (12.18)$$

The estimated fault in the output layer propagates through the network in this process to amend the connection weights. A gradient technique is used to calculate the amount to be added or deducted in each weight. The partial derivative of the non-linear activation function over each weight is used to determine the direction. A nonlinear boundary choice is "learned" in this procedure.

The fundamental benefit of using DL to process EEG data is that, unlike standard machine learning algorithms, it can handle unprocessed EEG data since it conducts feature extraction aka Automatic feature Engineering (FE). When creating a hand-crafted FE is not possible, new information can be extracted from data (raw) by an automatic FE, improving the classification result. In this way, CNNs can process the data by applying convolutions to the EEG data (input) in a variety of ways, such as 1-D convolutions to process individual channels or 2-D and 3-D convolutions to process neighbouring channels together.

In Anxiety Disorders like OCD, patients are not only provided with antipsychotic medication but also given the SSRI procedure for better results. The response of the drug is often examined at the genetic level to provide effective medication. Heavily built learning models are used to extract useful clues from genetic information of high dimensions. The drug-responsive coherent genetic indicators of OCD are detected using a filter ensemble fused feature selection approach. The prognostic nature of the model is further heightened by an unaided DL-based feature extraction technique. The feature extraction algorithm marks necessary information to find a rational representation of the input. The retrieved features are subsequently trained using supervised machine learning methods. The results of the aforesaid research revealed the critical gene markers of OCD that actively responded to medications. The potential of the treatment is improved dramatically when the medications are tailored to the individual's condition providing insights into a more personalized treatment of the affected persons [18].

An association of autoencoders and 3-D convolutional neural networks was used to build an algorithm predicting the stages of Alzheimer's disease in individuals. Experiments have proven that a 3D convolutions approach improves the classification performance by capturing local 3D patterns than a 2D approach in a CNN. In the experiment, the convolution layer trained with an autoencoder was used. MRIs (Structured Magnetic Resonance Imaging) of the brain used for a capable DNN technique help diagnose AD in individuals. This model offered a notable improvement for multi-class identification, while most other proposed research works accomplished binary identification. Early diagnosis and assessment of levels of the disease made the model efficient [19–21].

In the treatment of Bipolar Disorder using the Deep Learning algorithm, the Single Nucleotide Polymorphism (SNP) is taken under consideration. GWAS (Genome-wide association analysis) was obtained through the SNP as molecular genetic markers which were then conglomerated with the CNN to construct a detection model for bipolar disorder [22]. The recognition rate of the disorder was obtained as about 79% with this technique.

Feature selection becomes a necessity as most features in the data used for disease recognition are redundant and many times irrelevant. Therefore, only a few features align with the classification label. In the model using SNP, each sample was encoded

Table 12.4 Combinational pairing of bases

Digital code	Base pairing
0101	AA
0110	AT
0100	AC
0111	AG
1000	TC
1011	TG
1010	TT
0000	CC
0011	CG
1111	GG

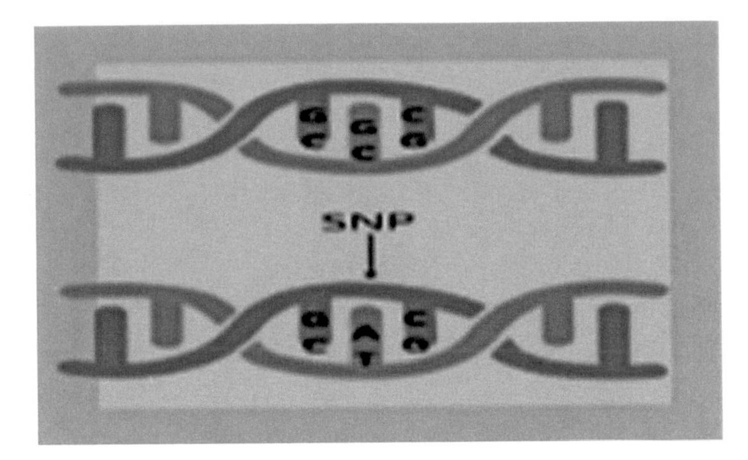

Fig. 12.5 Single nucleotide polymorphism

as a picture to reveal the combined pathogenic risk of the loci for bipolar disorder on different chromosomes by constructing resemblance in space of the sampled picture using the comparable genotypes at the same SNP for different samples and the limited genotypes at the same SNP. According to the combination of bases, the alleles of a certain site may have 10 cases (shown in Table 12.4). 0's and 1's were used to code instead of base [28].

One SNP is made of four pixels, therefore, the size of the convolution kernel must be maintained to learn the functional features of the sample. The performance of the CNN is also altered by the choice of hyperparameters, therefore adjusting their values based on the performance of the model is essential. The confusion matrix is used to get more information regarding model performance. The confusion matrix allows us to see if the model is "confused" when it comes to distinguishing between the two classes. It's a 2X2 matrix, as seen in the Fig. 12.5. Positive and Negative are the labels for the two rows and columns, which correspond to the two-class labels. The ground-truth labels are represented by the row labels, whereas the predicted labels are represented by the column labels in this case. This is something that could be altered [29] (Figs. 12.6 and 12.7).

Performance Measurement is given by [7, 11],

$$Precision_i = \frac{t_{p_i}}{t_{p_i} + f_{p_i}} \tag{12.19}$$

$$Recall_i = \frac{t_{p_i}}{Support_i} \tag{12.20}$$

$$F_1score_i = \frac{2}{Recall_i^{-1} + Precision_i^{-1}} \tag{12.21}$$

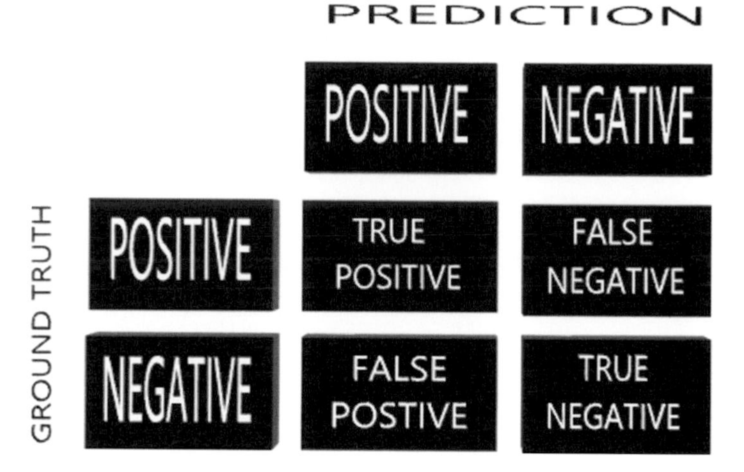

Fig. 12.6 Confusion matrix

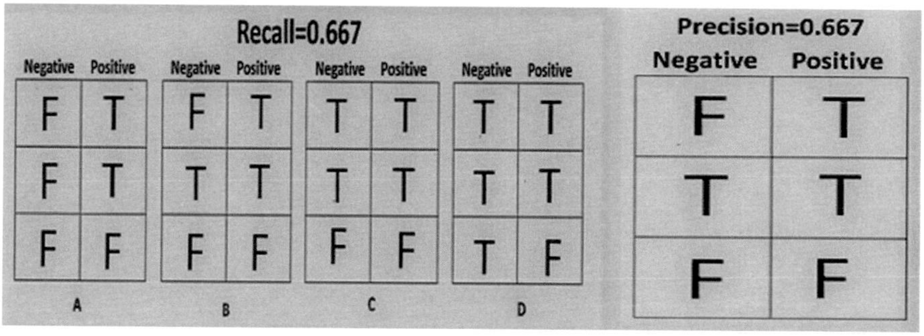

Fig. 12.7 Precision matrix and recall matrix

$$Accuracy = \frac{t_{p_i} + t_{n_i}}{t_{p_i} + f_{n_i} + f_{p_i} + t_{n_i}} \tag{12.22}$$

$$Averageaccuracy = \frac{\sum_{i=1}^{l} \frac{t_{p_i} + t_{n_i}}{t_{p_i} + f_{n_i} + f_{p_i} + t_{n_i}}}{l} \tag{12.23}$$

12.6 Challenges and Future Scope

The various techniques listed above indicate how the accurate analysis of neuroimages, records of patients, and genetic data by DL techniques detect the presence of a mental illness, further facilitating the diagnosis of the ailment. But, for the process to be efficient various shortcomings need to be eliminated. First, large data samples are required by DL architectures for the training of the models, hence posing a difficulty in neuroimaging analysis. Second, the imaging data is usually of a higher dimension, overfitting the DL model. Present studies suggest the utilization of the concept of MRI data pre-processing tools such as Data Processing Assistant for Resting-State fMRI, fMRI Pre-processing Pipeline, and the Statistical Parametric Mapping for the extraction of useful features before it is fed into the DL models. Although a natural attribute of DL is the capacity to learn meaningful patterns from input data, in the case involving analysis of neuroimages where small-sized samples and high dimensionality comes into play, feature engineering tools are highly required. The risk of overfitting of DL models with high-dimensional space gets mitigated with the use of feature engineering tools. DL models involving the use of FE techniques have been observed to outplay traditional ML models while predicting multiple conditions of depression, schizophrenia, and ADHD. However, extraction of features through such tools relies on prior knowledge; hence some information that may have been omitted would have been meaningful and useful for certain mental outcomes. Lastly, CNN remains to be one alternate way in which information is automatically extracted from raw data and is reported well in processing raw images data.

12.7 Conclusion

In the modern era, with rapid digitalization and industrial growth, mankind has found it extremely difficult to keep up with the extreme levels of stress and anxiety. Mental illness has become very common, but most times goes un-diagnosed. This creates other problems such as decrease in work capabilities and social issues. Tackling these issues requires us to dig up solutions from the highlight of the century: technology. Deep learning is an emerging field of study and can be wielded for a number of abnormal situations, mental health being no exception. With the rapidly growing field of bioinformatics, deep learning plays a pivotal role in detection by enabling speedy and scalable analysis of complex data like images of the brain. While this technique can be employed and worked at the practice for the early diagnosis of a mental illness, the same can be extended for real-time monitoring of the individual and improving the efficiency of treatment. Further research in this area could result in some outstanding and impactful outcomes: reducing suicide rates, improving the global peace index and elevating the quality of life.

References

1. Su, C., Xu, Z., Pathak, J., Wang, F.: Deep Learning in mental health outcome research: a scoping review 2020. Transl. Psychiatry. New York, USA
2. Kholifah, B., Syarif, I., Badriyah, T.: Mental disorder detection via social media mining using deep learning, pp. 309–316. Kinetik, Politeknik Elektronika Negeri Surabaya, Indonesia (2020)

3. Orabi, A.H., Buddhitha, P., Orabi, M.H., Inkpen, D.: Deep Learning for Depression Detection of Twitter Users, pp. 88–97. New Orleans, Louisiana Canada (2018)
4. Oha, J., Yunb, K., Maozb, U., Kima, T.-S., Chaea, J.-H.: Identifying Depression in the National Health and Nutrition Examination Survey Data Using a Deep Learning Algorithm. Elseivier, USA (2019)
5. Cho, G., Yim, J., Choi, Y., Ko, J., Lee, S.-H.: Review of machine learning algorithms for diagnosing mental illness. Psychiatry Investig. **16**(4), 262–269. Canada, USA, Korea (2019)
6. Heinsfeld AS, Franco AR, Craddock RC, Buchweitz A, Meneguzzi F.: Identification of autism spectrum using deep learning and ABIDE dataset. Brazil (2017)
7. Jain, R., Semwal, V.B., Kaushik P.: Deep ensemble learning approach for lower extremity activities recognition using wearable sensors. Expert Syst. e12743 (2021)
8. Challa, S.K., Kumar, A., Semwal, V.B.: A multibranch CNN-BiLSTM model for human activity recognition using wearable sensor data. Vis. Comput. (2021)
9. Min, S., Lee, B., Yun, S.: Deep Learning in Bioinformatics. Seoul National University, Seoul, Korea (2017)
10. Tyagi, N.: Understanding Bioinformatics as the Application of Machine Learning. India (2019)
11. Semwal, V.B., Lalwani, P., Mishra, M.K. et al.: An optimized feature selection using biogeography optimization technique for human walking activities recognition. Computing (2021)
12. Semwal, V.B., Gupta, A., Lalwani, P.: An optimized hybrid deep learning model using ensemble learning approach for human walking activities recognition. J. Supercomput. 1–24 (2021)
13. Bijalwan, V., Semwal, V.B., Mandal, T.K.: Fusion of multi-sensor based biomechanical gait analysis using vision and wearable sensor. IEEE Sens. J. (2021)
14. Sui, J., Jiang, R., Bustillo, J., Calhoun, V.: Neuroimaging-based individualized prediction of cognition and behaviour for mental disorders and health: methods and promises. Biol. Psychiatry (2020). Georgia and China
15. Semwal, V.B. et al.: Pattern identification of different human joints for different human walking styles using inertial measurement unit (IMU) sensor. Artif. Intell. Rev. 1–21 (2021)
16. Raj, M., Semwal, V.B., Nandi, G.C.: Bidirectional association of joint angle trajectories for humanoid locomotion: the restricted Boltzmann machine approach. Neural Comput. Appl. **30**, 1747–1755 (2018)
17. Rivera, M.J., Teruel, M.A., Juan Trujillo, A.M.: Diagnosis and prognosis of mental disorders by means of EEG and deep learning: a systematic mapping study. Springer, Spain (2021)
18. Dekaran, K., Sudha, M.: Predicting drug responsiveness with deep learning from the effects on gene expression of Obsessive–compulsive disorder affected cases, pp. 286–394. India (2020)
19. Payan, A., Montana, G.: Predicting Alzheimer's disease a neuroimaging study with 3D convolutional neural networks, pp. 355–362. Elsevier, Germany, USA (2015)
20. Islam, J., Zhang, Y.: Brain MRI analysis for Alzheimer's disease diagnosis using an ensemble system of deep convolutional neural networks. Brain Informatics. USA (2018)
21. Salehi, A.W., Baglat, P., Gupta, G.: Alzheimer's Disease Diagnosis using Deep Learning Techniques. Elsevier, Himachal Pradesh, India (2019)
22. Lin, E., Kuo, P.-H., Lin, W.-Y., Liu, Y.-L., Yang, A.C., Tsai, S.-J.: Prediction of probable major depressive disorder in the taiwan biobank: an integrated machine learning and genome-wide analysis approach. J. Pers. Med. **11**, 597 (2021)
23. Albawi, S., Mohammed, T.A., Al-Zawi, S.: Understanding of a convolutional neural network. In: 2017 International Conference on Engineering and Technology (ICET), pp. 1–6. (2017). https://doi.org/10.1109/ICEngTechnol.2017.8308186
24. Chien, Y.W., Hong, S.Y., Cheah, W.T., et al.: An automatic assessment system for Alzheimer's disease based on speech using feature sequence generator and recurrent neural network. Sci. Rep. **9**, 19597 (2019)
25. Yu, W., Lei, B., Ng, M.K., Cheung, A.C., Shen, Y., Wang, S.: Tensorizing GAN With High-Order pooling for Alzheimer's disease assessment. In: IEEE Transactions on Neural Networks and Learning Systems (2021)
26. Ranta, I., Teuho, J., Linden, J., Klén, R., Teräs, M., Kapanen, M., Keyriläinen, J.: Assessment of MRI-based attenuation correction for MRI-only radiotherapy treatment planning of the brain. Diagnostics **10**, 299 (2020)

27. Dua, N., Singh, S.N., Semwal, V.B.: Multi-input CNN-GRU based human activity recognition using wearable sensors. Computing (2021)
28. Gua, J., Wangb, Z., Kuenb, J., Mab, L., Shahroudyb, A., Shuaib, B., Liub, T., Wangb, X., Wangb, L., Wangb, G., Caic, J., Chenc, T.: Recent Advances in Convolutional Neural Networks. Singapore (2017)
29. Vakili, M., Ghamsari, M., Rezaei, M.: Performance Analysis and Comparison of Machine and Deep Learning Algorithms for IoT Data Classification (2020)

Part IV
Healthcare Informatics and Management

Chapter 13
Skin Disease Detection and Classification Using Deep Learning: An Approach to Automate the System of Dermographism for Society

Anurag Sinha, Sarvjeet Kumar Singh, Hassan Raza Mahmood, and Kshitiz Sinha

Abstract Skin disease is more normal than different sicknesses. Skin disease can be much inherited disease which directly or indirectly affects almost every age group we present an audit on profound learning strategies and their applications in skin infection conclusion. We first present a short prologue to skin infections and picture acquisition techniques in dermatology, and rundown a few freely accessible skin datasets for preparing and testing calculations. The fundamental thought of this undertaking is to work on the exactness of indicative frameworks by utilizing Image Processing and grouping methods. In the proposed framework, a picture caught on camera is taken as information. In this chapter, we propose an enormous scope; It contains 5660 clinical pictures, covering several sorts of skin infections. Each picture in this dataset is named by proficient specialists. IN this paper we have taken boundary of clinical traits which encoded in calculation and furthermore utilizing PCA we have optimitized the part factor of neural calculation. Thus we are employing deep learning to automate the function of skin disease detection and classification as it is considered boon of artificial intelligence and can benefit medical field on great level.

Keywords PCA · Neural network · Optimization · Skin disease dataset · Clinical attribute · Deep learning

---

A. Sinha (✉)
Department of Information Technology, Amity University Jharkhand, Ranchi, India
e-mail: anuragsinha257@gmail.com

S. K. Singh
Acsmch M.G.R. Deemed to be University, Department of M.B.B.S., Vellappanchavdi, Chennai, India

H. R. Mahmood
Fast Nukes Cfd Campus Fast-Nu, Fast Square, Faisalabad, Punjab, Pakistan

K. Sinha
Department of M.B.B.S, Shandong University, Shandong, China

A. Biswas et al. (eds.), *Artificial Intelligence for Societal Issues*, Intelligent Systems Reference Library 231, https://doi.org/10.1007/978-3-031-12419-8_13

13.1 Introduction

Skin infection among people has been a typical sickness, a huge number of individuals are experiencing different sorts of skin infections. Normally, these infections have stowed away perils which lead to not just absence of fearlessness and mental sorrow yet additionally lead to a gamble of skin malignant growth. Clinical specialists and undeniable level instruments are expected to analysis these skin infections due to non-accessibility of visual goal in skin sickness pictures. The proposed system incorporates deep learning methods, for example, CNN design and three predefined models called Alex Net, ResNet, Inception V3. A Dataset of pictures with seven illnesses has been taken for the Order of Skin infections. They incorporate infections like Melanoma, Nevus, Seborrheic Keratosis and so on The dataset was stretched out by adding pictures having cuts and consumes, which were named skin infection by a large portion of the current frameworks. The use of Deep Learning calculations has decreased the requirement for human work, for example, manual element extraction and information remaking for arrangement purposes [1].

Skin diseases happen for the most part among individuals. They are regularly achieved by factors like unmistakable living thing's telephones, an other eating schedule, and inside and outside factors, similar to the ever-evolving genetic social occasion of cells, synthetic substances, and invulnerable course of action of conditions. These parts might act together or in a course of action of skin disease. There are industrious and genuine contaminations, like skin irritation and psoriasis, and risky diseases like destructive melanoma. Late examiners have found the openness of solutions for these contaminations in the event that they are perceived in the first place stages. Atopic dermatitis, ordinarily called dermatitis, is a really long skin disease whose typical signs are dry and disturbed skin, rashes on the face, inside the elbows, behind the knees, and down on the ground. Hansen's disease, consistently called ailment, is achieved by lazy creating microorganisms and can impact the body and facial parts like nerves, skin, eyes and nose lining. Melanoma is outrageous and risky skin harmful development. The "Abcd's" of moles perceived on the skin are Asymmetry, Border, Color, and Diameter. Deviation recommends that the condition of one half doesn't arrange with the other half. Line suggests the edges of the mole are battered, darkened, or flighty. Concealing is unbalanced and may consolidate shades of dull, brown, and tan. The width of mole construes a change of size [1]. Ecological elements alongside various causes are the rule parts of skin diseases. Easily open acknowledgment plans are to be made for early assurance of skin contamination. Here the proposed paper gives a method for managing recognize various kinds of the infections like psoriasis, melanoma, not melanoma, ringworm and measles. The client gives commitment of the skin disease picture, which then the structure measures, features extraction using CNN computation and use softmax picture classifier to examine contaminations. Accepting no disease is found, that doesn't facilitate with the dataset ailment not found message is shown. Human skin shows wide combination of shades from the most dark brown to the lightest pinkish–white. Normal assurance is the inspiration driving why individuals show huge concealing assortments. Skin pigmentation in individuals are caused in view of the proportion of brilliant radiation (UVR) invading the skin, controlling its biochemical effects [2]. Melanin is the component that chooses the shade of human skin. The system perceives the shade of the skin to perceive the affliction. Complexion varies for various diseases. Considering the complexion and various properties like line, shape and surface the disorder can be found. In this paper, we present a total study of the new works on significant learning for skin ailment finding. We first give a succinct preamble to skin contaminations. Through composing research, we then present ordinary data procurement methodologies and once-over a couple of commonly used and transparently available skin disorder datasets for getting ready and testing significant learning models. From that point on, we depict the essential beginning of significant learning and present the notable significant learning plans. We then draw on the composition of purposes of significant learning in skin ailment analysis and familiarize the substance agreeing with different endeavors. Through inspecting the investigated composition, we present the challenges remained in the space of skin disease finding with significant learning and outfit rules to deal with these hardships later on. Considering the shortfall of all around enthusiasm for skin ailments and significant learning by more broad organizations, this paper could give the understanding of the critical thoughts related to skin disorder and significant getting at a legitimate level. It should be seen that the target of the review isn't to weaken the writing in the field [3].

Problem Definition and Contribution

One of the most continuous sicknesses among individuals all around the world is skin infection. Basal cell carcinoma (BCC), melanoma, intraepithelial carcinoma, and squamous cell carcinoma are instances of skin diseases (SCC). The event of skin malignant growth is as of now more noteworthy as the event of other new sorts of lung and bosom disease [1]. A few skin diseases have side effects that can consume a large chunk of the day to treat since they can develop for a really long time prior to being perceived. Thus, AI based finding becomes an integral factor since it can create an outcome in a short timeframe with more precision than human investigation using lab strategies. Profound Learning is the most generally involved innovation for skin sickness expectation. Profound learning models will utilize gathered information to distinguish and investigate highlights in unexposed information designs, bringing about huge proficiency even with low computational models. This study presents a strong component for precisely recognizing skin infections utilizing administrative methodologies that diminish analytic expenses. This has provoked the scientists to consider utilizing a profound learning model to arrange the skin infection in light of the picture of the impacted district [4].

Objective

The objective of the assignment is two-wrinkle; to make a convolution neural association to dissect diverse skin diseases like skin break out, psoriasis, moles, etc. that vanquishes the cons of existing methods and give better results and precision and to cultivate a useful and reliable system for dermatological sickness acknowledgment that can be used as a strong consistent appearance gadget for clinical understudies in the dermatology stream [5].

This chapter is organized in this chronological fashion:

I. Sections 13.1 and 13.2 contains overview and objective, scope of the chapter,
II. Section 13.3 contains rclated literature study done.
III. Section 13.4 contributes about the literature analysis and comparison of research gap and finding.
IV. Section 13.5 is all about what techniques are already available for covid-19 recognition.
V. In the Sect. 13.6 we have proposed our own method for analysis.
VI. Last section it discusses about result and conclusion.

Author contribution in this paper is as follows: author 1 and 2 contributed to the dataset preparation and background study and optimization of the algorithm with paper drafting. The third and fourth author of the paper has done all the introductory and theoretical literature investigates. Fourth Author has implemented and simulated the result in programming environment. This work is novel in terms of its algorithmic approach a layer of encoding value is added through model precision.

13.2 Background

13.2.1 Skin Disease Nature

Skin is our body's biggest organ, hair and nails are likewise sorts of skin and fall under this framework. Skin illnesses have been demonized for century's kin having skin conditions were projected out of the general public it were messy individuals with skin sicknesses needed to bear the obliviousness, bias and dread to fear they. Prior to understanding the skin conditions we should comprehend how skin is framed There are three essential microorganism layers in the undeveloped stage these are:

1. Endoderm.
2. Mesoderm.
3. Ectoderm.

The external hindrance of the skin is made of cells called keratinocytes. These cells are brought into the world in the basal or internal layer. They then, at that point, clear their path through numerous cell layers and move towards the surface. As keratinocytes travel, they make the fundamental parts required for the boundary. Each inch of skin is comprised of 19 million skin cells, 650 perspiration organs, 20 veins and 1,000 sensitive spot. The skin is bodies first line of safeguard against the rest of the world, including aggravations and allergens Specialized resistant cells live in skin and perceives interfering life forms and substance. They conveyed messages to animate a warrior reaction or hypersensitive reactions by enrolling particular white platelets from veins in the skin [7]. Skin infections are more normal than one could expect, consistently 1 out of 4 individuals manage skin conditions. Skin conditions influence 1.9 billion individuals consistently on account of obliviousness or deficiency of dermatologist. Skin sicknesses are a wide scope of conditions influencing the skin, and incorporate illnesses brought about by bacterial diseases, viral contaminations, contagious contaminations, unfavorably susceptible responses, skin malignant growth and parasites.

Skin diseases can be mainly classified into two different types:

1. Temporary skin diseases—numerous impermanent skin conditions exist including contact dermatitis and keratosispilaris.
2. Permanent skin disorder—Some persistent skin conditions are available from birth. While other show up out of nowhere further down the road.

The cause of skin disorders is not always known. Many permanent skin discascs have effective treatment while some are incurable [8]

Types of Skin Diseases.

Skin diseases can be categorised as either being caused by infections, allergies, autoimmune reactions, parasites or cancer (see Table 13.1).

13.2.2 Data Set Description

Data set is taken from https://archive.ics.uci.edu/ml/datasets/dermatology the dataset worked for this space, the family parentage incorporate has the value 1 expecting any of these sicknesses has been found in the family, and 0 anyway. The age incorporate essentially addresses the age of the patient. Every single other component (clinical and histopathological) was given a degree in the extent of 0 to 3. Here, 0 exhibits that the part was missing, 3 shows the greatest total possible, and 1, 2 show the general moderate characteristics. We have used two kinds of attributes of data for analysis of PCA optimization factor encoding of our paper. In Table 13.1, the data set is given with clinical attributes [8] (Table 13.2).

Table 13.1 Some common types of skin diseases are

Disease	Sample	Example
Bacterial infections	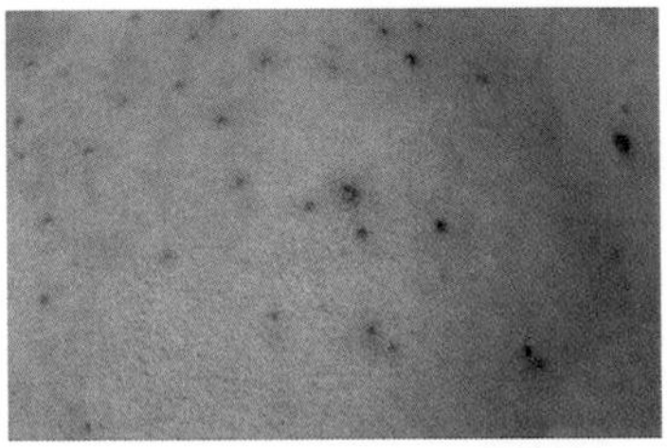	Boils an carbuncles, staph infections
Viral infections	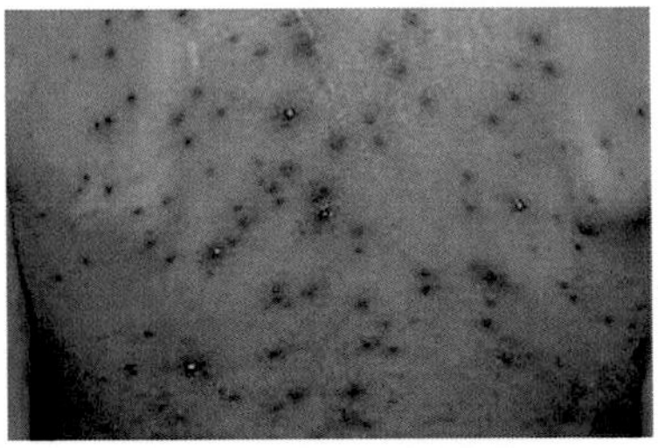	Warts, verucas, cold sores herpes, chickenpox, shingles
Fungal infections	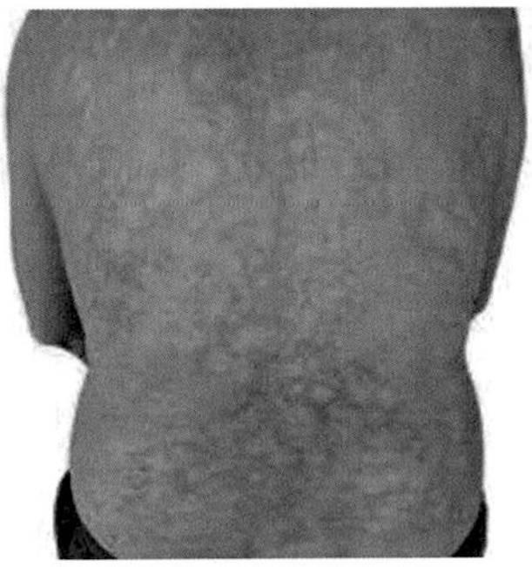	Ring worm, yeast infections
Allergic reactions		Eczema, hives, itching

(continued)

Table 13.1 (continued)

Disease	Sample	Example
Autoimmune diseases	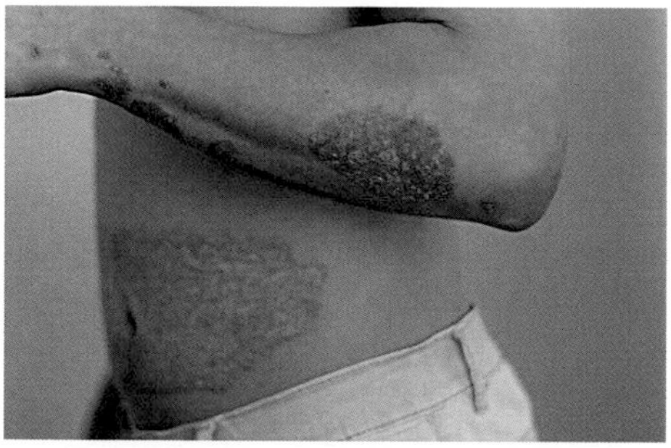	Psoriasis, eczema
Parasites	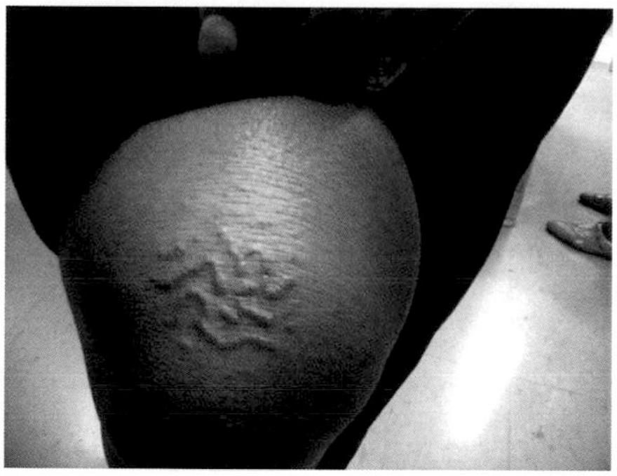	Scabies, bedbugs, head lice, mites

13.3 Literature Review

In paper [1], Author have utilized SVM and ANN information mining procedures, to group different sorts of erythema-squamous illnesses. They utilized a private weighted casting a ballot plan to join the two advances to accomplish the most noteworthy exactness of 99.25% in the preparation what's more, 98.99% in the testing stages. In paper [2], utilized data mining on utilized Best First Search highlight determination innovation method, and they eliminated 20 highlights from the dermatology dataset assortment gathered by the University of California Irving store and afterward utilized Bayesian innovation to accomplish 99.31% precision. In paper [3], anticipating different skin sicknesses utilizing the gullible Bayesian calculation. Programmed distinguishing proof of circulatory sickness dermatological highlights removed from Local Binary Pattern from impacted skin pictures and utilized for arrangement. In paper [4], Author have implemented an Artificial Neural Network ANN-based single level framework as well as a multi-model, staggered framework for dermatitis recognition. Also, ANN was applied in to recognize certain circulatory illnesses through the shade of the fingernails. A comparative execution was applied in illness recognition utilizing tongue pictures. A technique for diagnosing.

Table 13.2 Skin disease data set and attribute

State	Erythema	Scaling	Definite borders	Itching	Polygonal papules	Oral mucosal involvement	Family history, (0 or 1)	Knee and elbow involvement
Prosasis	4,464,356	−1.78	−0.02	0.69	14.41	10.28	869.21	130.79
Acne	634,892	−1.72	−0.24	2.09	15.95	4.64	941.95	58.05
Hives	5,307,331	14.25	−0.03	4.29	15.88	7.77	869.54	130.46
Rubeola	2,692,090	0.36	−0.01	1.07	14.35	10.51	861.06	138.94
Keloid	34,501,130	−2.01	−0.04	7.88	15.37	6.72	894.03	105.97
Seborehhic	4,417,714	9.32	−0.06	3.57	14.57	6.26	903.52	96.48
Curcuble	3,425,074	−2.37	−0.02	3.50	12.52	9.00	862.64	137.36
Iaison 1	796,165	5.39	−0.04	2.12	14.01	8.79	869.45	130.55
Syphiils	571,822	−7.77	−0.07	5.73	14.33	10.76	880.75	119.25
Seboric	16,396,515	12.52	−0.03	5.76	12.54	10.13	826.28	173.72
Shingle	8,383,915	7.07	−0.07	2.71	16.16	7.75	904.37	95.63
Vitilo	1,224,398	−2.50	−0.29	4.32	15.44	6.87	866.18	133.82
Necotriz	1,321,006	6.40	−0.03	2.24	15.00	7.34	887.18	112.82
Acne2	12,482,301	−7.28	−0.02	4.76	14.87	8.72	880.26	119.74
ca	6,114,745	−1.99	0.00	1.23	14.08	9.19	876.78	123.22
cb	2,923,179	−5.68	0.00	1.21	12.83	9.70	851.89	148.11
Seboric	2,694,641	−6.36	−0.05	2.28	14.53	9.35	868.32	131.68
Cold 20	4,065,556	0.15	−0.06	0.75	13.63	9.89	875.51	124.49
Vitigo	4,465,430	−7.98	−0.04	0.70	15.69	9.37	884.16	115.84
Syphilles	1,286,670	6.62	−0.02	0.58	10.31	9.79	856.33	143.67

(continued)

Table 13.2 (continued)

State	Erythema	Scaling	Definite borders	Itching	Polygonal papules	Oral mucosal involvement	Family history, (0 or 1)	Knee and elbow involvement
Necortix	5,375,156	2.18	−0.06	4.02	14.16	8.30	886.72	113.28
Acnw 3	6,379,304	−3.25	−0.01	3.24	12.72	9.02	865.26	134.74
Michigan	9,990,817	−2.70	0.00	1.93	13.63	8.88	877.62	122.38

In paper [5], framework was proposed utilizing artificial brain network as well as advanced picture handling in distinguishing BCC infection. The recognition depends on unique attributes of basal cell carcinoma. The framework will be able to accurately recognize the event of carcinoma utilizing the appropriate limit values with percent unwavering quality of 93.33%. In paper [6], framework included picture division, highlight extraction, and factual characterization to recognize and separate among gentle and serious dermatitis. When the dermatitis type was distinguished, a seriousness record was relegated to that picture. A few normal division strategies, similar to Otsu's, watershed, and district developing division, were carried out, and since none of these techniques gave right yields, shading based division utilizing k-implies bunching was utilized. Highlight extraction depended on shading highlights, surface highlights, and boundary highlights and the order was finished utilizing the SVM classifier technique. In paper [7], prepared the troupe techniques dependent on Mask RCNN and DeeplabV3 + strategies on ISIC-2017 division preparing set and assess the presentation of the gathering networks on ISIC-2017 testing set and PH2 dataset. Our outcomes showed that the proposed gathering strategies fragmented the skin sores with Sensitivity of 89.93% and Specificity of 97.94% for the ISIC-2017 testing set. The proposed group strategy Ensemble-A beat FrCN, FCNs, U-Net, and SegNet in Sensitivity by 4.4%, 8.8%, 22.7%, and 9.8% independently. Besides, the proposed assembling procedure Ensemble-S achieved an unequivocality score of 97.98% for clinically compassionate cases, 97.30% for the melanoma cases, and 98.58% for the seborrhea kurtosis cases on ISIC-2017 testing set, showing preferred execution over FrCN, FCNs, U-Net. In paper [8], Author have used Fuzzy interference system for for primary diagnosis and self assisted diagnosis based on own symptoms for covid—19. In paper [9], dataset is utilized and the proposed technique has beated different strategies with over 85% precision heartiness in perceiving the influenced area a lot quicker with practically 2 × lesser calculations than the traditional MobileNet model outcomes in negligible computational endeavors.

13.4 Proposed Method

13.4.1 Data Pre-processing

In this section we will describe our we proceeded in order to create our dataset that:

1. Collection of Skin Disease Data: We Tried to Collect Around 700.
2. Extraction of features from collected data: We checked and analyzed every data and we generated a list of pertinent features that we will use as inputs for our supervised model. We extracted around 28 features and developed a javascript file that acts as the middle-man between the add-on and the transfer learning method that extracts each feature value to pass it to the ANN. A description of these different features is given in the Table 13.1.

3. Preprocessing of the dataset: As early-stage dataset preprocessing may positively affect the project outcomes, we have used the data mining technique to transfer our data into a machine-understandable format, where we have constructed the measurable features, avoid any irrelevant/redundant data. by Data Quality Assessment. To ensure the quality of our dataset we have determined the values manually and automatically, compare both results.
4. Data normalisation and conversion to CSV format: We ensured the normalisation of our dataset in order to guarantee the accuracy of the intended results. All the values of all features where designed as 0/1 values. The file was then converted to CSV format to be readable and extractable for our model [10, 11].

13.4.2 Performance Metrics

Training presents a prime potential to expand our system's knowledge and learning, as our dataset is small and the training set is 70% of our dataset, while testing set is 30% which equal to 18 attributes, i.e. means we will test only using 17 attributes, which is not really sufficient for reaching high performance. Therefore, we will perform two training phases for our model to join the supervised the reinforcement learning. This means that we will use the available training dataset for performing a pre-processing for our model. Then for each new attribute, we will retain our model using the feedback from users. Each time a user discover an error in the given result, for example a attributes predicted as infected and the user discover is not and contrary. In this case the user will report this error. These faced problems will be used later to retrain again the model and increase its performance [12, 13].

13.4.3 Implementation

Image pre-processing—Because photographs come in a variety of game plans, such as conventional, false, grayscale, and so on, we must consider and standardize them before incorporating them into a cerebrum network. Grayscale is the process of converting a toned-down image into one that is starkly distinct. In AI computations, the complex character of diminishing estimation is commonly used. Because most photographs don't need to care about variety to be remembered, it's a good idea to utilise grayscale, which reduces the number of pixels in an image and so reduces the number of computations necessary. It is the approach connected with projecting image data pixels (force) to a preset range (often (0, 1) or (−1, 1), and is sometimes referred to as data re-scaling. This is commonly utilised on a variety of data sets, and you'll need to normalise all of them in order to do comparable calculations [14, 15] (Fig. 13.1).

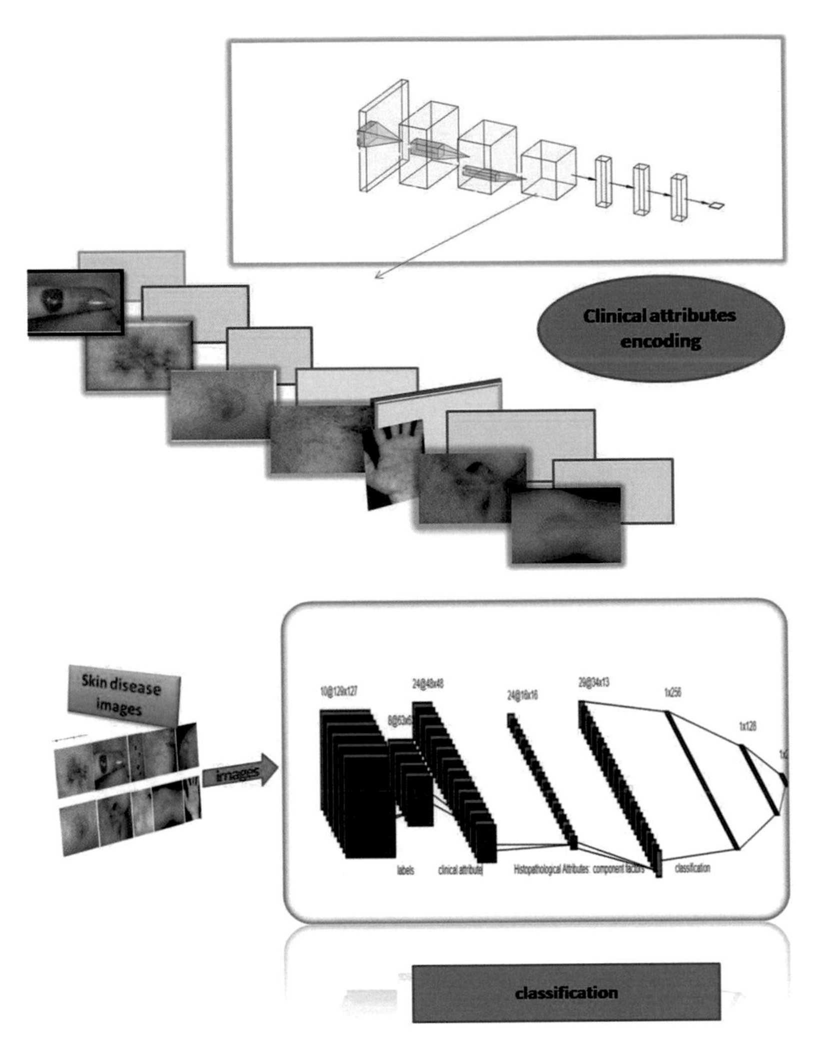

Fig. 13.1 Optimized classification based on clinical attributes

Division Thresholding—Thresholding is a type of image split in which the pixels of an image are changed to make the image easier to study. Thresholding is the process of converting a grayscale or assortment image into a two-dimensional image, which is essentially extremely distinguishing.

Worldwide Thresholding—Thresholding divides an image into items and foundations in general. It is the quickest of all thresholding methods. The division is then done by filtering each pixel and stepping it as a foundation or forward looking region based on the pixel's dull level.

Highlight Extraction—This portion extraction is the association in which we will address a raw image in a reduced development to aid dynamic model various evidence or orders. The cycle entails a reduction in the size of the assets that are expected to represent that in the more notable information designs. In AI, this connection is commonly used. This is the broad name for techniques in which we may create variable mixes to harmonies with these challenges while displaying the data with great precision. As a result, we've split this communication into two systems to improve component extraction.

GLCM (Gray Level Co-event Matrix)—The dull level co-occasion cross section (GLCM), also known as the dim level spatial dependency organisation, is a measurable approach for looking at surfaces that examines the spatial interaction of pixels. The GLCM limits depict the picture's outer layer by calculating how frequently sets of pixel with unambiguous properties and in a preset spatial connection occur in an image, creating a GLCM, and then extracting true measurements from this structure. (The surface channel limitations shown in Calculate Statistical Measures of Texture don't provide information about shape or possible spatial relationships of pixels in a picture.

Image Quality Assessment—One of the quality appraisal methodologies is picture quality assessment. The Full reference approach distinguishes them. Features of Image Quality Assessment.

Full-reference strategy—The whole reference methodology Metric will make a continual effort to learn more about the test photos by enrolling and distinguishing them from the suggested images, which will then have the option of being guaranteed to have the most ascribes. By averaging the squared power separations of twisted and reference picture pixels, as well as the corresponding proportion of peak signal-to-commotion degree (PSNR), the mean squared blunder (MSE) is determined [16].

Enhancers/Activation Functions—In essence, an inception work is a limitation placed on a fake mental organisation to assist it in learning complex models from data. While distinguishing and a neuron-based model that is to us, the order work on what has to be done to the accompanying neuron is nearing completion [17].

Max-Pooling—Pooling is the process of picking a value from a window while keeping a certain goal in mind, such as retaining data and lowering dimensionality. The most shocking component of that pixel is set using max-pooling, which selects the best value from all open attributes in the window. A depiction's dimensionality is reduced when it is pooled. For instance, if a 3*3 window size is preferred over a 10*10 part alliance, the most all-around basic value out of the available 9 credits is chosen to address that highlight window. Average pooling is another structure in which, instead of choosing the greatest value, the average of all credits in a certain window is utilised to keep an eye on that segment [18] (Fig. 13.2).

Dropout—Dropout is a regularization framework for discovering the best way to avoid overfitting in the first place. To deal with neuronal interdependency, the regularization approach is applied. As a result of this dependency, over fitting arises. An insignificant portion of neurons selected by the dropout rate are excused at each configuration step in dropout for a variety of reasons fitting arises. An insignificant portion of neurons selected by the dropout rate are excused at each configuration step in dropout for a variety of reasons.

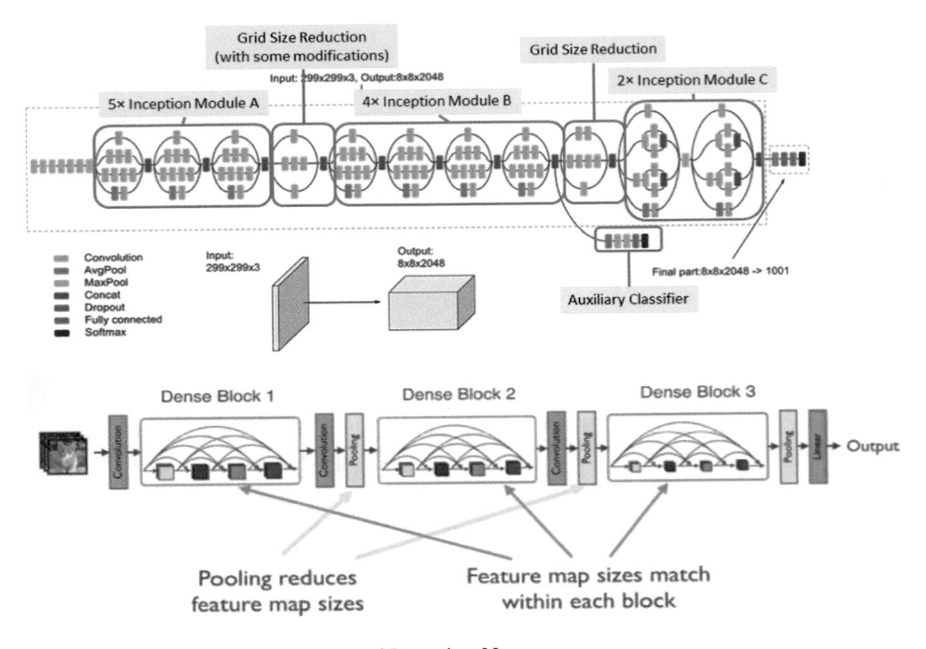

Fig. 13.2 Illustration of Dense-net and Inception-Net

Different layers that were nearer to the result highlights were prepared to dispense with additional data from the later convolution layers. We added an additional three layers to the most raised sign of each model, to be express, a completely related layer (FC2) with the eventual outcome of 512, a dropout layer, and another completely related layer (FC1) with a softmax classifier as portrayed in Fig. 13.3. A dropout layer was added to forestall covering, passing with 0.5 for every frontal cortex network utilized in this audit. The affiliation was prepared with a softmax classifier for 15 ages utilizing a RMSprop streamlining expert [18], with a learning velocity of 0.10111 and a bunch size of 33. At the same time, the consecutive presentation utilized as a classifier with the last three layers merges the softmax classifier.

Transfer Learning is the reuse of a pre-prepared model on another issue. It is at present exceptionally well known in the field of Deep Learning since it empowers you to prepare Deep Neural Networks with relatively little information. There are a some pre-prepared Machine Learning models out there that turned out to be very well known. One of them is the DenseNet-161 model, which was prepared for the ImageNet "Huge Visual Recognition Challenge". Regardless of the articles it was prepared to characterize are very unique contrasted with skin infections pictures, the elements identification part of such retrained model are frequently reused to order totally various pictures. I at first chose DenseNet-161, Resnet-V2, inception-net as a decent model to reuse in light of the fact that it has (had) further developed execution over others on ImageNet and it presents a fascinating engineering with each layer taking all former element maps as information. Besides notwithstanding it has a ton

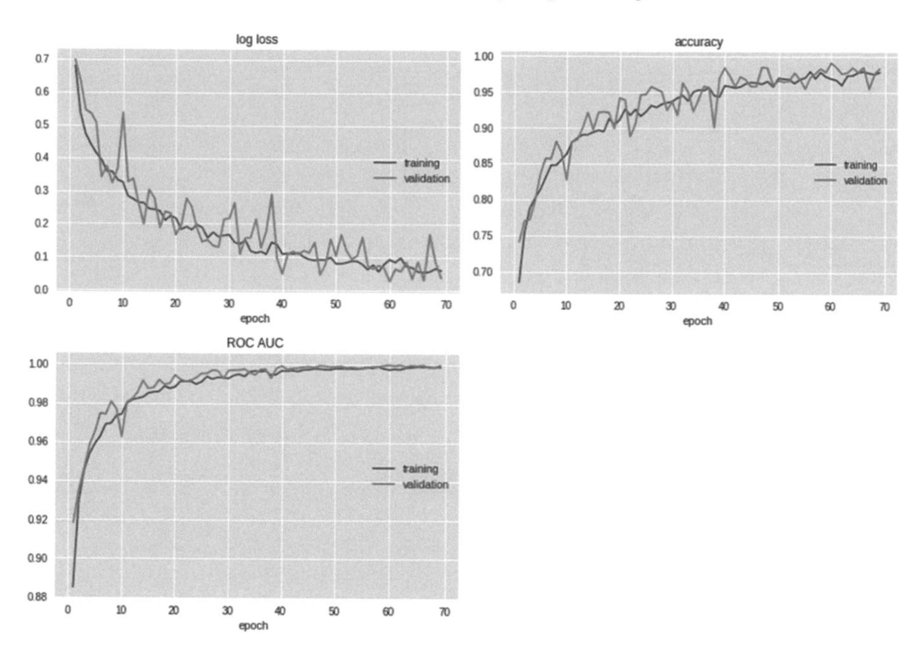

Fig. 13.3 Model training history

of layers, an opportunity to prepare a solitary age is basically the same as easier retrained models. The illustration of proposed one is given below.

13.5 Results and Discussion

We will at first see which the orders that the model by and large confused with one another. We will endeavor to check regardless of whether what the model expected was reasonable. For the present circumstance the misunderstandings look reasonable (none of the mistakes has all the earmarks of being plainly honest). This is a marker that our classifier is working really. Additionally, when we plot the chaos organization, we can see that the dispersal is strongly skewed: the model messes up the same way over and over yet it only from time to time perplexes various classes. This suggests that it basically feels that it's difficult to isolate a few specific classes between each other; this is customary lead.. The technique learn.lr_find() assists with observing an ideal learning rate. It utilizes the strategy created in the 2015 paper Cyclical Learning Rates for Training Neural Networks where we essentially continue expanding the gaining rate from a tiny worth, until the misfortune begins diminishing. Presently, do likewise, yet feature the pictures with a heatmap to see what portions of each picture prompted some unacceptable grouping [19, 20].

In Fig. 13.3 the model history plot is shown where loss plot, accuracy plot is given.

The result of prêt trained result of VGG-16, RES-net v2, Inception-Net model result is as shown in Fig. 13.4 where ROC curve, training validation graph and, confusion matrix is given.

We observe that the ResNet18 fluctuates significantly more than the ResNet50 when we put it up with a modest learning rate. Despite the fact that the learning rate was nearing zero, this method was employed. I repeated the modifying test #2 with the ResNet-152, this time increasing the learning rate reach to max lr = slice(1e-4,1e-5) and resetting the learning rate reach. In the provided situation, the ResNet18 had an underwriting accuracy of 0.904643. This looks to be the plan of action. To cope with shifting, flexible learning rates, the fit one cycle strategy employs a bend in which the rate is initially inflated and then dropped. If you interrupt a 20-year status in age #10 and then continue for another 9 years, the outcome will be different than if you planned consistently for 20 years. You should provide the option to note

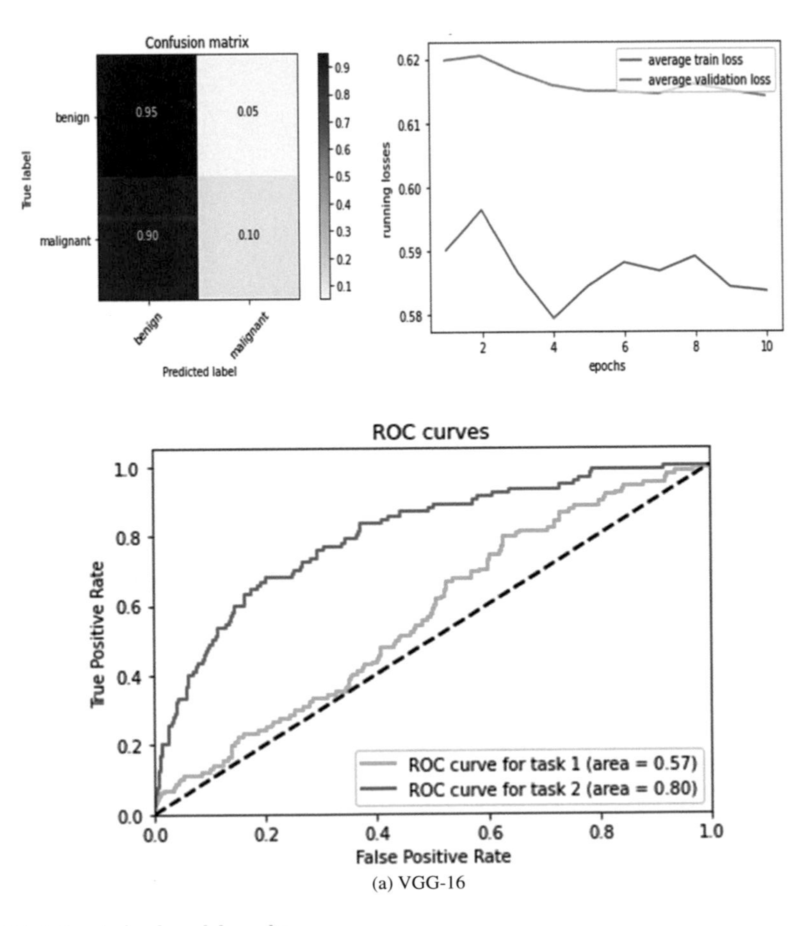

(a) VGG-16

Fig. 13.4 Pre trained model results

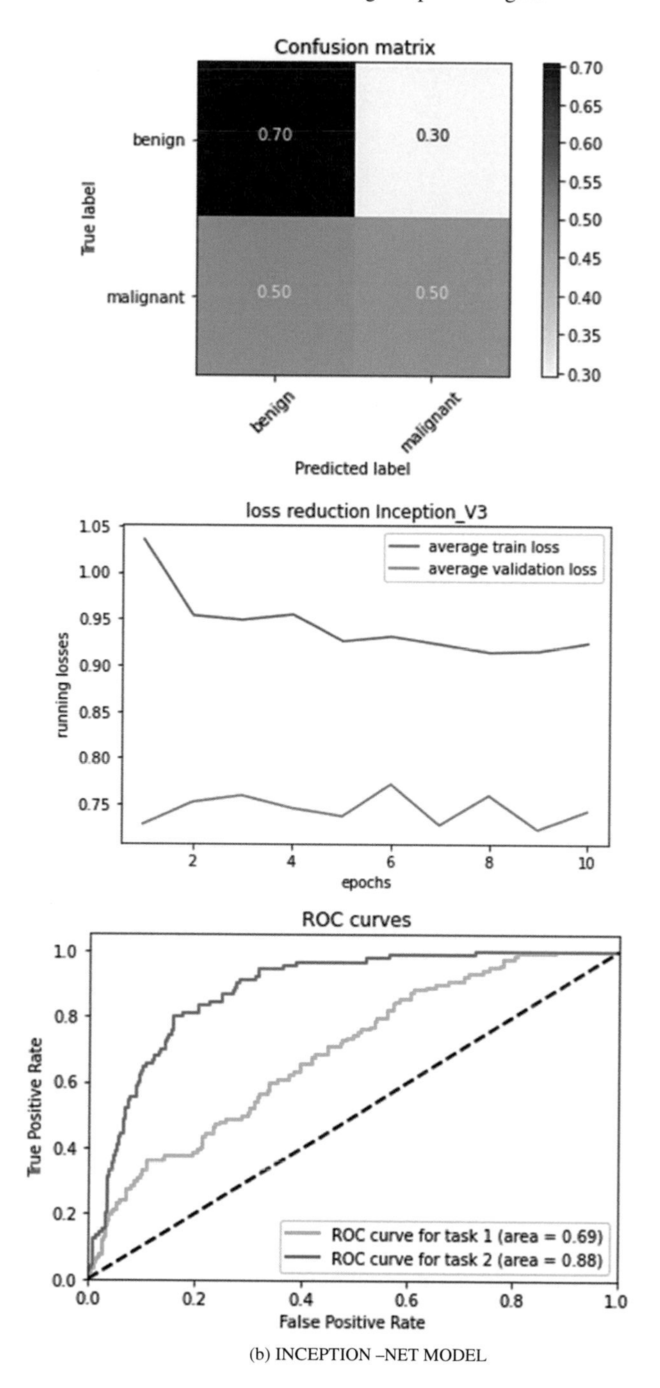

(b) INCEPTION –NET MODEL

Fig. 13.4 (continued)

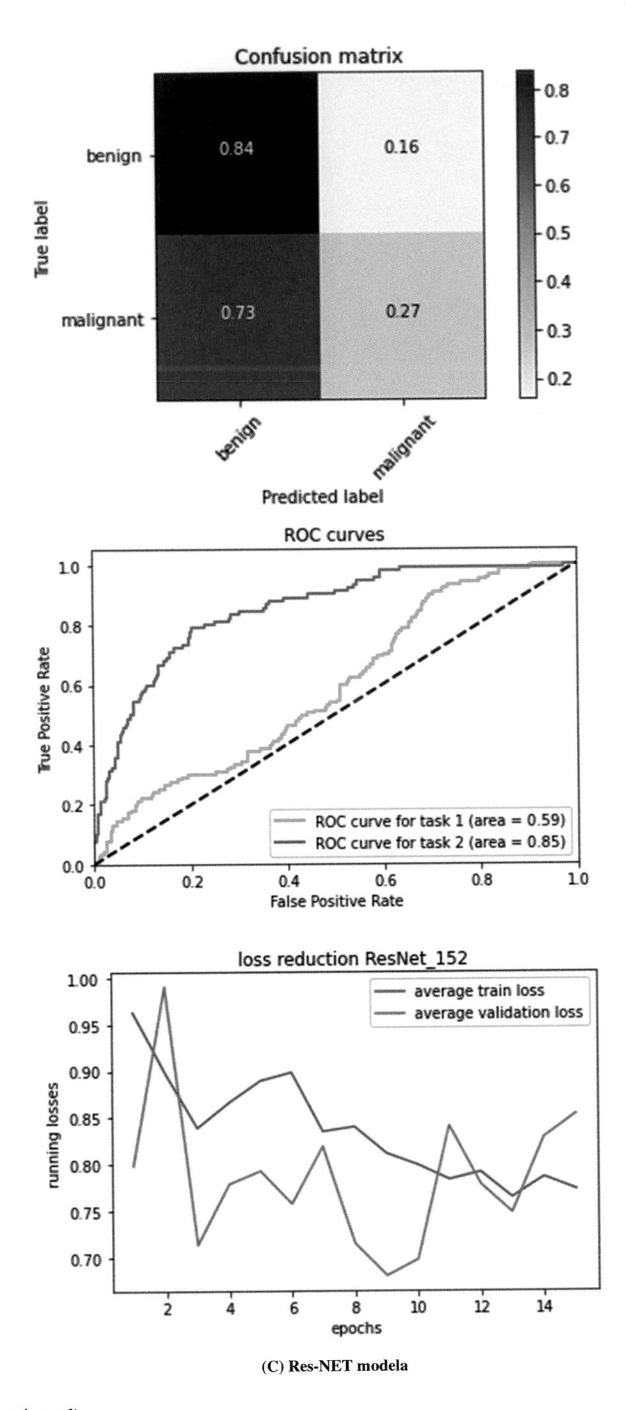

(C) Res-NET modela

Fig. 13.4 (continued)

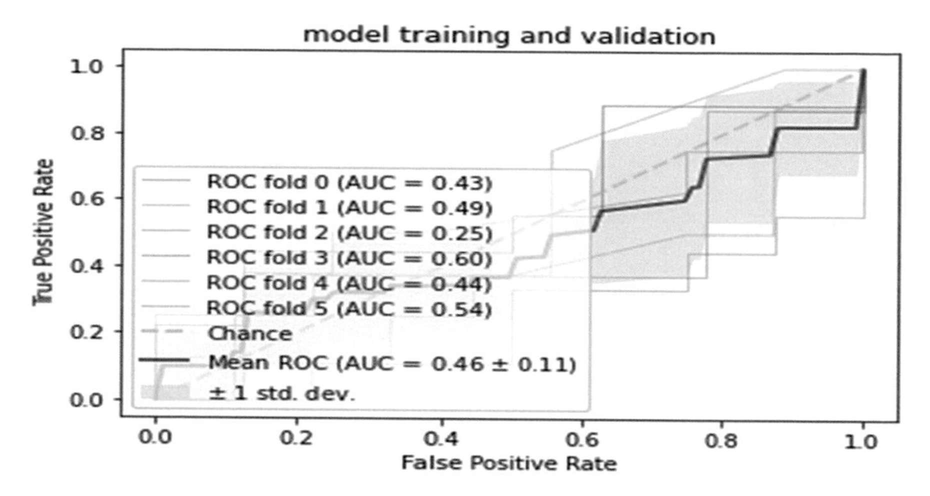

Fig. 13.5 Epoch curve graph of model layer wise negative and positive rates

where you stopped and then restart the plan cycle with the proper hyper limits for that section of the cycle [22–24] (Fig. 13.5).

RES-NET 152, InceptionV3, and VGG -16 have more FN occurrences, according to the state-of-the-art comparison. The models with the fewest FN are the most optimal when it comes to a certain sickness. Overall, a FN situation will contribute to the false perception that the patient is unaffected, perhaps leading to the spread of diseases among the healthy community. As a consequence, among the four models considered,, VGG16 dominated in accomplishing values up to 95.72%, 91.59%, 94.43%, 88.69%, and 94.78% for Acc, F1 score, PPV, Spc, and Sen, individually shown in Table 13.3 [25–27].

This exploration has assisted with fostering an assortment strategy for foreseeing skin sicknesses. This exploration is the most recent disclosure, on the grounds that to date, controllers and clinical foundations have never had a thorough arrangement for creating data frameworks. This might be because of restricted human asset limit with mastery in line innovation and lacking HR for data frameworks. Here, we see that exactness of the relative multitude of strategies expanded in contrast with without scaled information mining procedures. Presently, we gathering every one of the five strategies as one and play out the investigation and results are displayed in Table 13.4.

The attribute wise classification is shown in Table 13.4.

Table 13.3 Model accuracy comparison

Classifier	ACC	F1	PPV	SPC	SEN
VGG-16					
Res-NET 152	95.72%	91.59%	94.43%	88.69%	94.78%
Inception v3					

Table 13.4 Attribute wise classification accuracy

Classification_report		*Precision*	*Recall*	*F 1-score*	*Support*
Classification _report					
	Prosasis	1.11	1.00	1.00	24.00
	Acne	0.81	1.00	0.95	10
	Hives	1.09	1.00	1.00	11.0
	Rubeola	1.07	0.93	0.96	14.0
	Keloid	1.80	1.00	1.00	11
	Seborehhic	1.00	1.00	1.00	4.0
	Curcuble	0.99	0.99	0.99	74.0

PCA is a statistical procedure that uses an orthogonal transformation to turn a set of observations of possibly correlated variables into a set of values of linearly uncorrelated variables called principle components. The number of various main components equals the smaller of the number of original variables or the number of observations minus one. The first principal component has the greatest possible variance (that is, it accounts for as much data variability as possible), and each successive component has the greatest variance possible given the constraint that it be orthogonal to the preceding components. The resulting vectors constitute an uncorrelated orthogonal basis set. The probabilistic models PCA and Factor Analysis are both. As a consequence, model parameters and covariance may be determined using the probability of incoming information. Low position data contaminated with clinical characteristics (commotion fluctuation is quite comparable for each component) or heteroscedastic commotion (clamour change is distinct for each element), i.e. a demographic feature, is used to compare PCA and FA, as well as cross-approval. In the second phase, we compare the model probability to the probabilities derived using shrinkage covariance assessors.\. Both FA and PCA gain in terms of retrieving the size of the low position subspace when homoscedastic clamour is employed. PCA has a higher chance of succeeding in this situation than FA. PCA fizzles and overestimates the location when heteroscedastic commotion is available. Under fitting conditions, low position models are more frequent than shrinkage models, as seen in Fig. 13.6 [28, 28].

In Fig. 13.7 , the separated class of data set is shown where first outlined cluster depicts the reduced and binned cleaned test data and another segment shown result opposes data.

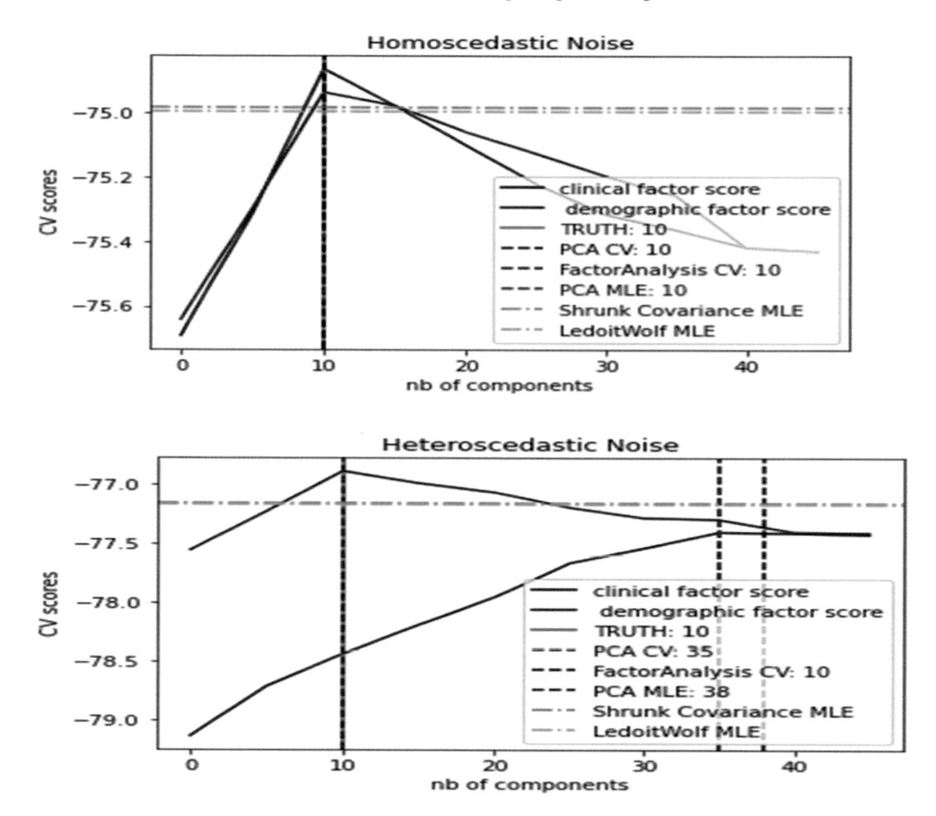

Fig. 13.6 Factor analysis

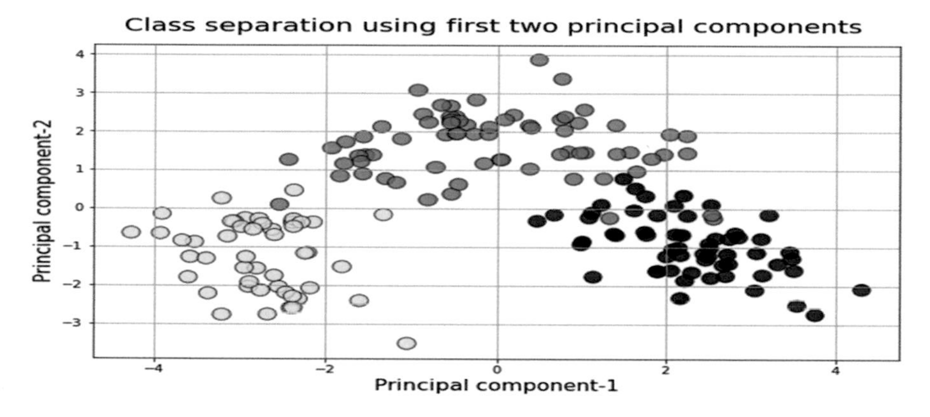

Fig. 13.7 Separated class of data cluster based on function kernel

13.6 Conclusions and Future Scope

In this chapter we have shown skin disease classification based on transfer learning method which consist clinical attribute with the maximum accuracy of 97% which has been evaluated on several performance metrics such as speciavity, sensitivity, precision, recall and support. In future more hybrid algorithm must be developed for organizing and gathering Neural organizations have numerous applications in clinical field that assistance in early conclusion and avoidance of infections. Convolutional Neural Networks have demonstrated that enormous number of datasets can be prepared inside a limited capacity to focus time and give more noteworthy exactness. Utilizing progressed computational procedures and enormous dataset, the framework can coordinate with the consequences of a dermatologist subsequently working on the quality guidelines in the space of medication and exploration.

References

1. Alkolifi Alenezi, N.S.: A method of skin disease detection using image processing and machine learning. Procedia Comput. Sci. **163**, 85–92 (2019). https://doi.org/10.1016/j.procs.2019.12.090
2. Kanani, P., Padole, M.: Deep learning to detect skin cancer using google colab. Int. J. Eng. Adv. Technol. **8**(6), 2176–2183 (2019). https://doi.org/10.35940/ijeat.F8587.088619
3. Chen, M., Zhou, P., Wu, D., Hu, L., Hassan, M.M., Alamri, A.: AI-Skin: Skin disease recognition based on self-learning and wide data collection through a closed-loop framework. Inf. Fusion **54**, 1–9 (2020). https://doi.org/10.1016/j.inffus.2019.06.005
4. Dosovitskiy, A. et al.: An Image is Worth 16x16 Words: Transformers for Image Recognition at Scale (2020). http://arxiv.org/abs/2010.11929
5. Li, H., Pan, Y., Zhao, J., Zhang, L.: Skin disease diagnosis with deep learning: a review,. Hongfeng Li (2020). http://arxiv.org/abs/2011.05627
6. Brock, A., De, S., Smith, S.L., Simonyan, K.: High-Performance Large-Scale Image Recognition Without Normalization. http://arxiv.org/abs/2102.06171
7. Tolstikhin, I. et al.: MLP-Mixer: An all-MLP Architecture for Vision, pp. 1–16 (2021). http://arxiv.org/abs/2105.01601
8. Patnaik, S.K., Sidhu, M.S., Gehlot, Y., Sharma, B., Muthu, P.: Automated skin disease identification using deep learning algorithm. Biomed. Pharmacol. J. **11**(3), 1429–1436 (2018). https://doi.org/10.13005/bpj/1507
9. Shoieb, D.A., Youssef, S.M., Aly, W.M.: Computer-Aided Model for Skin Diagnosis Using Deep Learning. J. Image Graph. **4**(2), 122–129 (2016). https://doi.org/10.18178/joig.4.2.122-129
10. Lian, D., Yu, Z., Sun, X., Gao, S.: AS-MLP: An Axial Shifted MLP Architecture for Vision, pp. 1–14 (2021). http://arxiv.org/abs/2107.08391
11. Garg, I., Panda, P., Roy, K.: A Low Effort Approach to Structured CNN Design Using PCA. IEEE Access **8**, 1347–1360 (2020). https://doi.org/10.1109/ACCESS.2019.2961960
12. Duan, S., Chen, K., Yu, X., Qian, M.: Automatic Multicarrier Waveform Classification via PCA and Convolutional Neural Networks. IEEE Access **6**, 51365–51375 (2018). https://doi.org/10.1109/ACCESS.2018.2869901
13. Romero Lopez, A., Giro-I-Nieto, X., Burdick, J., Marques, O.: Skin lesion classification from dermoscopic images using deep learning techniques. In: Proc. 13th IASTED Int. Conf. Biomed. Eng. BioMed 2017, pp. 49–54 (2017). https://doi.org/10.2316/P.2017.852-053

14. Sriwong, K., Bunrit, S., Kerdprasop, K., Kerdprasop, N.: Dermatological classification using deep learning of skin image and patient background knowledge. Int. J. Mach. Learn. Comput. **9**(6), 862–867 (2019). https://doi.org/10.18178/ijmlc.2019.9.6.884
15. Ahmad, B., Usama, M., Huang, C.M., Hwang, K., Hossain, M.S., Muhammad, G.: Discriminative Feature Learning for Skin Disease Classification Using Deep Convolutional Neural Network. IEEE Access **8**, 39025–39033 (2020). https://doi.org/10.1109/ACCESS.2020.2975198
16. Xie, B. et al.: XiangyaDerm: A clinical image dataset of asian race for skin disease aided diagnosis. Lect. Notes Comput. Sci. (including Subser. Lect. Notes Artif. Intell. Lect. Notes Bioinformatics), Vol. 11851 LNCS, pp. 22–31 (2019). https://doi.org/10.1007/978-3-030-33642-4_3
17. Souahlia, S., Bacha, K., Chaari, A.: MLP neural network-based decision for power transformers fault diagnosis using an improved combination of Rogers and Doernenburg ratios DGA. Int. J. Electr. Power Energy Syst. **43**(1), 1346–1353 (2012). https://doi.org/10.1016/j.ijepes.2012.05.067
18. Li, Y., Shen, L.: Skin lesion analysis towards melanoma detection using deep learning network. Sensors (Switzerland) **18**(2), 1–16 (2018). https://doi.org/10.3390/s18020556
19. Srinivasu, P.N., SivaSai, J.G., Ijaz, M.F., Bhoi, A.K., Kim, W., Kang, J.J.: Networks with MobileNet V2 and LSTM, pp. 1–27 (2021)
20. Goyal, M., Oakley, A., Bansal, P., Dancey, D., Yap, M.H.: Skin Lesion Segmentation in Dermoscopic Images with Ensemble Deep Learning Methods. IEEE Access **8**, 4171–4181 (2020). https://doi.org/10.1109/ACCESS.2019.2960504
21. Polat, K., Koc, K.O.: Detection of skin diseases from dermoscopy image using the combination of convolutional neural network and one-versus-all. J. Artif. Intell. Syst. 2(1):80–97 (2020). https://doi.org/10.33969/ais.2020.21006
22. Lu, Y., Xu, P.: Anomaly Detection for Skin Disease Images Using Variational Autoencoder (2018). http://arxiv.org/abs/1807.01349
23. Roslan, R., Razly, I.N.M., Sabri, N., Ibrahim, Z.: Evaluation of psoriasis skin disease classification using convolutional neural network. IAES Int. J. Artif. Intell. **9**(2), 349–355 (2020). https://doi.org/10.11591/ijai.v9.i2.pp349-355
24. Naronglerdrit, P., Mporas, K.: Evaluation of Big Data based CNN Models in Classification of Skin Lesions with Melanoma (2020)
25. Hemavathi, S., Jayasakthi Velmurugan, K.: Skin disease prediction and provision of medical advice using deep learning. J. Phys. Conf. Ser., **1724**(1) (2021). https://doi.org/10.1088/1742-6596/1724/1/012048
26. Purnama, I.K.E. et al.: Disease classification based on dermoscopic skin images using convolutional neural network in teledermatology system. In: 2019 Int. Conf. Comput. Eng. Network, Intell. Multimedia, CENIM 2019 - Proceeding, Vol. 2019, pp. 1–5 (2019). https://doi.org/10.1109/CENIM48368.2019.8973303.
27. Akmalia, N., Sihombing, P.: Skin diseases classification using local binary pattern and convolutional neural network. In: 2019 3rd International Conference on Electrical, Telecommunication and Computer Engineering (ELTICOM), February 2021, pp. 168–173. https://doi.org/10.1109/ELTICOM47379.2019.8943892.
28. Mahajan, K., Sharma, M., Vig, L.: Meta-DermDiagnosis: few-shot skin disease identification using meta-learning. IEEE Comput. Soc. Conf. Comput. Vis. Pattern Recognit. Work, 3142–3151 (2020). https://doi.org/10.1109/CVPRW50498.2020.00373
29. Naeem, A., Farooq, M.S., Khelifi, A., Abid, A.: Malignant Melanoma classification using deep learning: datasets, performance measurements, challenges and opportunities. IEEE Access **8**, 110575–110597 (2020). https://doi.org/10.1109/ACCESS.2020.3001507

Chapter 14
A Deep Learning Techniques for Brain Tumor Severity Level (K-CNN-BTSL) Using MRI Images

M. Saravanan, Suseela Sellamuthu, Saksham Bhardwaj, Chakshusman Mishr, and Rohit Parthasarathy

Abstract Brain tumor is revealed due to uncontrolled rapid growth of cells. If not found at an infinity stage, it causes death. Early stage of prediction of diseases of Human's Brain overcomes a lot of issues. Still classification and segmentation of Brain tumor is very challenging even though having existing promising solutions. Magnetic resonance imaging (MRI) is one forefront technique which is suitable for identifying disease for those who are sufferers. This research focuses on Deep learning techniques for identifying the risk of Brain Tumor and also explores the Brain Tumor detection, segmentation, classification and prediction of severity level of Brain tumor. In this work, MRI The training and testing is done by the CNN from MRI images. The Convolution Neural Network Model is used to classify the images and K-means algorithm is applied to segment the image. Feature extraction applied through Discrete Wavelet Transform and Principal Component Analysis to reduce the dimensional for better accuracy and computational power management. The experiment with deep learning algorithms indicates the effectiveness of the proposed work. A novel method of hybridization of K-Neighboring Algorithm and CNN (K-CNN-BTSL) predicts the severity level of a brain tumor either benign or malignant at an early point with 93% accuracy.

M. Saravanan (✉) · S. Bhardwaj · C. Mishr · R. Parthasarathy
Department of Networking and Communications, School of Computing, SRM Institute of Science and Technology, Kattankulathur, Chennai, India
e-mail: saravanm7@srmist.edu.in

S. Bhardwaj
e-mail: sa4456@srmist.edu.in

C. Mishr
e-mail: rs1253@srmist.edu.in

R. Parthasarathy
e-mail: cn2553@srmist.edu.in

S. Sellamuthu (✉)
School of Computing Sciences and Engineering, Vellore Institute of Technology, Chennai, India
e-mail: suseela.s@vit.ac.in

A. Biswas et al. (eds.), *Artificial Intelligence for Societal Issues*, Intelligent Systems Reference Library 231, https://doi.org/10.1007/978-3-031-12419-8_14

Keywords K-means algorithm · CNN model · Principal component analysis · Discrete wavelet transform · Mri images

14.1 Introduction

The human brain is the most complex system and biological processing power house which has sky rocketing capabilities. We are all still unaware of the power of the brain. The brain is also susceptible to a variety of diseases and has lost its divine power. Brain cancer is one of them and also has a wide variety of symptoms including confusion, sleepiness, seizures, and behavioral changes Headache, Nausea, Speech alterations, Problems in maintaining balance, Memory problems, Twitches and fits, Numbness, and tingling in arms or legs. Early-stage detection [1] of cancer can avoid fatality but is always not practically feasible. A tumor is classified as pre-carcinoma, benign, or malign. Benign tumors do not spread or unfold the effect to different organs, and hence the affected tissues can be surgically removed.

Figure 14.1 shows tumors types.Gliomas are a general term used in tumor identification that starts from tissues of the brain other than blood vessels and nerve cells. Meningiomas grow from the membranes that cover the brain and its s rounding central system. A pituitary tumor is present as a lump that resides within the brain. The impact of tumors varies from benign to malignant. Investigate the types of tumors that are mandatory for the clinical diagnosis of patients. The symptoms of the growth of a brain tum r depend upon factors such as tumor type, size, and placement. Tomography is one of the foremost techniques for the diagnosis of human brain tumors. Still, the traditional methodology for detection and classification of brain tumor is carried from the expertise of radiologists, the review of WHO's, and analyzing the characteristics of images. In Medical Imaging, various disease are identified using the recent advancements of deep and machine learning technique [2]. K-CNN is a subset of machine learning [3] that is mostly used for image detection and classification [4]. In this paper, a K-CNN classification is proposed to classify the impact of brain tumors such as benign or malignant.

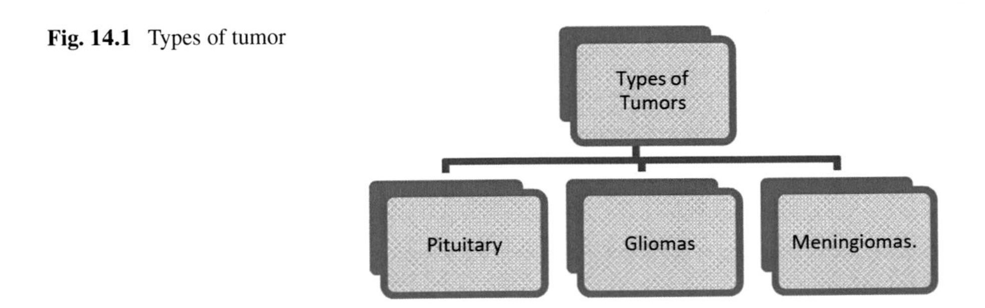

Fig. 14.1 Types of tumor

14.2 Related Work

The authors [4] focused on using watershed algorithms to separate the tumor area from the MRI scans of the brain. Extraction is dependent on several feature combinations like orientation, color, texture, and edge. The outcomes of the experiment have been compared with the ground truth images which contain labeled data. The tumor region is segmented using marker-based watershed algorithm. To identify the external and internal markers, the authors have used an atlas-based approach. In this approach, some early information about the image is needed for identifying the marker. The marker-based watershed algorithm has been implemented using an interactive tool kit, Ilastik. The proposed algorithm is not able to carry out automatic identification of markers and is not able to separate the tumor region based on the markers and color which are identified automatically. The orientation feature cannot be used as a factor for segmentation through the proposed method. The authors [5] focused on providing a solution for the low contrast of the MRI images. Low contrast images make it difficult to analyze the image for identifying any anomalies. The authors have proposed a strategy for enhancing the contrast of the MR Image and skull stripping through the use of the Morphological image processing technique. The data set consists of 120 MRI scans of the brain with each image having several sequences. FLAIR, T1, and T2 sequences of each image are available. The basic strategy for enhancing the MRI brain scans using mathematical morphology for separating parts of an image will provide useful information such as the portrayal of regions, shapes like edges, and borders. The author has not mentioned any limitations. The results obtained show that the methods proposed in this paper are efficient in enhancing the contrast of MRI images and their skull striping.

The authors [6] have proposed a technique for identifying brain tumor using MRI images [7, 8]. The image data set contains details of 120 subjects with brain tumors. The overall technique is split into the following steps: preprocessing, feature extraction, segmentation using local independent projection-based classification (LIPC) method, and p st-processing. A web analysis tool has been used to analyze the segmentation results obtained from the testing data. The projected classification technique needs no specific regularization as a result of the patch feature containing the contextual information of the neighborhood region within the image. The author focused on the segmentation and classification [9] of brain tumors. The overall procedure is as follows: The input image is segmented to obtain the tumor region. Segmentation is done using the watershed algorithm. Features are extracted from the tumor region using DAPP (Dynamic Angle projection pattern). Finally, normal or abnormal tumor is classified by the CNN classifier using data sets namely BRATS. The Watershed algorithm is successful in efficiently segmenting the tumor region, which also improves the process of feature extraction

The proposed method can classify tumors with an accuracy of 93 however, more features can be used and experimented with to get better accuracy. The probabilistic Neural Network (PNN) model proposed by authors [4], which depends on learning vector quantization (LVQ) methods with data and image analysis procedures to automatically classify brain tumors through MRI images. The algorithm is implemented using MATLAB. One limitation of the proposed method is that the accuracy results are different for different data sets. supervised machine learning [7] methods proposed by authors [4] to classify brain MR images into the following 5 types: normal, Meningioma, Anaplastic Astrocytoma, Cystic Oligodendroglioma, Ependymoma, and Lymphoma. The ML techniques used to automatically classify the image includes C4.5 decision tree algorithm and multi-layer perceptron (MLP). The features which are considered are area, solidarity, major axis length, minor axis length, Euler number, circularity. Experimental results show that classification using the C4.5 algorithm has a precision of 91% and the MLP algorithm has a precision of 95% which is the maximum precision achieved. The precision can be increased by using a larger data set and considering other features like those based on texture and intensity.

The segmentation of Brain tumor [6] is significant method for early tumor analysis and radiotherapy planning. The author addressed LIPC(local independent projection-based classification) method to classify the tumor of brain. The performance of the classification is improved by using softmax regression model. Further, this paper explores the adequacy of convolutional neural organizations when joined with different classifiers to expand the precision of cerebrum tumor division [7]. This intends to utilize the MRI picture of the mind to examine the technique for a more clear perspective on the tumor. SVM classifier is used for isolating the influenced region. Anisotropic filtering for commotion evacuation, are the vital stages in the strategy. Test outcomes showed that SVM accomplished 83% precision in the stage. At long last, the isolated locale of the plant is remembered as the primary picture for an alternate distinguishing proof. An MRI filter picture may contain an assortment of commotion. Legitimate detachment and morphological execution of the input pictures ought to be liberated from the noise. That is the reason we have utilized an anisotropic channel for better execution. SVM classifier fragments the pixels into two classes. Since the framework is intended for any MRI input SVM is utilized with part for solo learning. To extricate the tumor from the fragmented district, morphological activities are utilized. The framework henceforth can distinguish the tumor all the more precisely. Different classification approaches such as hidden Markova random field [8], Machine Learning [9], KNN [10], SVM [11], FCM [12, 13] are used for testing and training the data from unknown to known samples.

14.3 Problem Statement

Magnetic Resonance Imaging (MRI) [14], is the commonly used method for identifying and diagnosing the presence and growth of the tumor. However, this process is vulnerable to human subjectiveness, because handling such a huge volume of information is a challenging task for human's identification and observation. Detecting brain-tumors [15] at an early stage is based on the expertise of the radiologist. Medical analysis for the growth of the tumor is not complete without determining whether it is malignant or benign. To have confirmation about the severity of the tumor, it is typically required to perform a diagnostic test. Contrasting to tumors in other parts of the body, the biopsy of the brain tumor isn't sometimes obtained before definitive operation. To get a precise medical specialty, and to avoid subjectiveness and surgery, it's necessary to develop a good medical specialty tool that can detect and classify tumors with the help of MRI images. In this work, proposed a method that uses convolution neural networks, further detecting and classifying a tumor into benign or malignant. Using the various steps involved. All the features would help doctors in efficiently diagnosing their diseases. This would also help us removing the subjectivity that comes into play while; a doctor examines the MRI scan. By removing the subjectivity in determining the type of tumor that exists, helps us a lot in removing the errors that pertain to the diagnosis which would help a lot of patients that are suffering from the disease and thus would therefore help in the well- being of all the ones who are affected by it.

14.4 Proposed Work: K-CNN-BTSL (Brain Tumor Severity Level)

The proposed work K-CNN-BTSL consists of three modules: Pre Processing, Image segmentation, Feature Extraction and Classification (CNN). Figure 14.2 shows the system overview of the proposed work. In preprocessing, a picture for tumor classification is chosen. The selected image is then changed to grayscale format from RGB format. In the image segmentation, a median filter is used to remove any possible abnormalities or disturbances, to improve accuracy in the later stages. A filtered image is observed and segregated to find the tumor region using K-means clustering algorithm. From the identified tumor region, a few texture-related properties and features are identified using the options of the brain image square PCA (Principal Component Analysis) and Discrete Wavelet rework (DWT). The features obtained through the above-mentioned steps are compared with training data using the CNN [12] model. Finally, the result will highlight the type of tumor.

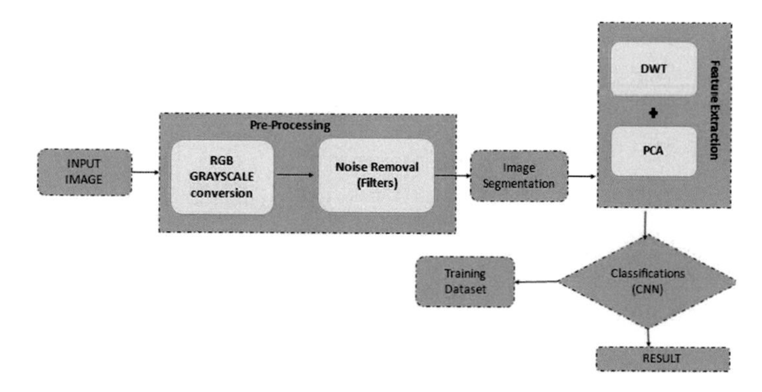

Fig. 14.2 System architecture of K-CNN-BTSL

14.4.1 Preprocessing

14.4.1.1 Data Set

One of the most commonly used imaging techniques to identify and assess the impact of brain tumors is MRI (Magnetic Resonance Imaging [14, 15]. Due to the huge amount of data reports generated by MRI, it becomes difficult to perform segmentation manually in an acceptable operational time. This limits the use of precise quantitative measurements in clinical practice. In this work, Algorithm 1 shows the Brain Tumor Identification using CNN.

Algorithm 1: K-Means Algorithm for segmentation

```
     Input: MIMG = {MIMG1, MIMG, MIMG2...., MIMGn} (set of Images to
     beclustered), k =Set of Clustered Images), Iters_Max: Maximum Iteration
     Output: CIMG = {CIMG1, CIMG2,....... CIMGk} (#clusters)
     CIMGLabel = {CIMG (cl) | cl=1,2..., n} (set of cluster labels of MIMG)
1    foreach CIMG (i) ε CIMG do
2             CIMG (i)  ←  CIMGLabel ε CIMG
3    end
4    foreach CIMG ε MIMG do
5             CIMG (cli) ← Dist_Min(cli, CIMG (i) i ε {1......k}
6    end
7    checked ←  false;
8    Iters_Max ←  0;
9    repeat
10         foreach CIMGLabel ε CIMG do
11            UpdateCluster(cli);
12         end
13         foreach cli ε MIMG do
14            Dist_Min ← (cli, CIMGLabel) i ε {1.....k};
15                If Dist_Min ≠ CIMGLabel (cli) then
16                   CIMGLabel (cli) ←  Dist_Min;
17                   checked ← true;
18                end
19            end
20         iters_Max ++;
21    until Checked = true and Iters_Max;
```

The data set contains MRI images as shown in Fig. 14.3. The table1 shows the data set used in the proposed system.

Preprocessing of data is a significant task [15] to extract the exacted damaged region. Depending upon the typical distinctive conditions fixed, multi-scale, linear, nonlinear [16] are used in distinctive conditions [17, 18] Find out the normal and abnormal tissues are very difficult because of noise and artifacts [16] exist in the image is very difficulty in analysis of an Image [19] Thus without human interaction, automated techniques for segmentation are suggested in [20]. In this work, preprocessing is used to perform certain operations like RGB to grayscale conversion

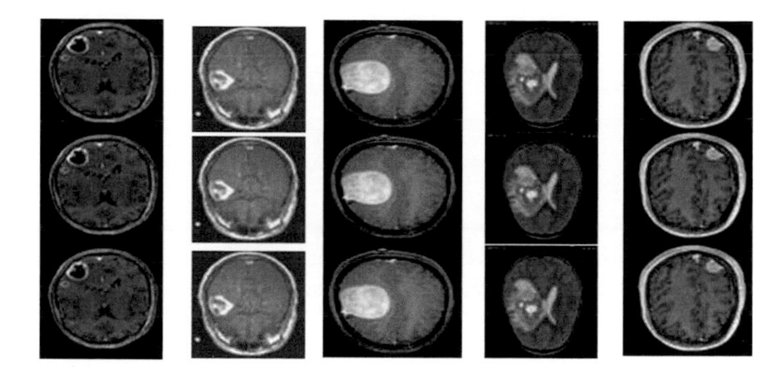

Fig. 14.3 Data set (MRI Images)

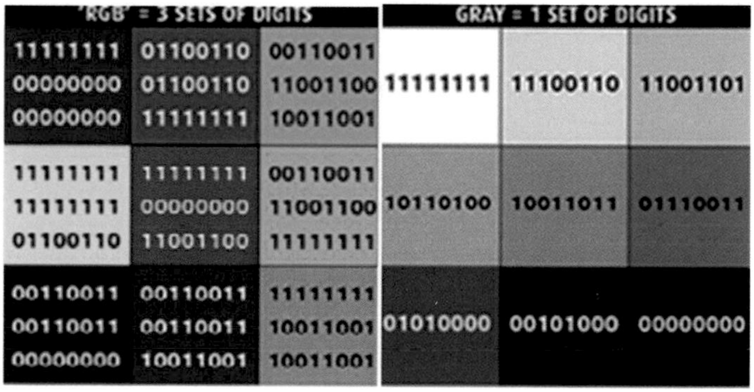

"RGB" = 3 SETS OF DIGITS			GRAY = 1 SET OF DIGITS		
11111111 00000000 00000000	01100110 01100110 11111111	00110011 11001100 10011001	11111111	11100110	11001101
11111111 11111111 01100110	11111111 00000000 11001100	00110011 11001100 11111111	10110100	10011011	01110011
00110011 00110011 00000000	00110011 00110011 10011001	11111111 10011001 10011001	01010000	00101000	00000000

Fig. 14.4 RGB to gray conversion

and median filtering. This is used to improve the clarity of the image. In preprocessing, RGB to grayscale conversion is shown in Fig. 14.4 used to change the RGB images to grayscale by dispensing the shade and saturation data while preserving the luminance. The median filter [21] is useful to the grayscale representation for additional noise removal. It is a nonlinear method that helps diminish the imprudent or salt-and-pepper commotion from the image. It is also useful in preserving the edges of the image, and at the same time decreasing the irregularities in the image. The median filter considers every component within the image successively and examines the close-by components to decide if it is similar to its surroundings or not. In median filtering, a pixel is interchanged with the median of the values of pixels surrounding it rather than changing it to the mean of the surrounding pixel values is shown in Fig. 14.5.

Displayed equations are centered and set on a separate line. The Eq. 14.1 is used to evaluate the median filtering values

123	125	126	130	140
122	124	126	127	135
118	120	150	125	134
119	115	119	123	133
111	116	110	120	130

Neighbourhood values:

115, 119, 120, 123, 124, 125, 126, 127, 150

Median value: 124

Fig. 14.5 Median filtering

$$M_filter(x, y) = \sum_{i=1}^{\infty} \sum_{j=1}^{-\infty} M_filter(x - i, y - j) * Con_filter(i, j) \tag{14.1}$$

where $M_filter(x - i, y - j)$ is preprocessed image and Con_filter is covolve filter.

14.4.2 Image Segmentation

K-means algorithm [22] is employed to perform image segmentation is shown in Fig. 14.6. It is an unsupervised clustering algorithm [23] which groups the input data given to it into multiple categories depending on their intrinsic separation from one another. The algorithm assumes that the information provided is in the form of a vector space and attempts to discover natural groups in the given input. The algorithm 1 is applied to segment the filtered image. This method is used to highlight the segmented and affected part of the tumor in the brain.

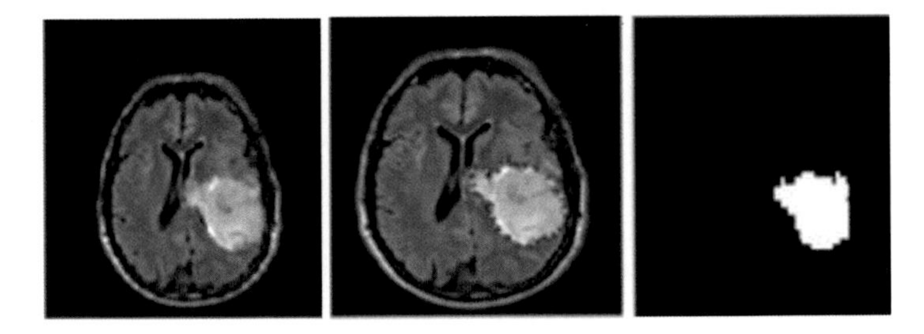

Fig. 14.6 K-means segmentation

14.4.3 Feature Extraction

The process of feature extraction [24] begins with the information which is initially available from the segmented area of the tumor. The focus is on extracting values of features which are non-repeating and will help in the generalization and learning steps in the later stages. One of the main purposes of feature extraction [25] is to reduce the dimensionality of the available information. In this step important features which are needed for image classification are extracted. This is done by extracting features from the tumor segmented region using Discrete Wavelet Transform and PCA. The wavelet transformation [26, 27] the wavelets are sampled discretely. When compared with other types of wavelet transformation, one of the advantage is that it captures both frequency and location information. This is beneficial than Fourier transformation, which has temporal resolution. The high level feature extraction is applied to categorize the Benign Tumor and Malignant. The main purpose of using PCA [28, 29, 36] is to reduce the dimensional of the available information in order to increase efficiency of processing. The calculations are achieved by using basic matrix operations along with statistics. GLCM (Gray Level Co-occurrence Matrix) [30] is applied to extract the texture properties of segmented region like standard deviation, correlation, mean, skewness, variance, energy, kurtosis, homogeneity, contrast and smoothness. The GLCM is generated from the given input image I using the gray comatrix (I) function. This matrix is also known as gray-level spatial dependence matrix. This function generates the GLCM by computing how often does a pixel having a gray-level (grayscale intensity) value of x appears in close proximity to a pixel with the value of y. Each component (x, y) present in GLCM represents frequency of occurrences wherein a pixel with value x occurred in close proximity to a pixel with value y. The function graycoprops (GLCM, properties) is used to calculate features which is mentioned in the properties using GLCM. This function normalizes the entries of the GLCM in order to make the sum of the elements as unity. Every component (i, j) present in the normalized GLCM represents the combined likelihood of occurrence of a couple of pixels with a exact correlation with gray level values as i and j in the image. Graycoprops() calculates the required properties using the normalized GLCM. The contrast of pixel intensity values of MRI image is defined by Eq. 14.2

$$Contrast = \sum_{i=0}^{w} n^2 \sum_{l=1}^{w} \sum_{m=1}^{h} \{T_r(l, m)\} |l - m| = n \tag{14.2}$$

where Tr(l, m) is the tumor region segmented image with width w and height h, respectively. The complexity of the brain tumor region is defined as Entropy in Eq. 14.3

$$Entropy = \sum_{i=0}^{w-1} \sum_{m=0}^{h-1} T_r(l, m) * log(T_r(l, m)) \tag{14.3}$$

The energy of the segmented image is defined as Eq. (14.4). Energy is the intensity of each pixel with other pixel in brain tumor region in segmented image.

$$Energy = \sum_{i=0}^{w-1}\sum_{m=0}^{h-1} T_r(l,m) * log(T_r(l,m)^2) \quad (14.4)$$

Skewness is used to find the symmetry of brain tumor image. It is lies between –0.5 and 0.5. Statistical measure of Kurtosis is used to represents the pixel intensity are light tailed or heavy tailed in the affected region.

14.5 K-CNN-BTSL

The CNN requires large number of brain images. To achieve optimum accuracy of brain images, CNN needs classifier [37–39]. Figure 14.7 shows the architecture of K-CNN-BTSL. The first layer performs the convolution between the given brain tumor images [31] and a small filter or kernel of a specific size. This output is known as a feature map. Different kernels are used to perform different operations like sharpening [32] the image, edge and corner detection etc. The feature map is given to the pooling layer is to diminish the size of the convoluted feature map [33] so as to decrease the computational expenses. Fully connected layers: In this layer, each neuron in one layer is associated with each neuron in another layer [34]. Here the input from the pooling layer is compacted out into a column vector and fed to the fully connected layer [34]. A K-CNN-BTSL model is trained with a labeled data set of images along with the corresponding output. During the training process, the network will repeatedly tune the weights for each neuron as it processes each image in the data set. The classification process begins in this layer. Algorithm 2 shows the Brain Tumor Identification using CNN [37–39].

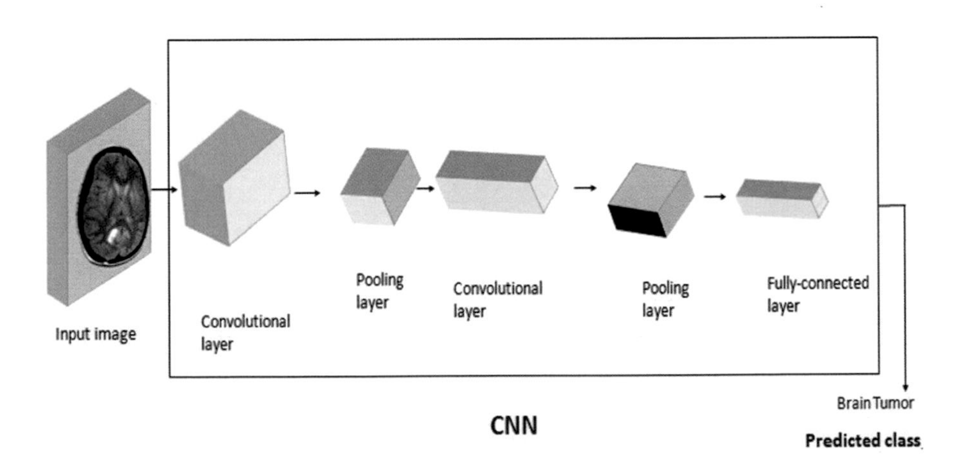

Fig. 14.7 Architecture of K-CNN-BTSL

Algorithm 2: Brain Tumor Identification using CNN

```
1.  Input: MRI Image MIMGn // n MRI Images
2.  Output: Classified Image CIMGn
3.  CIMAo ← ϕ
4.  while | CIMAo | ≤ MIMGn do
5.    Random Image= random (0,1)
6.      List Img ← linked list with RandomImage
7.        for all MIMGn € List Img do
8.          Rcheck= random (0,1)
9.          if Rcheck ≤ α0
10.             MIMGn . type = 'Layer_Skip'
11.             MIMGn .f1=random (0,1)
12.             MIMGn .f2=random(0,1)
13.         else
14.             MIMGn .type= 'Layer_Pool'
15.             poolCheck = random (0,1)
16.               if poolCheck ≤ 0.5 then
17.                 MIMGn . CIMGn = "benign"
18.               else
19.                 MIMGn . CIMGn = "malignant"
20.             end if
21.       end if
22. end for
23. end while
24.     CIMG0=CIMG0∪ List Img
25. return CIMG0
```

14.6 Results and Discussion

The proposed work is simulated using MATLAB 2016(a). It permits grid control, data and task scheduling, calculations execution, UI creation, and joining with programs written in different languages. CNN algorithm is used to classify the severity levels as "Benign Tumor and Malignant". The GLCM matrix is used to represent the features of the tumor extracted from the MRI Image is shown in Table 14.2. Table 14.1. DataSet(BRATS2015), 70 Normal Images and 60 From the Abnormal Images are trained by K-CNN-BTSL tested 155 Normal Images, 100 Abnormal Images tested. The entropy, Variance, Skewness, Energy, Contrast and Correlation are calculated and shown in Fig. 14.8.

Table 14.1 Data set (BRATS2015)

Mode	Normal	Abnormal
Training images	70	60
Testing images	155	100
Total images	225	166

Table 14.2 Values of statistical features of images

Values of statistical features of images								
Features	Contrast	Corelation	Energy	Enropy	RMS	Variance	Kurtosis	Skewness
Img1	0.2088	0.199	0.7621	3.1735	0.0898	0.0080	7.3232	0.4690
Img2	20.2271	0.0931	0.7686	3.2698	0.0898	0.0024	5.2231	0.3202
Img3	0.2161	0.1382	0.7548	3.3156	0.0898	0.0025	5.1431	0.3124
Img4	0.2414	0.1065	0.744	3.255	0.0898	0.003	5.1234	0.3222
Img5	0.2251	0.0991	0.7691	3.3182	0.0898	0.0021	5.221	0.3204
Img6	0.2033	0.1126	0.7554	3.4149	0.0898	0.0019	5.1934	0.3301
Img7	0.2341	0.1321	0.755	3.1562	0.0897	0.0035	5.2042	0.3452
Img8	0.2155	0.0951	0.7378	3.6286	0.0898	0.00898	5.3238	0.323
Img9	0.2122	0.1444	0.7498	3.1868	0.0243	0.0023	5.1877	0.3211
Img10	0.2342	0.1848	0.7978	3.3999	0.09876	0.00687	5.2376	0.3244
Img11	0.2412	0.1867	0.7664	3.188	0.05674	0.00564	5.1587	0.3155
Img12	0.2314	0.1049	0.7767	3.256	0.05622	0.00475	5.1463	0.3401
Img13	0.2054	0.1876	0.7853	3.264	0.01573	0.00436	5.0435	0.3301

Tables 14.3 and 14.4 shows the comparison result of K-CNN-BTSL. The proposed results compared with other existing algorithms. K-CNN-BTSL provided the 93% with contrast 0.2155, Entropy 3.6286, Variance 0.0089, Skewness o.469.

14.6.1 Testing with Benign Input

The severity level of Brain Tumor is identified and analyzed through algorithm 1 and algorithm2. From the data set , among the 391 images, segregated as Benign Images as shown from Figs. 14.9 and 14.10.

14.6.2 Testing with MALIGNANT Input

The Fig. 14.11 represents the malignant images identificd thc proposed work. Through the implementation parts the proposed system K-CNN-BTSL identified the severity level of the Brain Tumor effectively.

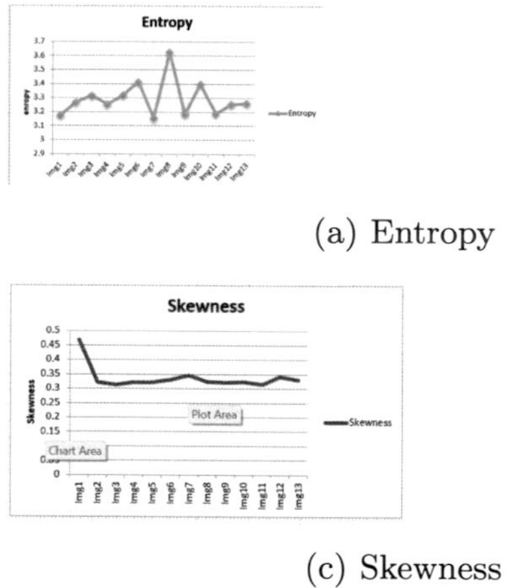

(a) Entropy

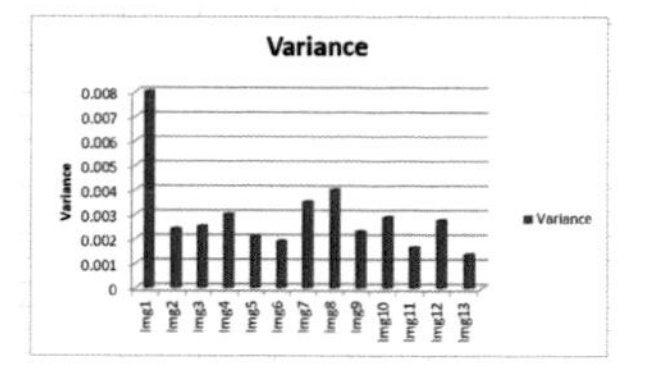

(b) variance

(c) Skewness

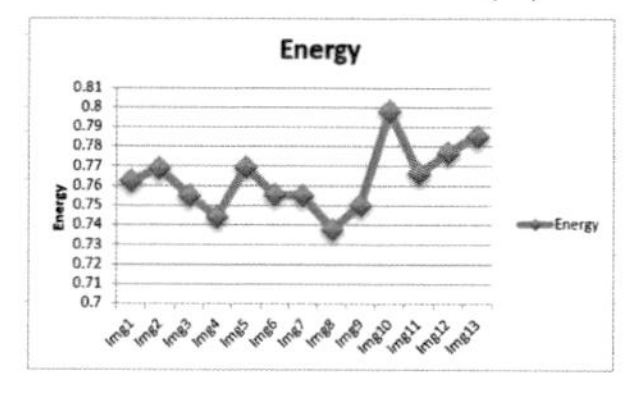

(d) energy

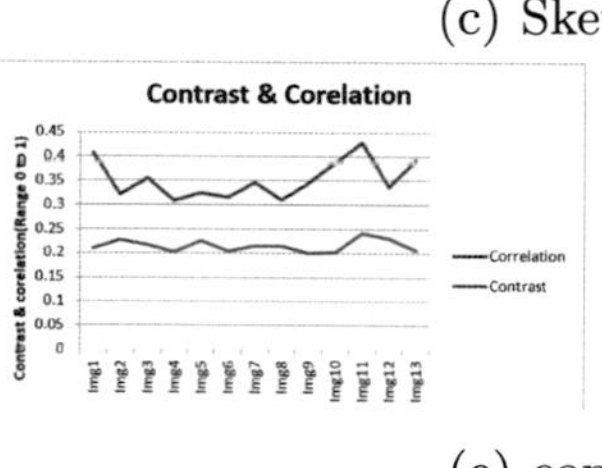

(e) contrast

Fig. 14.8 K-CNN-BTSL results

Table 14.3 Result of K-CNN-BTSL

Result of K-CNN-BTSL		
Segmentation methods	Data sets	Outcomes
Otsu algorithm [40], 2017	BRATS 2013 Synthetic	0.93±0.04 DSC on HG,0.90±0.02 DSC on LG0.87±0.06 Jaccard Index on HG,0.82±0.04 JaccardIndex on LG
Non-negative matrix factorization (NMF) [41], 2017	21 HGG patients	0.80 complete DSC, 0.74 core DSC and 0.65 active DSC tumor
HCSD [42], 2017	BRATS2012 Challenge	0.9102±0.0627 DSC, 0.9501±0.0518 SE,0.9980±0.0023 SP
Improved thresholding Method, [43], 2018	Harvard and Private collected images	0.948 Jaccard index on clinical and 0.961 Jaccard index on Harvard
Novel saliency method [44]	2018 BRATS 2013 Challenge	0.86±0.06 HG DSC,0.85±0.07 LG DSC

Table 14.4 K-CNN-BTSL

Result of K-CNN-BTSL

Segmentation methods	Data sets	Outcomes
BA and RG [45]	2018 BRATS 2015 Challenge	0.8741 Jaccard index,0.9036 DSC, 0.9827sensi tivity, 0.9772 specificity, 0.9753accuracy and 0.9585 precision
EM and FODPSO [46]	2019 192 MRI scan	0.93.4 ACC
Adaptive threshold and morphological operations [47]	2019 1340 Clinical MR images	0.85 0.85 DSC, 0.89 Jaccard index
3D semantic segmentation, [48]	2020 BRATS 2019	challenge 0.826 enhance, 0.882 complete, 0.837 core tumor
CNN model	2021 FLAIR, (T1T1C, and T2)weighted	0.957 ACC
K-CNN-BTSL	BRATS2015	0.93 ACC Contrast 0.2155, Enropy 3.6286, Variance 0.0089, Skewness 0.469

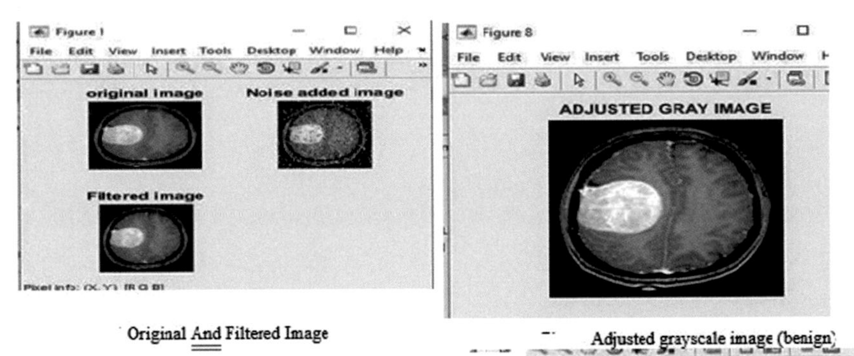

Original And Filtered Image

Adjusted grayscale image (benign)

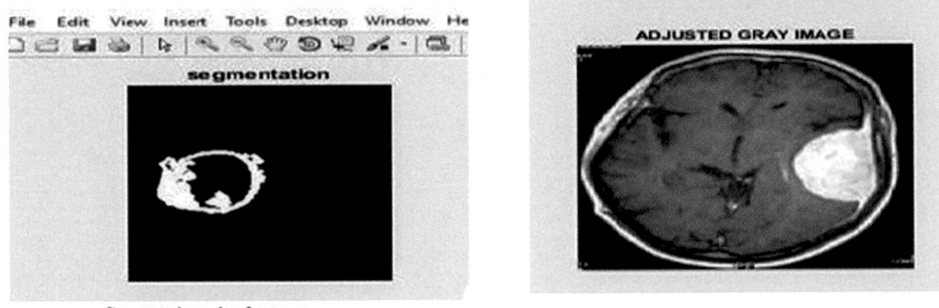

Segmentation using k-means

Feature Extraction

Fig. 14.9 Testing with Benign input

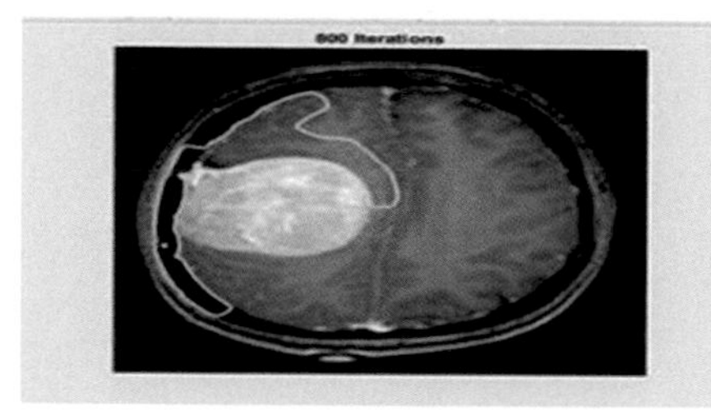

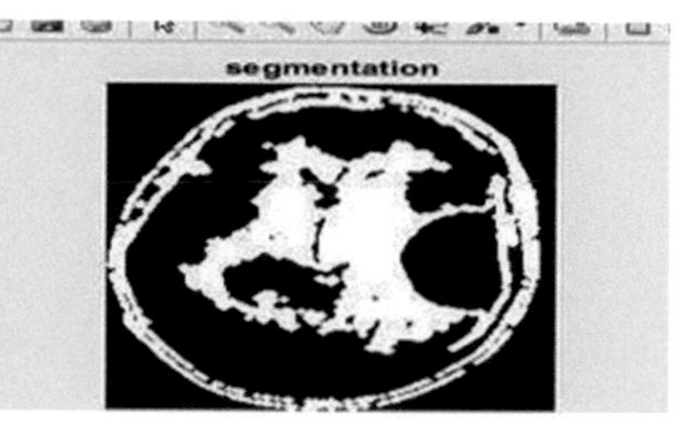

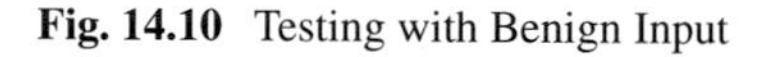
Tumor Border

Segmentation using K-Means ((Malignant)

Fig. 14.10 Testing with Benign Input

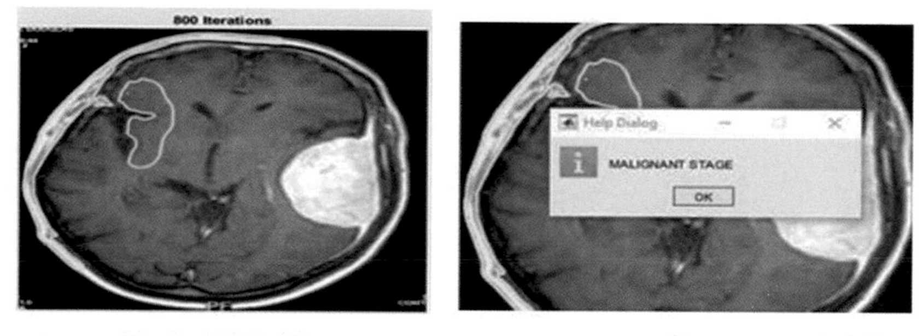

a.Feature Extraction

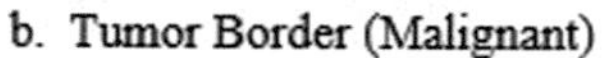
b. Tumor Border (Malignant)

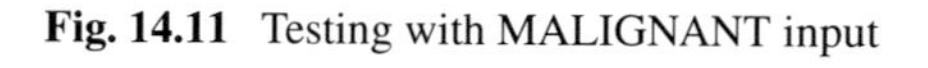
Fig. 14.11 Testing with MALIGNANT input

14.7 Conclusion

Technological advancement is being increased many folds for Brain tumor detection. Still researchers face many challenges for detection of brain tumor due to appearance of tumor like size, structure and life span. Among the existing tumor segmentation methods for detecting and analyzing the tumor using MR images, still challenges are there segmentation of unhealthy image and healthy images. In this proposed novel method of hybridization of K-Neighboring Algorithm and CNN (K-CNN-BTSL) is predicted the severity level of a brain tumor either benign or malignant at an early point with 93% accuracy. In Future work, improvements can be made in this work by classifying the tumors further into more specific types, like are gliomas, meningiomas, and pituitary tumors.[35]

References

1. Challa, S.K., Kumar, A. Semwal, V.B.: A for human activity recognition using wearable sensor data. Vis. Comput. (2021)
2. Jain, R., Semwal, V.B., Kaushik, P.: Deep ensemble learning approach for lower extremity activities recognition using wearable sensors. Expert Syst. e12743 (2021)
3. Semwal, V.B., Lalwani, P., Mishra, M.K. et al.: An optimized feature selection using biogeography optimization technique for human walking activities recognition. Computing (2021)
4. Benson, C.C., Lajish, V.L., Rajamani, K.: Brain tumor extraction from MRI brain images using marker based watershed algorithm. In: 2015 International Conference on Advances in Computing, Communications and Informatics (ICACCI), pp. 318–323. IEEE (2015)
5. Benson, C.C., Lajish, V.L.: Morphology based enhancement and skull stripping of MRI brain images. In: 2014 International Conference on Intelligent Computing Applications, pp. 254–257. IEEE (2014)
6. Huang, M., Yang, W., Yao, W., Jiang, J., Chen, W., Feng, Q.: Brain tumor segmentation based on local independent projection- based classification. IEEE Trans. Biomed. Eng. **61**(10), 2633–2645 (2014)
7. Semwal, V.B. et al.: Speed, cloth and pose invariant gait recognition-based person identification. In: Machine Learning: Theoretical Foundations and Practical Applications (2019)
8. Sapra, P., Singh, R., Khurana, S.: Brain tumor detection using neural network. Int. J. Sci. Mod. Eng. (IJISME) (2013). ISSN: 2319-6386
9. Jemimma, T.A., Jacob Vetharaj, Y.: Watershed algorithm based DAPP features for brain tumor segmentation and classification. In: 2018 International Conference on Smart Systems and Inventive Technology (ICSSIT), pp. 155–158. IEEE (2018)
10. George, D.N., Jehlol, H.B., Oleiwi, A.S.A.: Brain tumor detection using shape features and machine learning algorithms. Int. J. Adv. Res. Comput. Sci. Softw. Eng. **5**(10), 454–459 (2015)
11. Cui, B., Xie, M., Wang C.: A deep convolutional neural network learning transfer to SVM-based segmentation method for brain tumor. In: 2019 IEEE 11th International Conference on Advanced Infocomm Technology (ICAIT), pp. 1–5. IEEE (2019)
12. Rashid, M.H.O., Mamun, M.A., Hossain, M.A., Uddin, M.P.: Brain tumor detection using anisotropic filtering, SVM classifier and morphological operation from MR images. In: 2018 International Conference on Computer, Communication, Chemical, Material and Electronic Engineering (IC4ME2), pp. 1–4. IEEE (2018)
13. Abdulbaqi, H.S., Mat, M.Z., Omar, A.F., Mustafa, I.S.B., Abood, L.K.: Detecting brain tumor inmagnetic resonance images using hidden Markov random fields and threshold techniques. In: 2014 IEEE Student Conference on Research and Development, pp. 1–5 (2014)
14. Ali, A.H., Al-hadi, S.A., Naeemah, M.R., Mazher, A.N.: Classification of brain lesion using K-nearest neighbor technique and texture analysis. J. Phys. Conf. Ser. **01** (2018); Havaei, M., Davy, A., Warde-Farley, D., Biard, A., Courville, A., Bengio, Y., Pal, C., Jodoin, P.M., Larochelle, H.: Brain tumor segmentation with deep neural networks. Med. Image Anal. **35**, 18–31 (2017)
15. Nabizadeh, N., Kubat, M.: Brain tumors detection and segmentation in MR images: gabor wavelet versus statistical features. Comput. Electr. Eng. **45**, 286–301 (2015)
16. Supot, S., Thanapong, C., Chuchart, P., Manas, S.: Segmentation of magnetic resonance images using discrete curve evolution and fuzzy clustering. In: 2007 IEEE International Conference on Integration Technology, pp. 697–700 (2007)
17. Fletcher-Heath, L.M., Hall, L.O., Goldgof, D.B., Murtagh, F.R.: Automatic segmentation of non-enhancing brain tumors in magnetic resonance images. Artif. Intell. Med. **21**, 43–63 (2001)
18. Singh, A.: Detection of brain tumor in MRI images, using combination of fuzzy c-means and SVM. In: 2015 2nd International Conference on Signal Processing and Integrated Networks (SPIN), pp. 98–102. IEEE (20150
19. YasminM, Mohsin S., Sharif, M., Raza, M., Masood, S.: Brain image analysis: a survey. World Appl. Sci. J. **19**, 1484–1494 (2012)

20. Irum, I., Sharif, M., Raza, M., Yasmin, M.: Salt and pepper noise removal filter for 8-bit images based on local and global occurrences of grey levels as selection indicator. Nepal J. Sci. Technol. **15**, 123–132 (2014)
21. Pinto, A., Pereira, S., Correia, H., Oliveira, J., Rasteiro, D.M., Silva, C.A.: Brain tumour segmentation based on extremely randomized forest with high-level features. In: 2015 37th Annual International Conference of the IEEE Engineering in Medicine and Biology Society (EMBC), pp. 3037–3040. IEEE (2015)
22. Semwal, V.B., Mondal, K., Nandi, G.C.: Robust and accurate feature selection for humanoid push recovery and classification: deep learning approach. Neural Comput. Appl. **28**(3), 565–574 (2017)
23. Irum, I., Sharif, M., Yasmin, M., Raza, M., Azam, F.: A noise adaptive approach to impulse noise detection and reduction. Nepal J. Sci. Technol. **15**, 67–76 (2014)
24. Lu, X., Huang, Z., Yuan, Y.: MR image super-resolution via manifold regularized sparse learning. Neurocomputing **162**, 96–104 (2015)
25. Sawakare, S., Chaudhari, D.: Classification of brain tumor using discrete wavelet transform, principal component analysis and probabilistic neural network. Int. J. Res. Emerg. Sci. Technol. **1**(6), 2349–2761 (2014)
26. Mathew, A.R., Babu Anto, P.: Tumor detection and classification of MRI brain image using wavelet transform and SVM. In: 2017 International Conference on Signal Processing and Communication (ICSPC), pp. 75–78. IEEE (2017)
27. Sarhan, A.M.: Brain tumor classification in magnetic resonance images using deep learning and wavelet transform. J. Biomed. Sci. Eng. **13**(06), 102 (2020)
28. Arizmendi, C., Vellido, A., Romero, E.: Binary classification of brain tumours using a discrete wavelet transform and energy criteria. In: 2011 IEEE Second Latin American Symposium on Circuits and Systems (LASCAS), pp. 1–4. IEEE (2011)
29. Pratiwi, Mellisa, Harefa, Jeklin, Nanda, Sakka: Mammograms classification using gray-level co-occurrence matrix and radial basis function neural network. Procedia Comput. Sci. **59**, 83–91 (2015)
30. Badža, M.M., Barjaktarović, M.: Classification of brain tumors from MRI images using a convolutional neural network. Appl. Sci. **10**(6), 1999 (2020)
31. Simonyan, K., Zisserman, A.: Very deep convolutional networks for large-scale image recognition (2014). arXiv:1409.1556
32. Sawakare, S., Chaudhari, D.: Classification of brain tumor using discrete wavelet transform, principal component analysis and probabilistic neural network. Int. J. Res. Emerg. Sci. Technol. **1**(6), 2349–2761 (2014)
33. Mathew, A.R., Babu Anto, P.: Tumor detection and classification of MRI brain image using wavelet transform and SVM. In: 2017 International Conference on Signal Processing and Communication (ICSPC), pp. 75–78. IEEE (2017)
34. Sarhan, A.M.: Brain tumor classification in magnetic resonance images using deep learning and wavelet transform. J. Biomed. Sci. Eng. **13**(06), 102 (2020)
35. Arizmendi, C., Vellido, A., Romero, E.: Binary classification of brain tumours using a discrete wavelet transform and energy criteria. In: 2011 IEEE Second Latin American Symposium on Circuits and Systems (LASCAS), pp. 1–4. IEEE (2011)
36. Semwal, V.B. et al.: Pattern identification of different human joints for different human walking stylesusing inertial measurement unit (IMU) sensor. Artif. Intell. Rev. 1–21 (2021)
37. Dua, N., Singh, S.N., Semwal, V.B.: Multi-input CNN-GRU based human activity recognition using wearable sensors. Computing (2021)
38. Bijalwan, V., Semwal, V.B., Gupta, V.: Wearable sensor-based pattern mining for human activity recognition: deep learning approach. Ind. Robot.: Int. J. Robot. Res. Appl. (2021)
39. Abbasi, S., Tajeripour, F.: Detection of brain tumor in 3D MRI images using local binary patterns and histogram orientation gradient. Neurocomputing **219**, 526–535 (2017)
40. Sauwen, N., Acou, M., Sima, D.M., Veraart, J., Maes, F., Himmelreich, U., et al.: Semi-automated brain tumor segmentation on multi-parametric MRI using regularized non-negative matrix factorization. BMC Med. Imaging **17**, 29 (2017)

41. Ilunga-Mbuyamba, E., Avina-Cervantes, J.G., Garcia-Perez, A., de Jesus Romero-Troncoso, R., Aguirre-Ramos, H., Cruz-Aceves, I., Chalopin, C.: Localized active contour model with background intensity compensation applied on automaticMRbrain tumor segmentation. Neurocomputing **220**, 84–97
42. Ilunga-Mbuyamba, E., Avina-Cervantes, J.G., Garcia-Perez, A., de Jesus Romero-Troncoso, R., Aguirre-Ramos, H., Cruz-Aceves, I., Chalopin, C.: Localized active contour model with background intensity compensation applied on automaticMRbrain tumor segmentation. Neurocomputing **220**, 84–97
43. Akbar, S., Akram, M.U., Sharif, M.,Tariq, A., Khan, S.A.: Decision support system for detection of hypertensive retinopathy using arteriovenous ratio. Artif. Intell. Med. **90**, 15–24 (2018)
44. Banerjee, S., Mitra, S., Shankar, B.U.: Automated 3D segmentation of brain tumor using visual saliency. Inf. Sci. **424**, 337–353 (2018)
45. Raja, N.S.M., Fernandes, S.L., Dey, N., Satapathy, S.C., Rajinikanth, V.: Contrast enhanced medical MRI evaluation using Tsallis entropy and region growing segmentation. J. Ambient. Intell. Human Comput. 1–12 (2018)
46. Subudhi, A., Dash, M., Sabut, S.: Automated segmentation and classification of brain stroke using expectation-maximization and random forest classifier. Biocybern. Biomed. Eng. **40**, 277–289 (2020)
47. Gupta, N., Bhatele, P., Khanna, P.: Glioma detection on brain MRIs using texture and morphological features with ensemble learning. Biomed. Signal Process Control **47**, 115–125 (2019)
48. Myronenko, A., Hatamizadeh, A.: Robust semantic segmentation of brain tumor regions from 3D MRIs (2020). arXiv:2001.02040

Chapter 15
COVID-19 Detection in X-Rays Using Image Processing CNN Algorithm

Tilak Raj and Suraj Arya

Abstract The most generally utilized strategy for Covid-19 discovery is a continuous Polymerase Chain Response (RT-PCR). These kits for covid-19 detection are over-priced and they require an estimated time of hours or days to verify the disease. Because of the smaller affectability of the test, it gives higher deceptive outcomes. To detect this problem, X-rays of the chest and CT scan methods are utilized to evaluate Covid-19. Here chest X-Rays are chosen for model development. The reason to choose them is that the majority of clinics have X-ray equipment. The computerized investigation of Coronavirus was carried by using 371 X-ray images gathered from the Kaggle dataset to apply the Convolutional Neural Network algorithm. Some images were of infected patients of the Covid19 and others were of normal people. There was an acceptable prediction accuracy of 97.22%. In light of our discoveries, the model can help in making decisions for early Covid-19 identification.

Keywords Covid-19 · X-ray (chest) · Convolutional neural network · Machine learning · Deep learning

15.1 Introduction

Covid-19 infection rose in the year 2019 and was initially distinguished in the Wuhan city, China in December 2019 as having symptoms of pneumonia; an obscure virus spread a major symptom. Afterward, the worldwide council on Scientific Classification of Infections distinguished the main spreading specialist of Covid-19 as a unique form of Coronavirus, naming the variant as Severe Acute Respiratory Syndrome Coronavirus2 (SARS-CoV-2) [1].

The Coronavirus outbreak flared up quickly not only in China, but this virus engulfs the whole world. Hence, the World Health Organization reported it as a global epidemic on 12th March, 2020. The assessed number of affirmed cases are more than 519,748,323 and the number of deaths is more than 6 million, respectively,

T. Raj (✉) · S. Arya
Department of CS & IT (MCA), Central University of Haryana, Jant 123029, India
e-mail: tilakraj34048@gmail.com

A. Biswas et al. (eds.), *Artificial Intelligence for Societal Issues*, Intelligent Systems Reference Library 231, https://doi.org/10.1007/978-3-031-12419-8_15

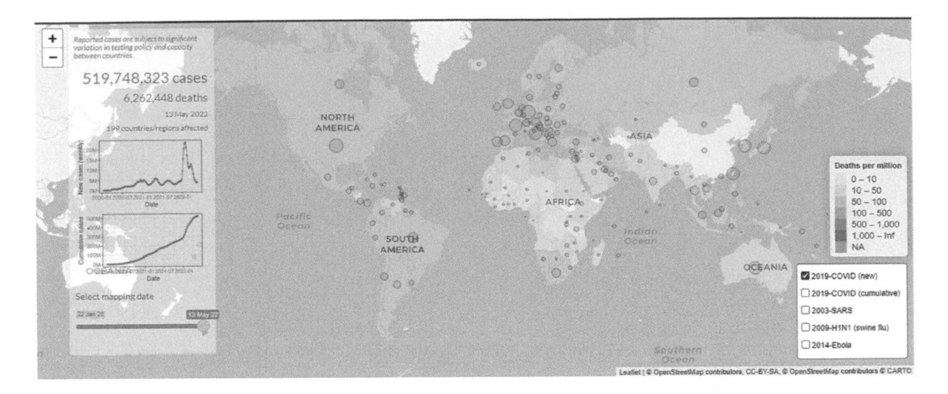

Fig. 15.1 Covid data worldwide

in 197 nations as of May 13, 2022 see Fig. 15.1 below for reference and for live data check citation [2, 3].

Covid-19 spread when a person inhale in contaminated air holding drops and minimal airborne molecule having the infection. The risk of inhaling these molecules is most critical when people are in closeness, however they can still breath these molecules over a distance, especially when they are indoors. Transmission of this virus can happen whenever the infected fluid is showered or sprinkled in the eyes, mouth or nose and sometimes through infected surfaces. A person can stay infectious for up to 20 days, and can spread the contamination whether or not they show side effects.

Few administrative majors were taken to control the risk of virus transmission. This includes travel restrictions, self-isolation, social distance restrictions, prohibitions on public transport, schools, and colleges. Some countries-imposed lock-down or curfew to halt the rapid spread of the corona infection. This had adverse effects on business, education, health, and the travel sector [1].

Effective screening of suspected infected individuals is one of the most important measures in containing the Covid-19 epidemic and preventing the contagion from spreading further among the population. The Nucleic Acid Detection test, Reverse Transcription Polymerase Chain reaction (RT-PCR) test, is a most precise evaluating approach to identify the new coronary pneumonia. Nasopharyngeal Swabs, Sputum, as well as blood and stool samples, and other lower respiratory tract secretions, are major samples used for covid detection. The RT-PCR test is an efficient approach for Covid-19 diagnosis. However, this is a difficult, time-consuming, and labor-intensive method [3, 4].

An alternate testing technique for COVID 19 screening was the radiography image assessment where x-ray images e.g. Chest x-ray (CXR) or Computed Tomography images are examined for a visual marker for SARS-CoV-2. Early investigations have revealed that patient chest x-beam irregularities that are common for people tainted with Covid-19 can be distinguished from regular person X-rays by an expert radiologist. The Fig. 15.2 below illustrates the chest X-ray depictions that have been

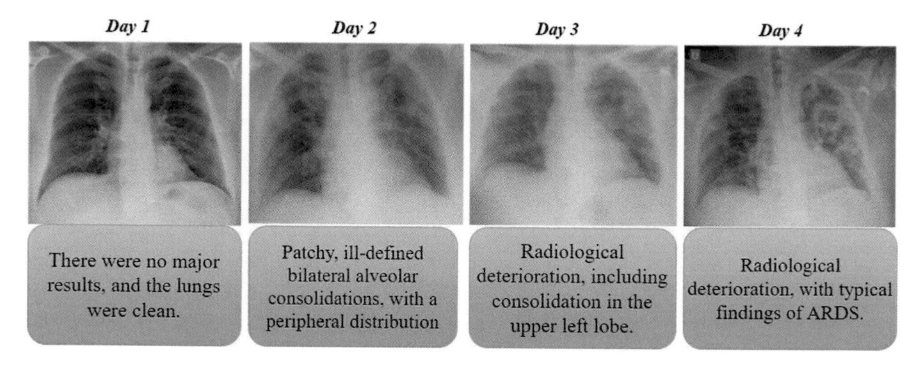

Fig. 15.2 Covid-19 patient who is aged 50 and is suffering from pneumonia

analyzed for over a week. It depicts a Covid-19 patient who is aged 50 and is suffering from pneumonia [5, 6].

Machine-Learning and Deep-Learning models are worked for the computerized finding of Covid-19 in this chapter. The trained model requires a raw image to deliver the result when passed through the well-built architecture. The proposed model is trained by using a total image of 371 chest X-rays, gathered promptly from trusted sources and in an unsorted structure.

15.2 Method and Materials

15.2.1 About X-Rays Dataset

For the research purpose in this study, X-ray images from the Kaggle source were utilized to diagnose Covid-19. Chest X-ray investigation is a form of the diagnostic procedures most cost-effective and accessible for detecting Covid-19. Public and indirect data gathering from the hospitals and clinicians is used to collect the data. Here, lung X-rays are used as it is the most influenced organ due to the virus. Fusic Fenta used images from various open and free platforms, that contained positive image and negative image cases. Fusic Fenta maintained the Covid-19 X-rays images database used in this research. There are currently 185 X-ray images in the database which are identified as Covid-19 confirmed cases and 186 X-ray images that are detected as a NORMAL case are present in the database.

The above Fig. 15.3 illustrate the Covid-19-pneumonia sample of a few pictures used from the dataset [7] for the training and validation of the deep learning CNN model. The covid X-rays might be viewed in a whitish tone or blurred in the scanned pictures. Lung gets affected due to the abundance of the angiotensin converting enzyme 2 (ACE2) enzyme [1].

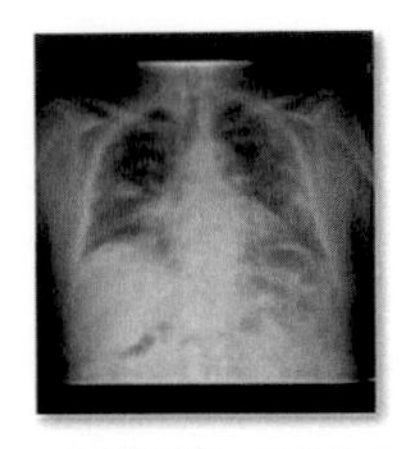

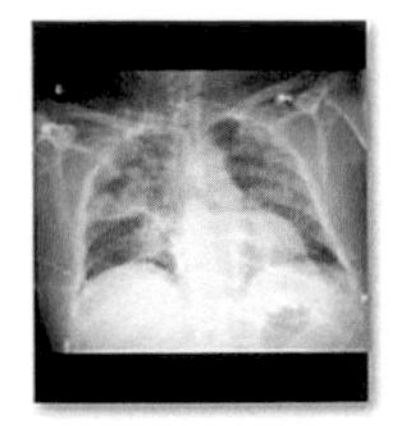

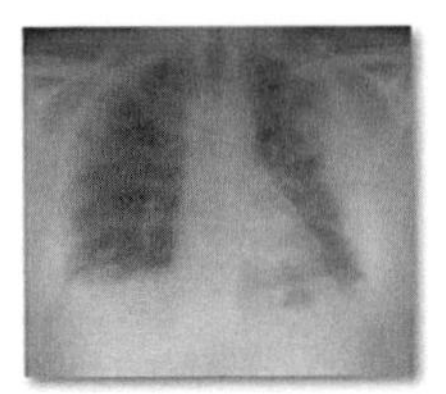

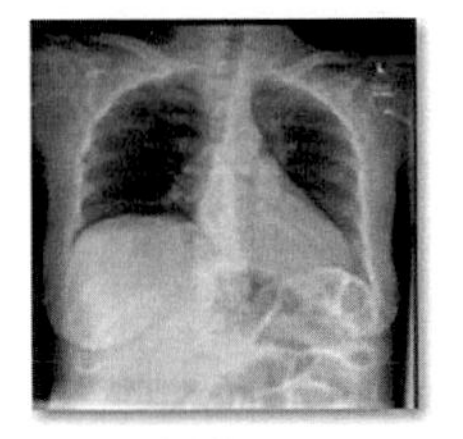

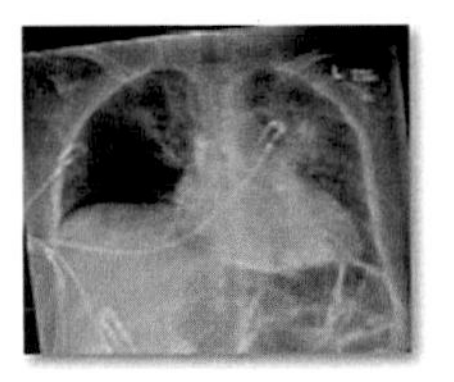

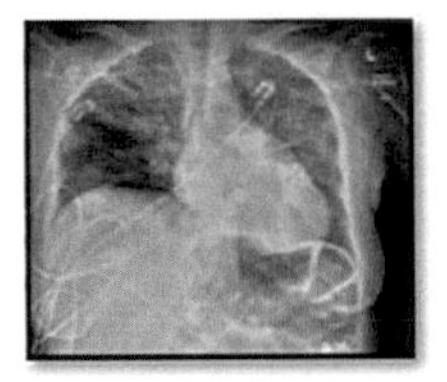

Fig. 15.3 Covid-19 image from the database

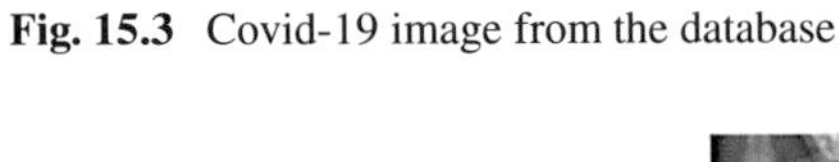

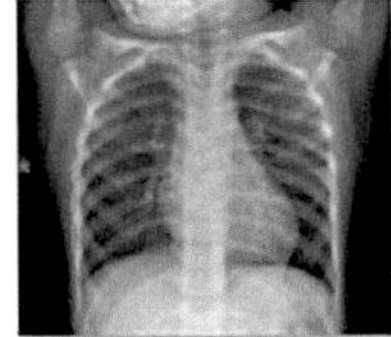

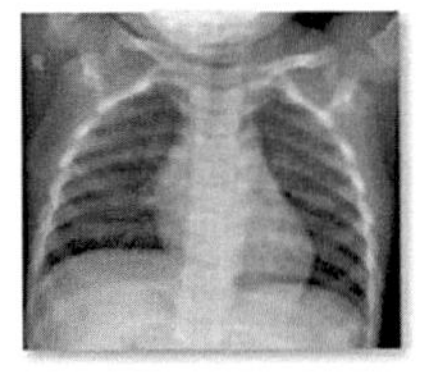

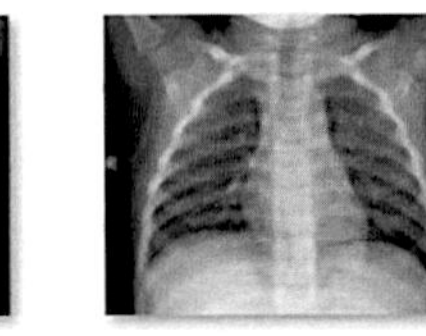

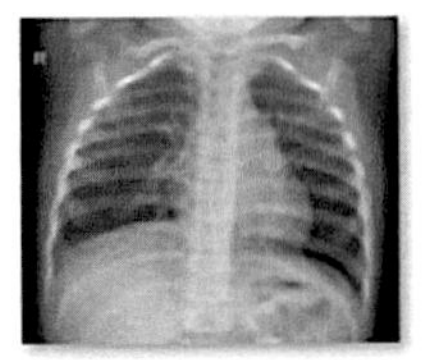

Fig. 15.4 Normal images from the used database

The Fig. 15.4 below exhibit the X-rays images of the normal person gathered from various sources and well-formatted by the Fusic Fenta over Kaggle [7] for the study purpose. Normal images have a clearer view of the lung rather than those infected by the Covid-19.

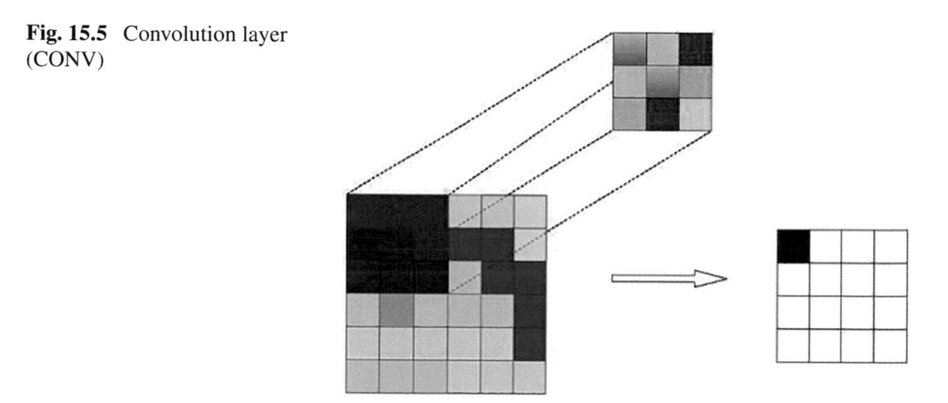

Fig. 15.5 Convolution layer (CONV)

15.2.2 CNN Architecture

Convolutional neural organizations (CNNs) are a kind of neural network architecture that works on deep learning methods. A CNN design consists of convolution and pooling layers at the alternate position, further connected with one or more fully connected last layers of the network [8, 9].

- **The Convolutional Layer**
  The convolutional layer is the best aspect of this architecture that identifies the presence of important features from the input. Figure 15.5 below depict the process where, we take a small matrix (kernel or filter), and move it over our image, and modify its values based on the value received from the filter. The following formula is used to calculate the feature value, where "f" and "h" is denoted as kernel and input images are written m and n are used to represent the index of the resultant matrix rows and columns [10].

$$G[m,n] = (f * h)[m,n] = \sum_j \sum_k h[j,k] f[m-j, n-k] \tag{15.1}$$

- **Pooling Layer**
  A Convolutional Layer is combined with a Pooling Layer. The main goal it exhibits is to diminish the extracted features, to cut down on computational expenses. Reducing relationships between layers and working freely on each characteristic helps achieving this. Some of the pooling techniques are explained below in Fig. 15.6.
  In Max Pooling, the more prominent number is brought from the feature matrix map. In average pooling, a predefined section is selected and average is determined. When the total sum of a marked area is calculated, it is called Sum Pooling. This layer acts as a link in between the Fully Connected Layer and the Convolutional Layer [11].

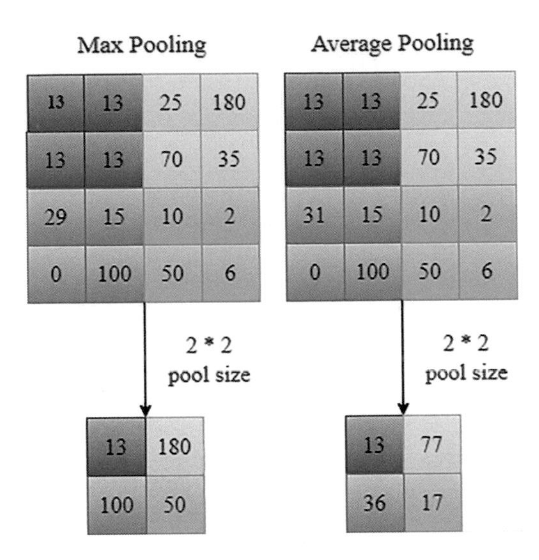

Fig. 15.6 Max pooling and average pooling

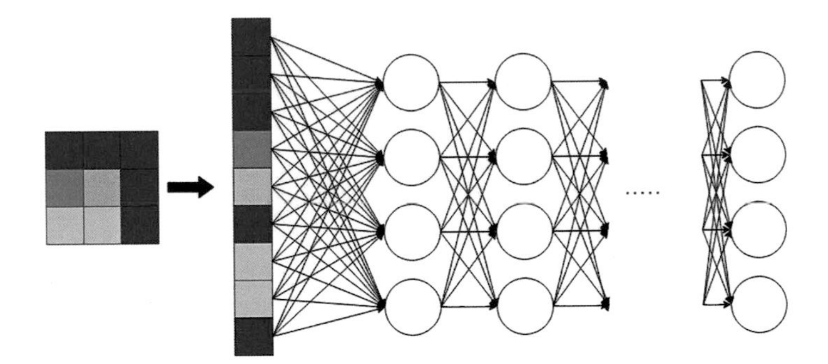

Fig. 15.7 Fully connected

- **Fully Connected Layer**
  The Fully Connected (FC) Fig. 15.7 comprises weights, biases together with the neuron which helps in connecting layers comprised of other neurons. The last few layers of CNN Architecture are built using the FC layers [12].
  The picture from the preceding levels is flattened in this layer and sent to the FC layer. Flattened vector then at that point goes through many more FC layers where the numerical operation on component happens. At this stage, the grouping of similar kinds of classification starts to occur.
- **Dropout**
  The training dataset used is liable to overfitting when all elements are connected to the FC layer. Overfitting occurs when a model's execution on the training data is great to such an extent that it influences the model's outcome when applied to new data.

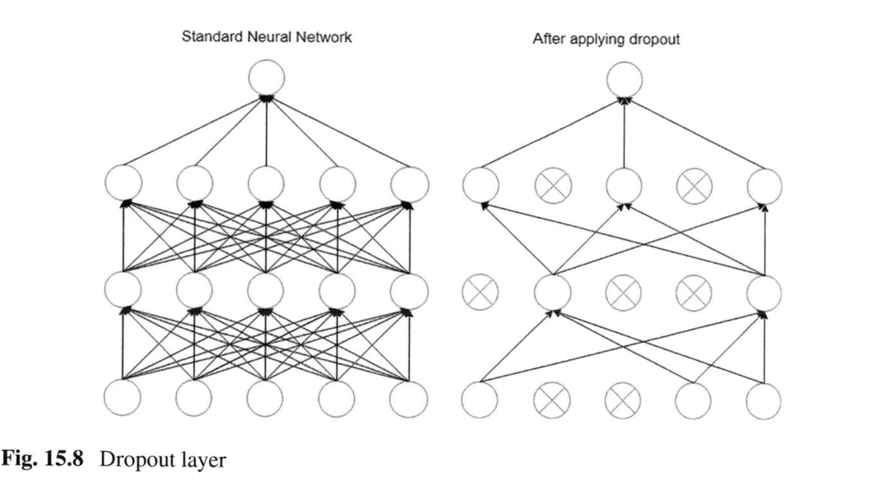

Fig. 15.8 Dropout layer

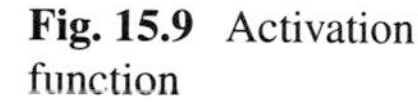

Fig. 15.9 Activation function

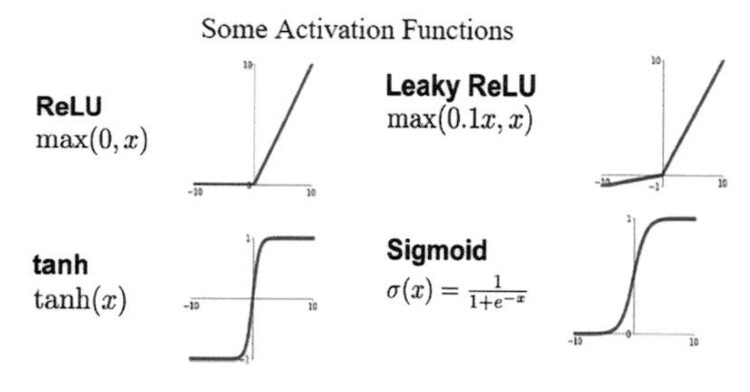

To overcome this difficulty, check Fig. 15.8 the dropout layer is utilized, which reduces the size of a model and depresses certain neurons from the neural network during training [12].

- **Activation Layer**
  The activation function is one of the key contributors to the CNN model. It is utilized for learning and approximating a wide range of complicated connections between network variables. Basically said, it picks which model data should be passed further and which should not pass at the conclusion of the network. It makes the network non-linear. Several popular activation functions are depicted below in Fig. 15.9 are ReLU, tanH and Sigmoid functions. Every one of these has a particular use. Here, Sigmoid functions for a CNN binary classification model are chosen [13].

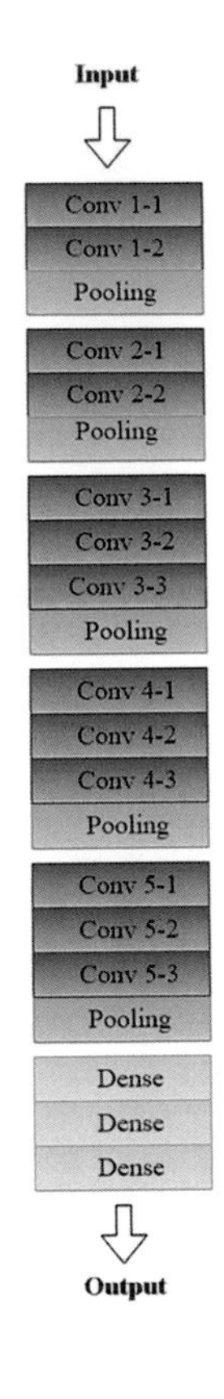

Fig. 15.10 VGGNET

15.2.3 Basic Requirement

Python, is a high-level programming language, which is utilized to train the proposed deep learning model. It comprises of countless libraries that can help us in this model building. Deep Learning can be code in python and it is the best technique to implement machine learning models.

This project is implemented on the local-machine with an i7-3520M Intel(R) Core(TM) processor clocked @ 2.90GHz, 8GB ram. Software: Window 10 pro version 20H2, Jupyter Notebook, Anaconda, and necessary required Python libraries like(Numpy, Pandas, Plotly, Matplotlib, Tenserflow, Scikit-learn, Keras, Seaborn and etc.).

15.3 Methodology

The "VGGNET" neural network for predicting Covid-19 from radiation is the Deep Learning approach utilized in this study [14]. The network's input is a picture of aspect (224, 224) RGB. The Fig. 15.10, initial convolutional layers 1 and 2 have

64 channels and filter size of 3*3 and same padding. Following to that a maximum pool layer of 2*2 stride, filter size 256 and of 3*3 filter size two convolution layers are added. After this there are three grouping of 3 convolution-layer each having 512 filter of 3*3 size with same padding together and a maximum pool layer. The test photo is then sent into a convolution layer heap of two. The filters used in this convolution and max pooling layers are 3*3. After this the heap of convolution-layer and max-pooling layer. Following that there are 3 fully connected-layer, which we will use to flatten the result to make it a feature-vector. The primary layer takes data from the previous feature vector and outputs a vector and so on. Then the output of third fully connected layer is passed to the softmax layer in order to normalize the classification vector.

Explaination of the workflow

This is the explanation of the workflow of the proposed model Fig. 15.3. X-ray chest pictures are obtained from kaggle, where images from numerous sources are combined to produce the dataset. The collection contains a total of 371 pictures combined of covid-19 positive and negative people. Training set and Validation set are divide by the ratio of 80%, 20% and 23 image of normal and positive peoples are collected separated for the further prediction and proper evaluation of our model.

At first, the model is trained and evaluate by using the necessary data. The data is pre-processed, so that it consume less computing power. The trained image and validation image are resized to the particular size here (244, 244) image size is chosen. Data augmentation is done where important features are extracted from the supplied images. These feature are extracted by the CNN model classifier used in the workflow. Here epoch value is chosen as "10", thus the entire data set is passed back and forth through the neural network (Fig. 15.11).

It can consume time according to your database size, if any error is encountered then the process of feature extraction has to be restarted by solving the encountered error and if the epoch step are completed successfully then our final working model is completely prepared and ready for the test purpose. Finally, the model is trained and now its time to test its efficiency and prediction result. The model efficiency, accuracy, true positive rate, false positive rate etc. can be calculated by using the confusion matrix extracted from the model. For making prediction via the proposed model, the image reside in the prediction data-set is utilized. One random image is chosen and then the image resizing is done making it suitable for the further use. The image get reduced of the desired size suggested by you for the model to work efficiently. The image is transfer to the CNN Covid-19 classifier. This classifier consist of various type of layers like Convolution2D, Max Pooling2D, Dropout layer and at the end these layers are further extended by using the Flatten, Dense layers which is the part of the fully connected layer.

After all this process the CNN model internally do lot of the processing and convert the user entered image to the array of one dimension. Further reduction of

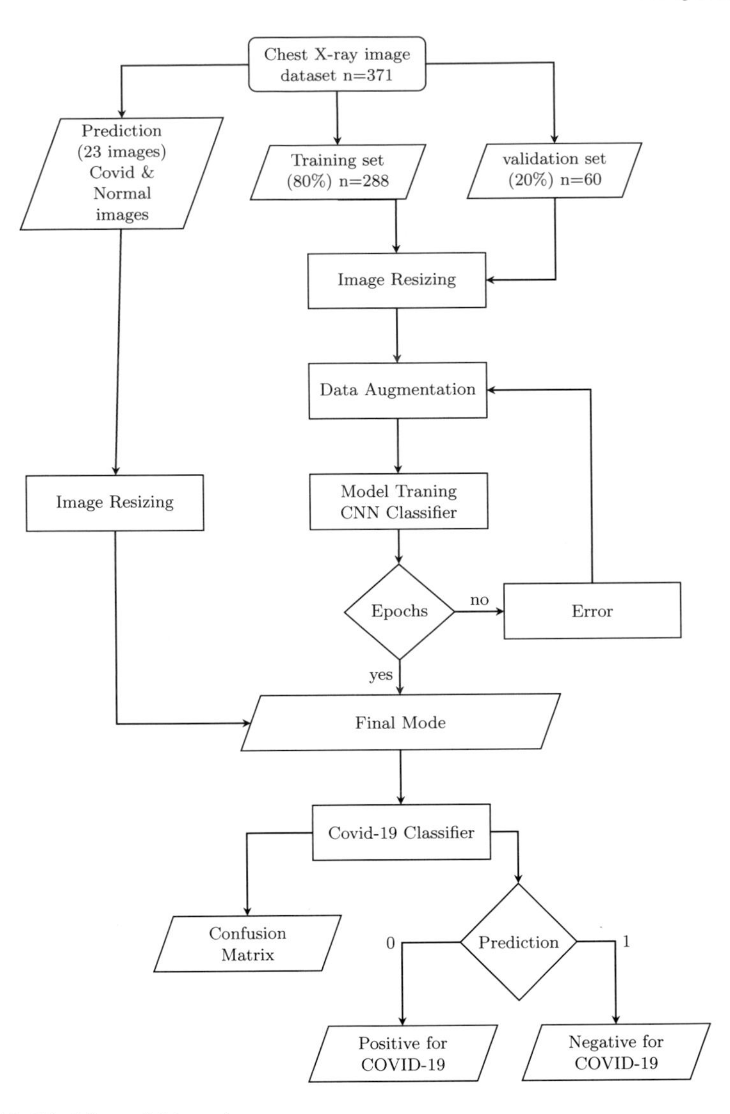

Fig. 15.11 Workflow of this study

the array the result of the solved array came out be in the binary form either "0" or "1". If the result of the array after solving came out to be "0" then the user entered image for the Covid-19 detection is the positive image and if the result of the outcome be "1" then the user entered image outcome to be Negative for the Covid-19(normal) (Table 15.1).

In the Convolutional Neural Network, the kernel acts as a filter used to gather the features from the images. A kernel is a form of a matrix that covers the input data and proceeds with the dot product, given data in a particular region and the result is

Table 15.1 Model summary of used trained CNN model

Type	Kernel size	Kernel	Shape
Convolution2D	3 × 3	32	222, 222, 32
Convolution2D	3 × 3	64	220, 220, 64
Max Pooling2D	2 × 2	–	110, 110, 64
Dropout (0.25)			110, 110, 64
Convolution2D	3 × 3	64	108, 108, 64
Max Pooling2D	2 × 2	–	54, 54, 64
Dropout (0.25)			54, 54, 64
Convolution2D	3 × 3	128	52, 52, 128
Max Pooling2D	2 × 2	–	26, 26, 128
Dropout (0.25)			26, 26, 128
Convolution2D	3 × 3	128	24, 24, 128
Max Pooling2D	2 × 2	–	12, 12, 128
Dropout (0.25)			12, 12, 128
Flatten			18432
Dense			64
Dropout			64
Dense			1

gathered as the matrix of the dot. The Stride value is used over the input data to move the kernel. Two columns of pixels follow the kernel shift in input matrix, in case the stride value is 2. In an easy term, a kernel is used to remove high-level attributes from the image. The images are prepared by using ".flow_from_directory()" to produce batches of image data with their labels directly from the given jpeg in their respective folders [15].

For Train_datagen: shear_range= 0.2, rescale= 1./255, horizontal_filp= True, zoom_range=0.2

For Train_generator: target_size=(224,224), class_mode='binary', batch_size=32

rescale is the value that is used to multiply the data before carrying any processing. The pictures used here contain 0–255 RGB coefficients, but this value is far more complicated to handle by our model. Thus cut the value down in between 0 and 1 by 1/255 scaling factor.

- Zoom_range is used for zooming the pictures.
- Shear_range is applied for shearing transformations.
- Horizontal_flip is applied for flipping half part of images horizontally [16].

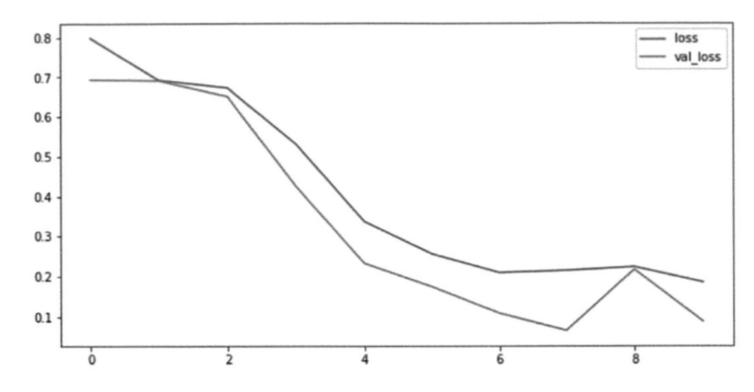

Fig. 15.12 loss versus val_loss after epoch

15.4 Experimental Analysis

Experiment Result

These generators may now be used to train our model. An epoch takes an estimate of 20–30s on GPU and an estimate of 300–400s on CPU. You can run this model on a CPU if you have the required processing time limit. Here, the Epoch value of 10 is processed [17].

One epoch is calculated as the training vectors once used for updating the weights. In this project, the use of sequential training where weights get updated after each training vector is successively gone through the preparation calculation. The accuracy of 0.9317 is attained, which is decent. This is partly because the model was trained on a dataset that already contains Covid-19 and normal images in a well-structured format.

When you train a model, the data related to loss and accuracy-related to our model. As loss and val_loss is tend to be nearly the same and also if the values are converging the are model is good.

The above plot Fig. 15.12 is obtained by training loss and validation loss after the epoch process is completed the obtained result of the Loss = 0.1397 and that of Val loss = 0.0874(approx). Now consider it to be a perfect fit as the value of loss and val_loss are nearly similar.

The above plot Fig. 15.13 is obtained by training accuracy and validation accuracy after the epoch process is completed the obtained result of the Accuracy= 0.9722 and that of Val_accuaracy = 1.0000.

The below graph Fig. 15.14 depicts the model accuracy and Fig. 15.15 model loss which was obtained after the epoch process was carried out. Model accuracy defines how well the trained model performs in the real scenario and how accurate its result can be output: train_accuracy = 1.0000, test accuracy = 0.97. The losses on training and validation are computed and interpreted on how well the model performs in the course of these training and test data sets output: train_loss = 0.2, test_loss = 0.1. The graph lines depict how well the accuracy and loss are implemented in

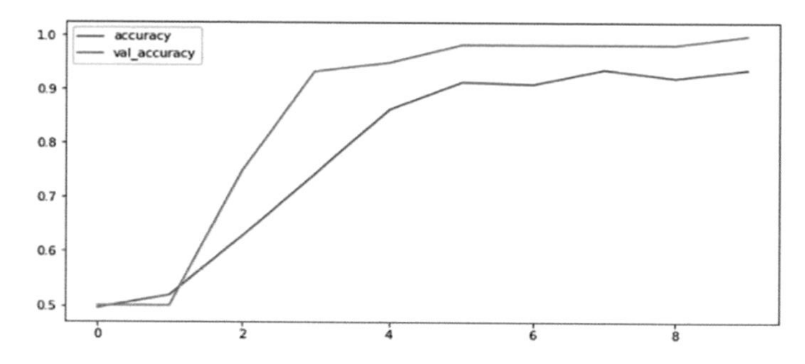

Fig. 15.13 loss versus val_loss after epoch

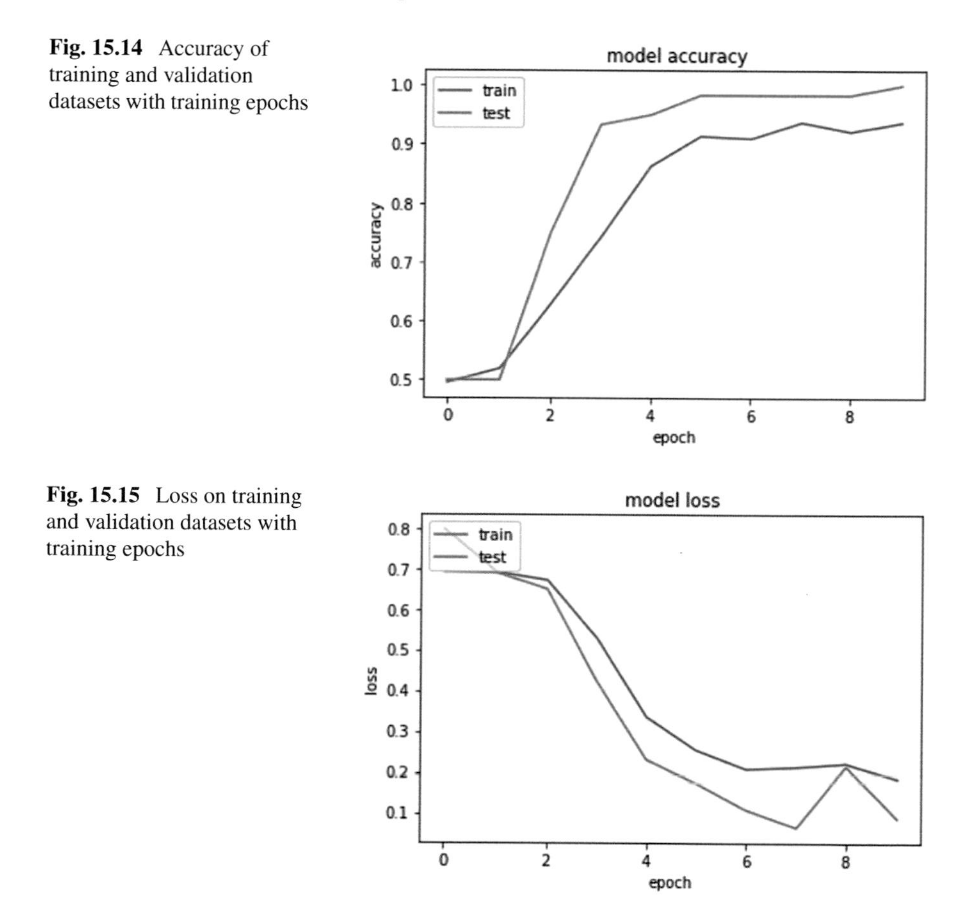

Fig. 15.14 Accuracy of training and validation datasets with training epochs

Fig. 15.15 Loss on training and validation datasets with training epochs

the model. As there is a gap between the line of train and test thus the overfitting is reduced. If these lines merge or coincide to a greater extent then our model can become overtrained and false results can be provided.

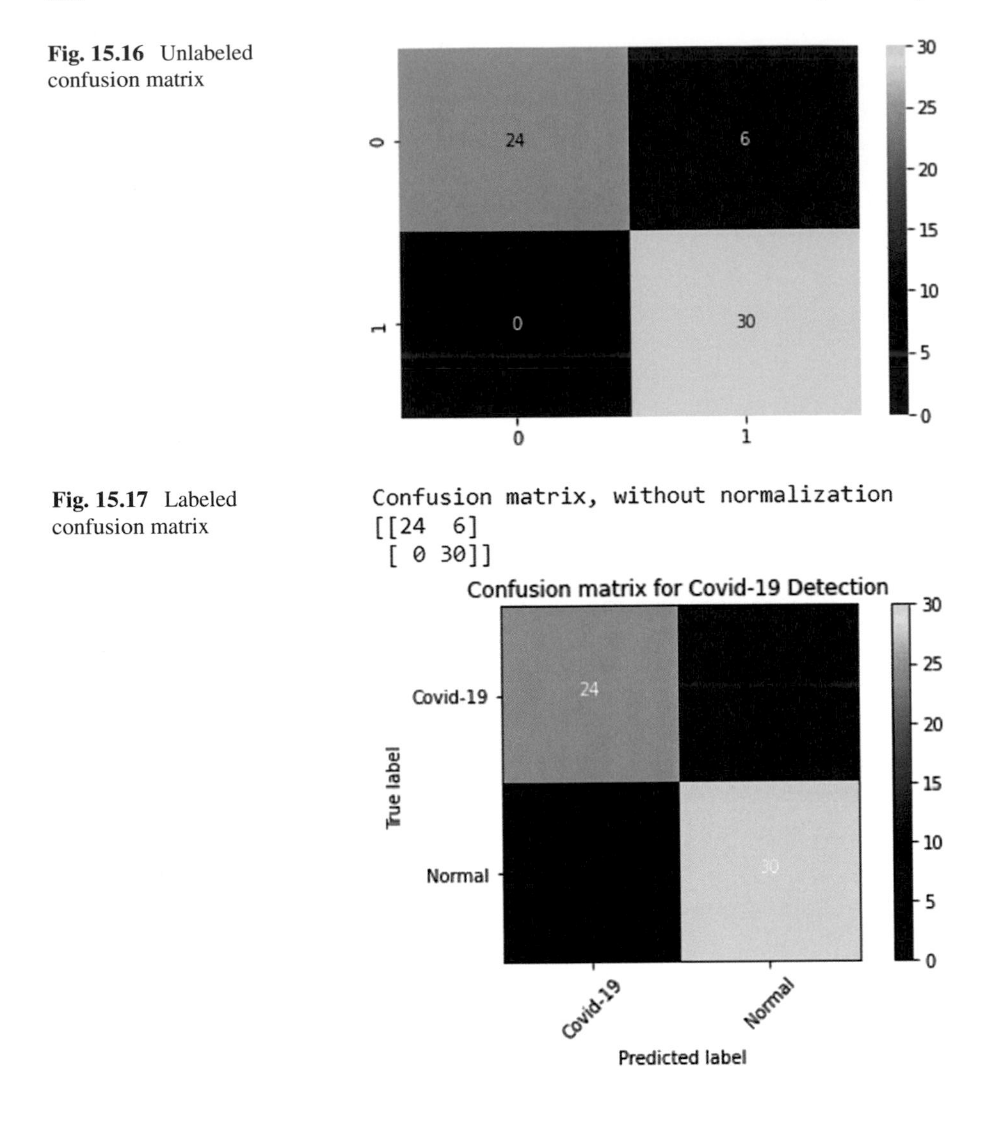

Fig. 15.16 Unlabeled confusion matrix

Fig. 15.17 Labeled confusion matrix

Performance Matrix

To measure the performance of our model, these matrices Fig. 15.16 are used: TP is represented as the accurate (Covid-19) case. The system misunderstands FP. It refers to the (normal cases). FN refers to Covid-19 instances wrongfully categorized as normal cases; TN refers to cases correctly labeled as normal cases [18].

That's the list of values that are commonly calculated from a binary classifier confusion matrix from Fig. 15.17 are-

Total= TP+FP+FN+FP= 24+6+0+30= 60

Actual Yes= TP+FN= 24+0= 24

Actual No= FP+TN= 6+30= 36
Predicted Yes= TP+FP= 24+6= 30
Predicted No= FN+TN= 0+30= 30

- Accuracy: (TP+TN)/total = (24+30)/60 = 0.9
- Misclassification Rate: (FP+FN)/total = (6+0)/60 = 0.1
- True Positive Rate: TP/actual yes = 24/24 = 1
- False Positive Rate: FP/actual no = 6/36 = 0.1666
- True Negative Rate: TN/actual no = 30/30 = 0.8333
- Precision: TP/predicted yes = 24/30 = 0.8
- Prevalence: actual yes/total = 24/60 = 0.4

Now that all the processing is done, and our model is done with the training, Now the pictures URL are provided in the field of image path as required by our model and load the image into our model for future processing and prediction. The image is loaded into the model with the particular size (224, 224). Then the user−entered image is converted into an array of numbers because understand an image is not an easy task for the computer so the array of numbers is used for their understanding.

The value of number at each index is reduced this is done by dividing the number at particular index with "224" our provided number in order to lower the value for better understanding and using resource and time efficiently. "img.shape" can retrive the shape of an image. It give back the tuple of the columns, number of rows, and channels.

Model.predict is a Keras class that can be used to make trends, predictions and it most probably gives the highest probability value. The obtained answer gets stored in the result variable. The result is in the binary form whether the input image is Covid-19 positive or negative and the association related to it is "0 for true" and "1 for false". After all, the processing is done the result is checked with the highest probability obtained. If the result of the array is equal to the "1" then the outcome of the model is "Negative for Covid-19" see Fig. 15.18 otherwise the result would be "Positive for Covid-19" see Fig. 15.19 when the array result is equal to "0" [19].

15.5 Discussion

15.5.1 Some Issues Handled by Deep Learning

Deep Learning is the main part of the artificial intelligent,and its the most fascinating technology at present. The approaches of ML and DL techniques have produced wide solutions in the medical field and are currently used to fight against covid-19 [20].

- Diagnosis of Covid-19
  - The team of researchers at Alibaba Research Academy has developed the AI-based resolution where they have achieved 96% of the accuracy. This network

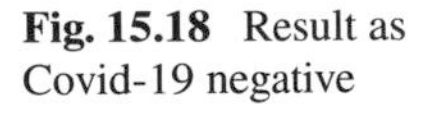

Fig. 15.18 Result as Covid-19 negative

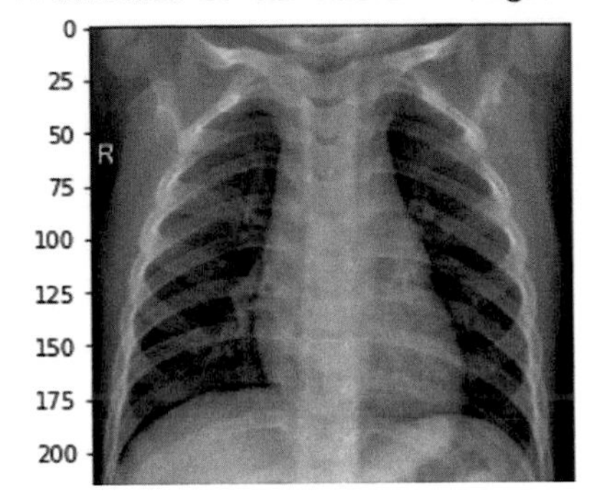

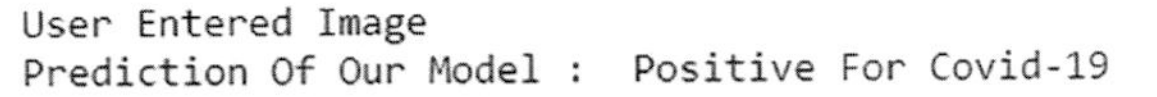

Fig. 15.19 Result as Covid-19 positive

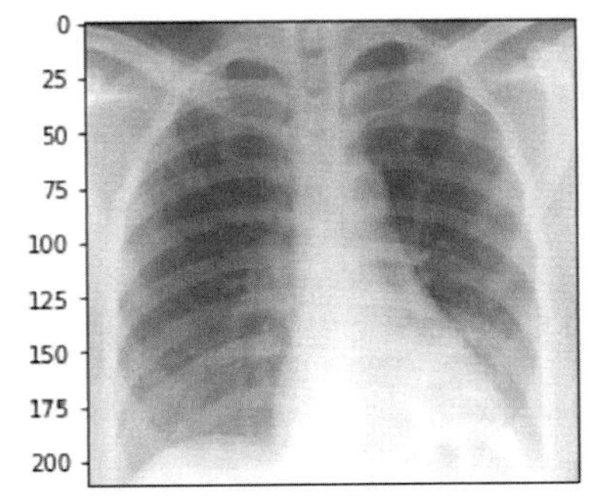

acquires the CT and X-rays scans of both the covid-19 positive and negative individuals. These systems are widely used by various hospitals in China already.

- AI can be used for drug screening. Many researchers in china find out how deep learning can assist in discovery of medicine and other investigation related to drug screening. Various investigators were in a position to discover possible covid-19 antibodies through the drug screening.

- Classification
  - The researcher in the field of health found that the people who experience this infected virus show some respiratory imperfection. This led the scientist and organizations to come together and create algorithms that can help in screening and predicting the infection based on breathing problems.

- Forecasting
  - The establishment of disease surveillance systems can help to restrict the illness's spread. The creation of this forecasting model can help the government to

predict the area where virus cases can rise, thus give an upper hand to control the spread of the covid by taking effective majors. Forecasting method for disease spread prediction is resolved by the use of Deep Learning.

15.5.2 Advantage of the Proposed Model

In deep learning, a model was trained directly from the noise, visual or text to complete tasks and may reach very high precision, often greater than performance on the human level. A system can learn by using the deep learning for evaluating COVID-19 kind of radio-logical image. Data processing minimizes the very negative errors and false-positive in the detection and identification of this sickness and gives a pretty unique potential to particularly serve patients with ease, rapid, affordable, and safe diagnostic options [21].

- RT-PCR kits are definitely costly hence a generally cheap and affordable system for Covid-19 detection basically is needed, thus this system literally helps to specifically detect Covid infection via X-rays images in a subtle way.
- Using chest X-ray pictures for diagnosis of COVID-19 should generally be advised as X-rays are rapidly obtained and at very low-cost particularly nearby hospitals or clinics in a fairly big way.
- The proposed model in this chapter can particularly help to kind of understand the CNN classifier over the X-rays image and by using this work various kind of other architectures can be implemented effectively.
- Anyone around the world can use this system (if available online publicly) for covid detection only computer and internet kind of is required, which is quite significant.
- The efficiency of these models can be increased over time by using sort of more feasible data for prediction which can easily essentially surpass the radiologist visual Covid-19 detection technique with the use of various scan in a major way.
- It can help us in the early detection of coronavirus, which can affect the covid chain and generally help us to literally break it easily in a big way.
- This model can reduce the work pressure on the clinical staff and they can focus on actually other seriously sort of ill patients.

15.6 Conclusion and Future Direction

As Covid-19 cases are expanding every day, numerous nations are confronting asset deficiencies. Henceforth, it is important to recognize every positive case during this crisis.

The accuracy of classification generally denotes the performance of the classification algorithm. The research shows that artificial neural networks and convolutional neural network algorithms may be employed and good results from classification algorithms VGG16 can be produced. In this study, it was noticed that Covid-19, which is a life-threatening disease, was classified with an accuracy of 97.22% using Convolutional Neural Network (VGG16) algorithms utilizing X-ray scans of the patients, thus depicting the performance of the algorithm for early recognition of Covid-19.The outcomes propose that the CNN based architectures have the ability for the correct recognition of Covid-19 infection.

This model can further be transformed in a precise way to make efficient and accurate prediction thus help to reduce the stress and responsibility on the clinical staff, this proposed model can be used to foster the device for the early discovery of Covid-19 disease. We can enhance the accuracy further by increasing the size of the positive and negative images in the database and different architectures of CNNs like: ResNet, ZFNet and GoogLeNet etc. can be used to test the reliable architecture which can provide better result. At, last we can use the model as back-end to develop the user interface and that application can be hosted online for public use across world-wide. In future we can work on developing the new architecture model based on the CNNs for the identification of COVID-19 and other diseases in the medical domain.

References

1. Wikipedia contributors. COVID-19. Wikipedia. Accessed July 2021, from https://en.wikipedia.org/wiki/COVID-19
2. Covid-19 tracker. (n.d.). Retrieved 9 Jan 2022, from https://vac-lshtm.shinyapps.io/ncov_tracker/
3. WHO Coronavirus (COVID-19) Dashboard. World Health Organization. https://covid19.who.int/
4. Giri, B., Pandey, S., Shrestha, R., Pokharel, K., Ligler, F.S., Neupane, B.B.: Review of analytical performance of COVID-19 detection methods. Anal. Bioanal. Chem. **413**(1), 35–48 (2021). https://doi.org/10.1007/s00216-020-02889-x
5. New-ai-diagnostic. Debuglies (2020). https://debuglies.com/2020/05/12/new-ai-diagnostic-can-predict-whether-someone-is-likely-to-have-covid-19-based-on-their-symptoms/
6. Bell, D.J.: COVID-19 | Radiology Reference Article | Radiopaedia.org. Radiopaedia. https://radiopaedia.org/articles/covid-19-4
7. Kaggle.com. 2022. Chest Xray for covid-19 detection. [online] Available at: https://www.kaggle.com/fusicfenta/chest-xray-for-covid19-detection/. [Accessed 20 Nov 2021]
8. Ozturk, T., Talo, M., Yildirim, E.A., Baloglu, U.B., Yildirim, O., Acharya, U.R.: Automated detection of COVID-19 cases using deep neural networks with X-ray images. Comput. Biol. Med. **121**, 103792 (2020)
9. Sun, Y., Xue, B., Zhang, M., Yen, G.G., Lv, J.: Automatically designing CNN architectures using the genetic algorithm for image classification. IEEE Trans. Cybern. **50**(9), 3840–3854 (2020)
10. Skalski, P.: Gentle Dive into Math Behind Convolutional Neural Networks (2019). Online: https://www.towardsdatascience.com/gentle-dive-into-math-behind-convolutional-neural-networks-79a07dd44cf9. Accessed 5 May

11. CS 230—Convolutional Neural Networks Cheatsheet. Stanford. https://stanford.edu/~shervine/teaching/cs-230/cheatsheet-convolutional-neural-networks
12. Brownlee, J.: Dropout Regularization in Deep Learning Models With Keras. Machine Learning Mastery (2020).https://machinelearningmastery.com/dropout-regularization-deep-learning-models-keras/
13. Zhao, W., Fu, H., Luk, W., Yu, T., Wang, S., Feng, B., Yang, G.: F-CNN: An FPGA-based framework for training convolutional neural networks. In 2016 IEEE 27Th International Conference on Application-Specific Systems, Architectures and Processors (ASAP), pp. 107–114. IEEE (2016)
14. Umer, M., Ashraf, I., Ullah, S., Mehmood, A., Choi, G.S.: COVINet: a convolutional neural network approach for predicting COVID-19 from chest X-ray images. J. Ambient Intell. Human. Comput. 1–13 (2021)
15. Rahimzadeh, M., Attar, A.: A modified deep convolutional neural network for detecting COVID-19 and pneumonia from chest X-ray images based on the concatenation of Xception and ResNet50V2. Inf. Med. Unlock. **19**, 100360 (2020)
16. Team, K.: Keras documentation: models API. Keras. https://keras.io/api/models/
17. Sarin, S.: Exploring data augmentation with keras and tensorflow. Medium (2021). https://towardsdatascience.com/exploring-image-data-augmentation-with-keras-and-tensorflow-a8162d89b844
18. Shiddieqy, H.A., Hariadi, F.I., Adiono, T.: Implementation of deep-learning based image classification on single board computer. In: 2017 International Symposium on Electronics and Smart Devices (ISESD), pp. 133–137. IEEE (2017)
19. Panwar, H., Gupta, P.K., Siddiqui, M.K., Morales-Menendez, R., Singh, V.: Application of deep learning for fast detection of COVID-19 in X-Rays using nCOVnet. Chaos, Solitons Fractals **138**, 109944 (2020)
20. Nayak, J., Naik, B., Dinesh, P., Vakula, K., Dash, P. B., Pelusi, D.: Significance of deep learning for Covid-19: state-of-the-art review. Res. Biomed. Eng. 1–24 (2021)
21. Ghaderzadeh, M., Asadi, F.: Deep learning in the detection and diagnosis of COVID-19 using radiology modalities: a systematic review. J. Healthcare Eng. (2021)

Chapter 16
Black Fungus Prediction in Covid Contrived Patients Using Deep Learning

Mohammad Abdul Hameed, Mohammad Safi Ur. Rahman, and Ayesha Banu

Abstract The COVID-19 has impacted many nations across the world, disrupting lives, economies and societies. The Post covid complications are much worse. Now a spike of black fungus is being observed among individuals hospitalized for or recovering from COVID19 infection. Black Fungus can even be life-threatening and fatal if not diagnosed and treated in its initial stages. Deep Learning has been revolutionizing in the Health sector and has path-breaking applications. The proposed research study aims at anticipating black fungus in covid affected patients majorly through deep learning. The Dataset used in this study has attributes related to covid19, the presence of any comorbidities like cancer or diabetes, medical treatment given to patient for covid, and the location of mucormycosis in the body. Popular deep learning methods such as RNN, CNN and ANN were used in this research. SVM (Support Vector Machine) which is also an efficient supervised Machine Learning Algorithm has also been used. The RNN (Recurrent Neural Networks) algorithm showed superior accuracy in predicting black fungus comparatively over other algorithms. This study provides a wide overview of the relative performance of different deep learning algorithms for black fungus prediction. Based on the covid treatment taken by the patient and his comorbidities, this study can help doctors by predicting patients who are more vulnerable to get infected by black fungus. Such Patients can be kept under supervision and special care can be taken. This relative performance of algorithms is often used to assist researchers in the selection of an appropriate deep learning algorithm for their studies.

M. Abdul Hameed (✉)
Department of Computer Science Engineering, Osmania University, Hyderabad, India
e-mail: hameed@gmail.com

M. S. Ur. Rahman
Department of Civil Engineering, National Institute of Technology, Warangal, India

M. S. Ur. Rahman · A. Banu
Department of Computer Science Engineering, Vaagdevi College of Engineering Warangal, Warangal, India

A. Biswas et al. (eds.), *Artificial Intelligence for Societal Issues*, Intelligent Systems Reference Library 231, https://doi.org/10.1007/978-3-031-12419-8_16

Keywords Deep learning · Machine learning · Black fungus · Covid 19 · Health care

16.1 Introduction

The COVID-19 coronavirus disease has been considered as humanity's deadliest calamity since World War II. For many years, the social, political and financial effects of this epidemic will be felt. It's been less than a year since the globe was struck by the terrible Covid-19 virus, and now another sickness has emerged that affects those who are infected with Covid19 or who have recovered from Covid. This Black Fungus appears to be more hazardous, with a far higher fatality rate than Covid-19. Mucormycosis, commonly known as Black Fungus, is a deadly fungal infection produced by moulds called mucormycetes [1]. Molds can be found all over the place. Mucormycosis is most common in those with health problems, comorbidities, or those who use drugs that impair the body's ability to fight infection and sickness. After inhaling due to fungus spores in the air, it commonly affects the sinuses or lungs. It can also appear on the skin after a cut, a burn, or any type of skin injury. There are 5 types of Mucormycosis classified on the basics of the body location they attack to namely Rhinocerebral (sinus and brain) mucormycosis [2, 3], Pulmonary (lung) mucormycosis, Gastrointestinal mucormycosis [4, 5], Cutaneous (skin) mucormycosis and Disseminated mucormycosis. A presumptive diagnosis is made based on the patient's medical history, medical examination, and hence the patient's factors associated for a fungal infection. It's tough to make a clear diagnosis. There are no tests for serology or blood that can help. In the presence of specific tissue stains looking for characteristic structural components, the growth of fungi from an infected tissue biopsy (tissue acquired by surgical removal or endoscopes with biopsy equipment) may assist identify the fungus and aid in the ultimate diagnosis. Although tests like CT or MRI scans can assist determine the amount of infections or tissue loss, they aren't specific for mucormycosis. Mucormycosis can cause severe blindness, organ malfunction, bodily tissue loss owing to infection, debridement and death. Mucormycosis treatment should be swift and aggressive, since by the time a tentative diagnosis is made, the patient has frequently sustained irreversible tissue damage. This simple fungus is visible in its early stages, but its lethal consequences appear quickly. When it comes to treating black fungus, medications have two major objectives. Antifungal drugs are used to stop the spread of the fungus, as well as medications to treat any underlying conditions that may be debilitating. Patients who have been using steroids should probably cease taking them since they promote the survival of fungus in the body. Antifungal medication is frequently required for a long time for patients infected with black fungus, depending on the severity of the condition. The examination of CT scan reports were used to try to detect the disease. This strategy necessitated expert doctors manually classifying the photos based on previous experiences. When the number of cases gradually increased, this strategy proved to be useful. However, due to the fungus's exponential growth, there is an urgent need for speedier options to determine the severity of the infection and begin treatment for the afflicted patient right once.

Deep Learning techniques have recently exploded in popularity, providing a solid foundation for a variety of healthcare applications. Deep learning in healthcare can help clinicians discover hidden opportunities and trends in clinical data, allowing them to treat their patients more efficiently and quickly. In the medical field, deep learning has game-changing applications. Some of the breathtaking applications include Drug Discovery, Medical Imaging, Alzheimer's Disease, Insurance Fraud and Genome [6]. Deep Learning is taught on vast volumes of data and makes predictions and outcomes using several processing units. Deep learning is a type of neural network that mimics the human brain by operating through a network of neurons, similar to how neural networks mimic the human brain by operating through a network of neurons. The goal of this study is to use prominent Deep Learning Algorithms to anticipate the Black Fungus. The approach employs ANN, CNN, and RNN algorithms. The outcome was also predicted using a supervised Machine Learning method. SVM (Support Vector Machine) is a sophisticated ML tool that can predict outcomes more accurately than other supervised machine learning models. RNNs (Recurrent Neural Networks) are utilized extensively in NLP activities because they specialize in processing sequences. Convolutional neural networks (CNNs) are image processing neural networks that have one or more convolutional layers.

As the covid virus was quite new to the mankind and black fungus illness following or as a result of covid attack being much newer, to the medical practitioner's as well, so there was a dilemma among the doctors regarding the treatment and diagnosis of black fungus infection. The major issue was with the identification of mucor present in the body. This delay and dilemma in treatment of black fungus could be lethal to the affected patients and hence an immediate requirement of proper diagnostics was needed. The major goal of this study is to assist medical practitioners in swiftly recognizing infections so that treatment can begin. This study examines the patient's covid medication, as well as his or her comorbidities, and predicts which patients are at risk of contracting black fungus infection. Because the condition was so new, there wasn't much research available in this domain, and this study can aid in this situation. Helping the medical practitioners in quicker identification of black fungus, which in turn will provide quicker treatment to the patient.

16.2 Dataset

The dataset used in this study was taken from the database of PubMed (24/28)and Google Scholar (4/28) [7–34]. The number of afflicted patients was low because it was the early days of the Black Fungus Outbreak after being infected with covid. The data collection used in this investigation consisted of medical records from 101 patients who had their covid19 status validated. There were around 82 samples of Indian people among the 101, with the other samples coming from various parts of the world. Age, Gender, Patient Comorbidities, Covid19 Treatment, Mucormycosis

Location, and Patient Mucormycosis Diagnosis were all important factors in the dataset. These characteristics were used to determine whether or not the patient would become infected with Black Fungus. Male patients accounted for 78.9% of the cases in 101 samples. When covid was still active in the patients, they were more susceptible to black fungus (60 patients were active with covid and 41 patients were affected during post covid state). Fungus was discovered to be present in the Nasal/Sinus in the majority of cases. Ninety-five patients were proven to have black fungus, while the remaining six were suspected of having black fungus. The dataset was severely skewed because the majority of patients had been diagnosed with Black Fungus. The model was at risk of becoming over fitted due to data scarcity and non-availability. To avoid this situation, numerous tools and procedures were used to keep the data from becoming unbalanced. Weight balancing, over and under sampling of data are some of the approaches used.

16.3 Machine Learning Architectures

Artificial Neural Networks (ANN), Convolutional Neural Networks (CNN), Recurrent Neural Networks (RNN), and Support Vector Machines are the models employed in this study (SVM). Deep Learning is a subset of Machine Learning that is built on the foundation of Artificial Neural Networks. Deep Learning is a sort of machine learning that is taught on vast volumes of data and makes predictions and outcomes using numerous processing units. It's a neural network with numerous hidden layers in between the output and input layers that has a particular level of complexity. Non-linear relationships can be modeled and processed using these algorithms.

16.3.1 Concurrent Neural Networks (CNN)

CNNs (convolutional neural networks) are neural networks with one or more convolutional layers that are mostly utilized for image processing. Convolution Layer, Rectified Layer Unit, Pooling Layer, and Fully Connected Layer are all important layers in a CNN. Figure 16.1 shows a typical CNN.

Convolution Layer. One of the key advantages of convolution is that the array size shrinks with each convolution. The smaller the array, the easier and faster it is to anticipate.

Relu Layer. Relu Layer stands for Rectified Layer Unit. The Relu is a technique that zeros out negative data input. The Relu Layer is notorious for increasing image nonlinearity. If the function receives any negative input, it returns 0. It delivers the same value for any positive x. Relu activation function is given by Eq. 16.1

$$f(x) = max(0, x) \tag{16.1}$$

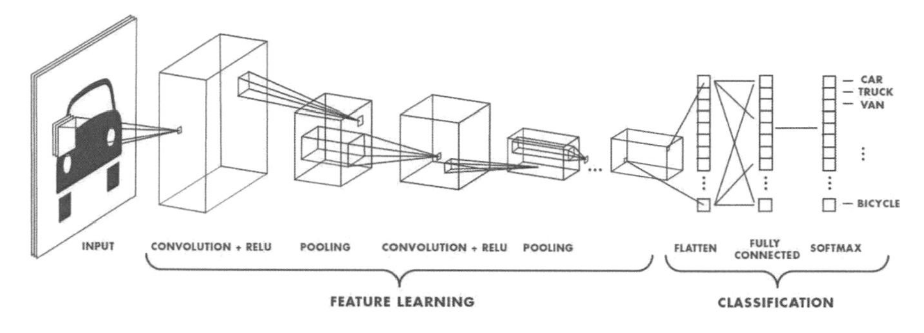

Fig. 16.1 Convolutional Neural Network [35]

Pooling Layer. The Max Pooling Layer is a pooling layer that is commonly utilized. The Max Pooling Layer holds the biggest input value and reject the rest within a filter. The data is passed to Flattening after the Pooling Layer. Before being transferred to the Fully Connected Layer, this flattening is required.

Fully Connected Layer. Depending on the features, there may be several fully connected layers. It takes the second most time to render, after the Convolution Layer. The Flattened vector is connected to a few fully connected layers that perform the same mathematical operations as Artificial Neural Networks.

16.3.2 Recurrent Neural Networks (RNN)

RNNs (Recurrent Neural Networks) are utilized extensively in Natural Language Processing (NLP) activities because they specialize in processing sequences. These methods are often used for ordinal and temporal applications and require sequential data or time series. The output of recurrent neural networks is dependent on the previous sections of the sequence. RNN's core and most essential feature is a hidden state that remembers some sequence information. In a conventional RNN, one layer's output is combined with the successive input, which is then recirculated back into the layer, yielding in contextual 'memory.' There are various applications of these RNNs that include speech recognition, language modeling and many more [36]. A typical RNN is shown in Fig. 16.2.

16.3.3 Artificial Neural Networks (ANN)

Artificial Neural Networks (ANN) are a processing system with great performance taking biological neural networks as inspiration. ANNs are parallel distributed processing technologies, and connectionist systems. ANN gathers a large number of

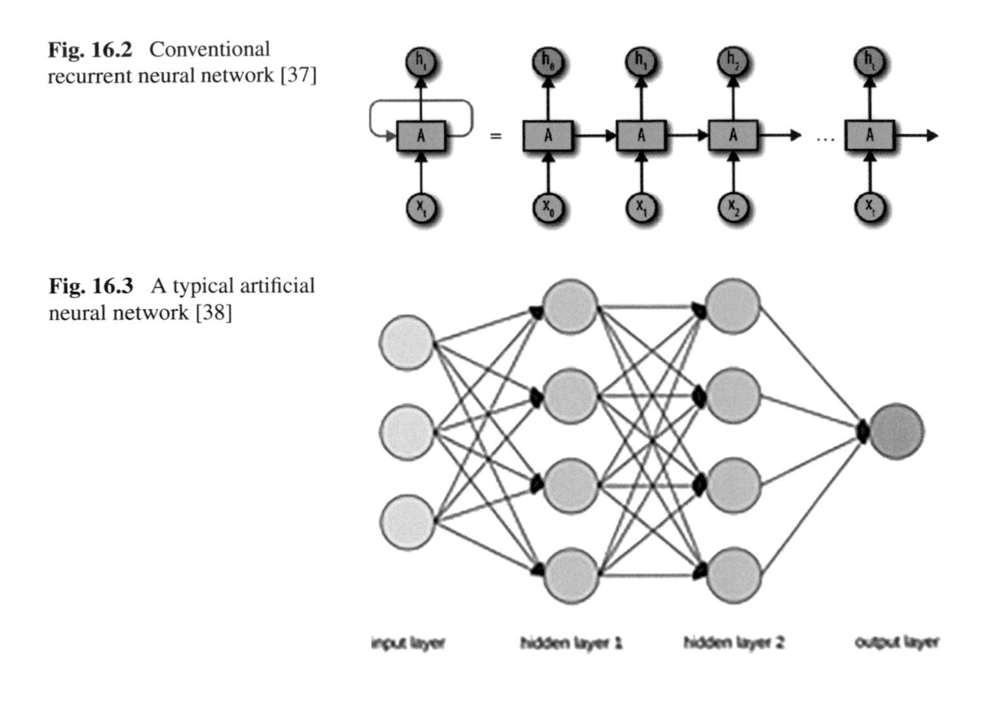

Fig. 16.2 Conventional recurrent neural network [37]

Fig. 16.3 A typical artificial neural network [38]

units that are linked in some way to allow them to communicate with one another. These rudimentary processors, also known as nodes or neurons, work in a parallel method. Every neuron is connected to other neurons via a connecting connection. Figure 16.3 shows ANN structure and Fig. 16.4 shows the structure of an activation function.

Activation Function in ANN. The output of an input function, or the output of a node given in inputs, is defined by the activation function. They merely decide whether to deactivate or activate neurons to achieve the desired result. Activation functions are vital in Neural Networks because they perform a nonlinear modification on the input in order to improve the outputs of a difficult neural network. ReLU, Softmax, Tanh, linear, and Sigmoid are the most prevalent activation functions. The structure of an activation function is shown in Fig. 16.4.

16.3.4 Support Vector Machine (SVM)

Among the many supervised machine learning models, the Support Vector Machine is one of the most popular. The support vector machine technique aims to find a separating hyperplane in an N-dimensional domain that classify data points. This plane should be as far away as possible from both data points' classes. The support vector machine is the ideal method because it achieves great accuracy while consuming

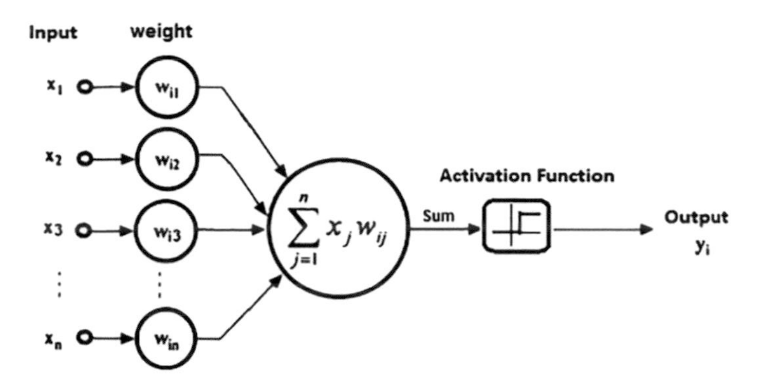

Fig. 16.4 Structure of an activation function [39]

minimal computing resources. In classification models, the Support Vector Machine is commonly utilized.

16.4 Experiment Methodology

16.4.1 Experiment Setup

The experiment was carried out using the black fungus dataset, which contains data on confirmed and probable black fungus cases. The dataset only had 101 cases in it, therefore it was quite little. Because Deep Learning Models necessitate a large dataset, our dataset was at risk of being over fit. By balancing the data, the problem was solved. We employed a variety of approaches to sample the data, including weight balancing and over and under sampling. It is critical to divide the data into training and testing sets; otherwise, the model would be over fit. As a result, we divided our dataset into 80% training data and 20% test data. The tensor package in Python was used to do the experimental assessments of CNN, RNN, ANN, and SVM.

16.4.2 Loss Function

In a machine learning model, the Loss Function is crucial. Machines learn by using a loss function. It's a way to see how effectively a particular algorithm models the data. The loss function would shoot an extremely big number if forecasts deviated too far from actual results. Eventually, the loss function learns to lower prediction error with the aid of some optimization function [40]. In our study, we have used binary cross-entropy loss function.

16.4.3 Evaluation Criteria

It is critical to assess the performance of any model. There are a number of parameters that may be used to assess the model's performance. We employed accuracy, F1 score, and recall in our research. The most commonly used evaluation parameter is accuracy, which is defined as the percentage of forecasts that are correct in a given number of predictions. Percentage of positive instances out of total projected positive instances is called precision. Percentage of positive cases out of total positive instances is called recall. The harmonic mean of precision and recall is the F1 score. Below are the equations for each of the evaluation parameters.

$$precision = \frac{TP}{TP + FP} \tag{16.2}$$

$$recall = \frac{TP}{TP + FN} \tag{16.3}$$

$$F1score = \frac{2 * precision * recall}{precision + recall} \tag{16.4}$$

$$accuracy = \frac{TP + TN}{TP + FN + TN + FP} \tag{16.5}$$

16.4.4 Experimental Results

Figure 11 shows the Performance comparison of different algorithms on test data based on individual classes. Highest Precision in Confirmed class was obtained in RNN and Highest Precision in Suspected Class was also obtained using both CNN and RNN algorithm. Highest F1 score for both confirmed and suspected class was obtained using RNN with F1 score of 0.97. Graphical Representation of Comparison of each evaluation metrics for different algorithms has been shown in Figs. 16.5, 16.6, 16.7, 16.8. Table 16.1 and comparison graphs clearly show that RNN algorithm has dominance over other models in terms of performance metrics. Table 16.1 and comparison graphs clearly show that RNN algorithm has dominance over other models in terms of performance metrics.

The below graphical representations, which compare several performance measures for different algorithms, indicate that RNN significantly outperforms the others. RNN predicts Black Fungus with the highest accuracy of 97.37%, fol-lowed by CNN, ANN, and SVM. When compared to other algorithms, RNN has the greatest F1 score. RNN has the highest precision, which is the percentage of positive cases out of the total projected positive instances, followed by CNN, ANN, and SVM. Because recall and accuracy are mutually exclusive, we witness a fall in recall value as precision improves.

Table 16.1 Performance metrics values for different algorithms used in the model

Method Used	Classes	Precision	F1 score	Recall
RNN	Confirmed	1.00	0.97	0.95
	Suspected	0.95	0.97	1.00
CNN	Confirmed	0.95	0.95	0.95
	Suspected	0.95	0.95	0.95
ANN	Confirmed	0.94	0.89	0.92
	Suspected	0.90	0.95	0.92
SVM	Confirmed	0.94	0.95	0.88
	Suspected	0.82	0.79	0.86

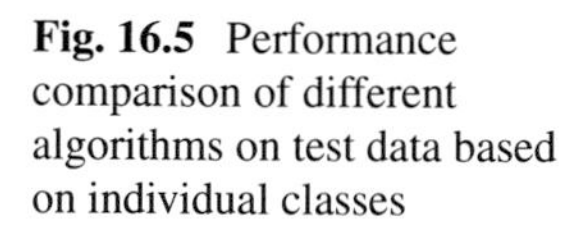
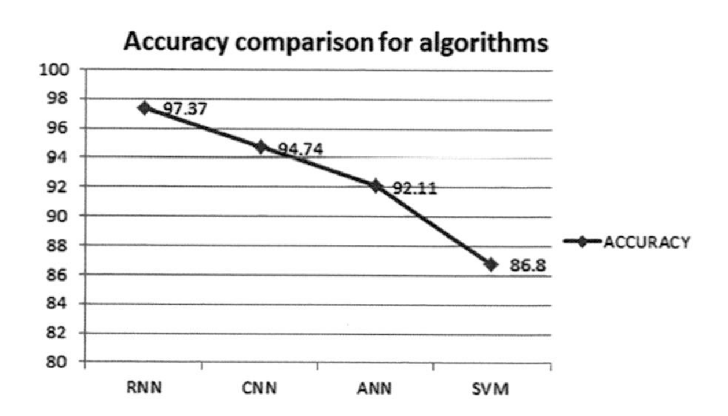

Fig. 16.5 Performance comparison of different algorithms on test data based on individual classes

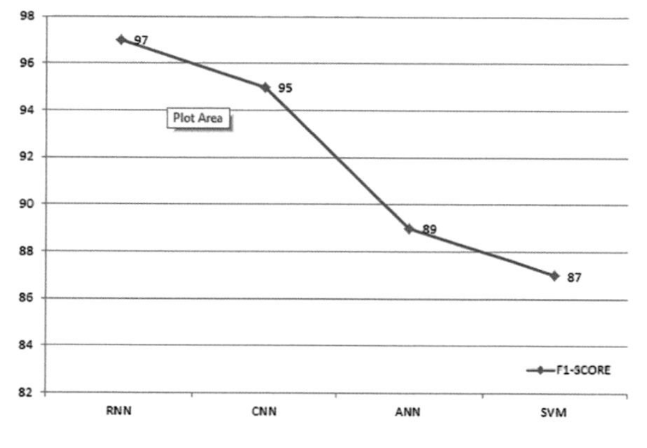

Fig. 16.6 F1-score comparison of all algorithms

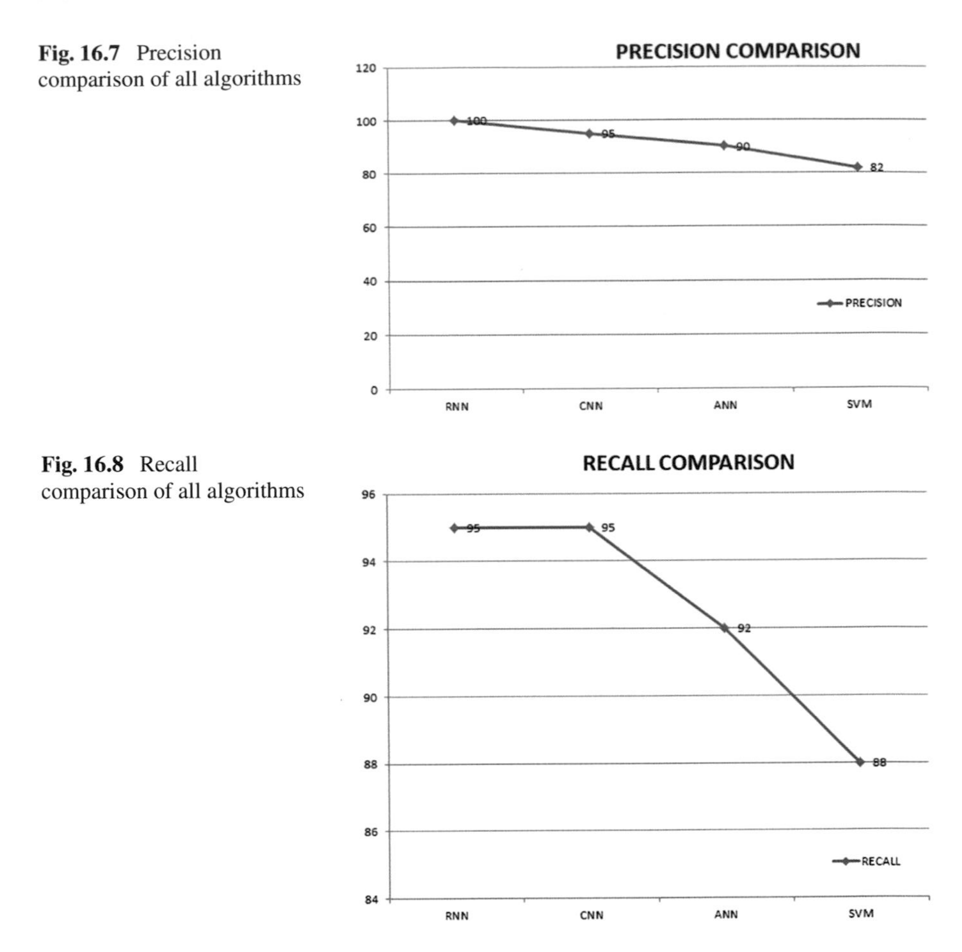

Fig. 16.7 Precision comparison of all algorithms

Fig. 16.8 Recall comparison of all algorithms

16.5 Conclusion

Based on our research and observations, we can conclude that RNN is capable of detecting the Black Fungus rather well. Better results can be obtained by fine-tuning the parameters. Scenarios with an 80:20 training-to-test ratio produce the best results, with a 97.37% accuracy and a 97% F1 score. To work efficiently and achieve better results, deep learning requires huge datasets, yet in our situation the dataset was little. Because black Fungus began infecting patients during the second wave of Covid 19, and this study began in the early stages of the second wave, the number of infected patients was limited, resulting in a tiny dataset. The study's other key problem was an unbalanced dataset. Because most classification techniques are based on the premise that each category has an equal number of samples, uneven classifications create a hurdle to prediction. As a result, models become skewed and biased. Various strategies such as oversampling (blind copy), oversampling (smote), and undersampling were employed to avoid this and improve test accuracy. This resulted in well-balanced data and a high level of test accuracy. The goal of this research was to assist doctors in identifying patients who have been infected with covid and are susceptible to black fungus. Patients with comorbidities such as diabetes who have been given steroids during their covid treatment are more prone to black fungus, according to the findings of the study. It may be deduced from the data that roughly 80% of the patients infected with black fungus were men. When covid was active in the body, over 60% of the patients were infected with mucormycosis. Approximately 30% of the individuals who were afflicted with the black fungus succumbed. This 30% indicates that black fungus is a fatal condition that requires prompt treatment. Some important conclusions can be obtained from the dataset analysis. Mucormycosis is more common in male patients. Patients with active covid virus are more likely to get mucormycosis. Doctors can identify patients who are more susceptible to mucormycosis based on their comorbidities and covid therapy. Such patients can be closely monitored and given specific attention. Medical practitioners should attempt to avoid using excessive steroids on these patients unless it is absolutely necessary. Machine learning is exploding in every industry, and health care is one of them. Machine learning technologies can considerably assist doctors by giving faster and more accurate answers. Digital diagnostics is one such application. Healthcare is becoming smarter as a result of machine learning. It can be difficult to preserve data about patients and their health difficulties, but once the data is available, it can be used to assess patient health using knowledge acquired from large data sets by aggregating millions of observations of diseases that a patient could have. This way machine learning can assist humanity in a variety of fields and will go a long way in future. Machine learning advancements will allow for the automatic detection of most diseases in their early stages. It will also improve disease detection efficiency and accuracy, easing the burden on clinicians.

References

1. Centers for Disease Control and Prevention, National Center for Emerging and Zoonotic Infectious Diseases (NCEZID), Division of Foodborne, Waterborne, and Environmental Diseases (DFWED). https://www.cdc.gov/fungal/diseases/mucormycosis/definition.html
2. Song, Y., Qiao, J., Giovanni, G., Liu, G., Yang, H., Wu, J., Chen, J.: Mucormycosis in renal transplant recipients: review of 174 reported cases external icon. BMC Infect Dis. **17**(1), 283 (2017). Apr
3. Abdalla, A., Adelmann, D., Fahal, A., Verbrugh, H., Van Belkum, A., De Hoog, S.: Environmental occurrence of Madurella mycetomatis, the major agent of human eumycetoma in Sudanexternal icon. J. Clin. Microbiol. **40**(3), 1031–1036 (2002). Mar
4. Vallabhaneni, S., Mody, R.K.: Gastrointestinal mucormycosis in neonates: a review external icon. Curr. Fungal Infect. Rep. (2015)
5. Francis, J.R., Villanueva, P., Bryant, P., Blyth, C.C.: Mucormycosis in children: review and recommendations for managementexternal icon. J. Pediatric Infect. Dis. Soc. **7**(2), 159–164 (2018). May 15
6. Top 5 applications of Deep Learning in healthcare. https://www.allerin.com/blog/top-5-applications-of-deep-learning-in-healthcare

7. Mehta, S., Pandey, A.: Rhino-orbital mucormycosis associated with COVID-19. Cureus **12**(9) e10726. [PMC free article] [PubMed] [Google Scholar]
8. Garg, D., Muthu, V., Sehgal, I.S., Ramachandran, R.: Coronavirus disease (Covid-19) associated mucormycosis (CAM): case report and systematic review of literature. Mycopathologia. **186**(2), 289–298. [PMC free article] [PubMed] [Google Scholar]
9. Maini, A., Tomar, G., Khanna, D., Kini, Y., Mehta, H., Bhagyasree, V.: Sino-orbital mucormycosis in a COVID-19 patient: a case report. Int. J. Surg. Case Rep. **82**, 105957. [PMC free article] [PubMed] [Google Scholar]
10. Saldanha, M., Reddy, R., Vincent, M.J.: Title of the article: paranasal mucormycosis in COVID-19 patient. Indian J. Otolaryngol. Head Neck Surg. **22**, 1–4. https://doi.org/10.1007/s12070-021-02574-0. [Online ahead of print] [PMC free article] [PubMed] [CrossRef] [Google Scholar]
11. Revannavar, S.M., Supriya, P.S., Samaga, L.: COVID-19 triggering mucormycosis in a susceptible patient: a new phenomenon in the developing world? BMJ Case Rep. **14**(4), e241663. [PMC free article] [PubMed] [Google Scholar]
12. Sen, M., Lahane, S., Lahane, T.P.: Mucor in a viral land: a tale of two pathogens. Indian J. Ophthalmol. **69**, 244–252. [PMC free article] [PubMed] [Google Scholar]
13. Sarkar, S., Gokhale, T., Choudhury, S.S., Deb, A.K.: COVID-19 and orbital mucormycosis. Indian J. Ophthalmol. **69**(4), 1002–1004. [PMC free article] [PubMed] [Google Scholar]
14. Mishra, N., Mutya, V.S.S., Thomas, A.: A case series of invasive mucormycosis in patients with COVID-19 infection. Int. J. Otorhinolaryngol. Head Neck Surg. **7**(5), 867–870. [Google Scholar]
15. Satish, D., Joy, D., Ross, A., Balasubramanya, B.: Mucormycosis coinfection associated with global COVID-19: a case series from India. Int. J. Otorhinolaryngol Head Neck Surg. **7**(5), 815–820. [Google Scholar]
16. Moorthy, A., Gaikwad, R., Krishna, S.: SARS-CoV-2, uncontrolled diabetes and corticosteroids-an unholy trinity in invasive fungal infections of the maxillofacial region? A retrospective, multi-centric analysis. J. Maxillofac. Oral Surg. **6**, 1–8.https://doi.org/10.1007/s12663-021-01532-1. [Online ahead of print] [PMC free article] [PubMed] [CrossRef] [Google Scholar]
17. Sharma, S., Grover, M., Bhargava, S., Samdani, S., Kataria, T.: Post coronavirus disease mucormycosis: a deadly addition to the pandemic spectrum. J. Laryngol. Otol. 1–6. https://doi.org/10.1017/S0022215121000992.[Online ahead of print] [PMC free article] [PubMed] [CrossRef] [Google Scholar]
18. Hanley, B., Naresh, K.N., Roufosse, C.: Histopathological findings and viral tropism in UK patients with severe fatal COVID-19: a post-mortem study. Lancet Microbe. **1**(6), e245-e253. [PMC free article] [PubMed] [Google Scholar]
19. Dallalzadeh, L.O., Ozzello, D.J., Liu, C.Y., Kikkawa, D.O., Korn, B.S.: Secondary infection with rhino-orbital cerebral mucormycosis associated with COVID-19. Orbit 1–4. https://doi.org/10.1080/01676830.2021.1903044. [Online ahead of print] [PubMed] [CrossRef] [Google Scholar]
20. Werthman-Ehrenreich, A.: Mucormycosis with orbital compartment syndrome in a patient with COVID-19. Am. J. Emerg. Med. **42**, 264. e5-264.e8, In press. [PMC free article] [PubMed] [Google Scholar]
21. Placik, D.A., Taylor, W.L., Wnuk, N.M.: Bronchopleural fistula development in the setting of novel therapies for acute respiratory distress syndrome in SARS-CoV-2 pneumonia. Radiol. Case Rep. **15**(11), 2378–2381. [PMC free article] [PubMed] [Google Scholar]
22. Mekonnen, Z.K., Ashraf, D.C., Jankowski, T.: Acute invasive rhino-orbital mucormycosis in a patient with COVID-19-associated acute respiratory distress syndrome. Ophthalmic Plast. Reconstr. Surg. **37**, e40-80 (2021). [PMC free article] [PubMed] [Google Scholar]
23. Alekesyev, K., Didenko, L., Chaudhry, B.: Rhinocerebral mucormycosis and COVID-19 pneumonia. J. Med. Cases **12**(3), 85–89 (2021). [PMC free article] [PubMed] [Google Scholar]
24. Johnson, A.K., Ghazarian, Z., Cendrowski, K.D., Persichino, J.G.: Pulmonary aspergillosis and mucormycosis in a patient with COVID-19. Med. Mycol. Case Rep. **32**, 64–67 (2021). [PMC free article] [PubMed] [Google Scholar]

25. Kanwar, A., Jordan, A., Olewiler, S., Wehberg, K., Cortes, M., Jackson, B.R.: A fatal case of Rhizopus azygosporus pneumonia following COVID-19. J. Fungi (Basel) **7**(3), 174 (2021). [PMC free article] [PubMed] [Google Scholar]
26. Khatri, A., Chang, K.M., Berlinrut, I., Wallach, F.: Mucormycosis after Coronavirus disease 2019 infection in a heart transplant recipient - case report and review of literature. J. Mycol. Med. **31**(2), 101125 (2021). [PMC free article] [PubMed] [Google Scholar]
27. Monte Junior, E.S.D., Santos, M.E.L.D., Ribeiro, I.B.: Rare and fatal gastrointestinal mucormycosis (zygomycosis) in a COVID-19 patient: a case report. Clin. Endosc. **53**(6), 746–749 (2020). [PMC free article] [PubMed] [Google Scholar]
28. Pasero, D., Sanna, S., Liperi, C.: A challenging complication following SARS-CoV-2 infection: a case of pulmonary mucormycosis. Infection 1–6 (2020). [PMC free article] [PubMed] [Google Scholar]
29. Bellanger, A.P., Navellou, J.C., Lepiller, Q.: Mixed mold infection with Aspergillus fumigatus and Rhizopus microsporus in a severe acute respiratory syndrome Coronavirus 2 (SARS-CoV-2) patient. Infect. Dis. News (2021). https://doi.org/10.1016/j.idnow.2021.01.010.S2666-9919(21)00030-0. [Online ahead of print] [PMC free article] [PubMed] [CrossRef] [Google Scholar]
30. Karimi-Galougahi, M., Arastou, S., Haseli, S.: Fulminant mucormycosis complicating coronavirus disease 2019 (COVID-19). Int. Forum Allergy Rhinol. (2021). https://doi.org/10.1002/alr.22785. [Online ahead of print] [PMC free article] [PubMed] [CrossRef] [Google Scholar]
31. Veisi, A., Bagheri, A., Eshaghi, M., Rikhtehgar, M.H., Rezaei Kanavi, M., Farjad, R.: Rhino-orbital mucormycosis during steroid therapy in COVID-19 patients: a case report. Eur. J. Ophthalmol. (2021) https://doi.org/10.1177/11206721211009450.11206721211009450. [Online ahead of print] [PubMed] [CrossRef] [Google Scholar]
32. Sargin, F., Akbulut, M., Karaduman, S., Sungurtekin, H.: Severe rhinocerebral mucormycosis case developed after COVID 19. J. Bacteriol. Parasitol. **12**, 386 (2021). [Google Scholar]
33. Waizel-Haiat, S., Guerrero-Paz, J.A., Sanchez-Hurtado, L.: A case of fatal rhino-orbital mucormycosis associated with new onset diabetic ketoacidosis and COVID-19. Cureus **13**, e13163 (2021). [PMC free article] [PubMed] [Google Scholar]
34. Zurl, C., Hoenigl, M., Schulz, E.: Autopsy proven pulmonary mucormycosis due to Rhizopus microsporus in a critically ill COVID-19 patient with underlying hematological malignancy. J. Fungi (Basel). **7**(2), 88 (2021). [PMC free article] [PubMed] [Google Scholar]
35. Saha, S.: Towards data science, convolutional neural network (2018). https://towardsdatascience.com/a-comprehensive-guide-to-convolutional-neural-networks-the-eli5-way-3bd2b1164a53. Accessed 19 May 2022
36. Karpathy, A., Fei-Fei, L.: Deep visual-semantic alignments for gen erating image descriptions. In: Proceedings of the IEEE Conference on Computer Vision and Pattern Recognition, pp. 3128–3137 (2015)
37. Kost, A., Altabey, W., Noori, M., Awad, T.: Applying neural networks for tire pressure monitoring systems. Struct. Durab. Health Monit. **13**, 247–266 (2019). https://doi.org/10.32604/sdhm.2019.07025
38. Arden Dertat, Towards Data Science, Artificial Neural Network (2017). https://towardsdatascience.com/applied-deep-learning-part-1-artificial-neural-networks-d7834f67a4f6. Accessed 19 May 2022
39. "Neural Networks," Umut's Tech Blog, 24-Jun-2009. http://hevi.info/tag/neural-networks/. Accessed 19 May 2022
40. Common Loss functions in machine learning. https://towardsdatascience.com/common-loss-functions-in-machine-learning-46af0ffc4d23

Chapter 17
AI-Driven Fuzzy Decision Making Framework for Efficient Utilization of COVID-19 Vaccination

Dalip and Deepika

Abstract These days, COVID-19 virus is spreading quickly at high rated speed in almost all over the world. The effected people are further spreading it become carrier and unknowingly at faster rate. So, it is most to educate the people about this communicable disease. To prevent from further spreading of COVID-19 disease there is a requirement of proper distribution of vaccines and also vaccinate the population on their importance. Artificial Intelligence (AI) plays a major role to prevent the COVID-19 pandemic disease through its efficient vaccination campaign. This chapter explains the Fuzzy Decision Making Framework for Efficient Utilization of COVID-19 Vaccination (FDMF-EUV) using Deep Learning which is designed for effective utilization of vaccination which further controls the spread of the coronavirus. There was no vaccine developed for COVID-19 till May, 2020, but now the number of vaccines has been developed by different countries. The prime motive of every country is to vaccinate their population hastily but the main challenge is the distribution of the vaccine and to vaccinate their population according to their importance. The major problem is facing by those countries that are planning to unlock their regions from the lockdown and how to vaccinate their population on a priority basis to prevent the virus. The proposed framework contributes to resolving the above-mentioned problems by assigning fuzzy weights to the classified populations for efficient vaccine utilization. The assigned weights help to set the priority of vaccination among the population. For experimental results, the concept of the FDMF-EUV framework is designed by considering the scenario of the classified population of India. Hence, it is concluded that FDMF-EUV is a fully utilized novel framework and works efficiently to vaccinate the population quickly on a priority basis.

Keywords Artificial intelligence · Deep learning · Covid-19 · Vaccination · Fuzzy weights · Experienced · FDMF-EUV

Dalip (✉)
M.M. Institute of Computer Technology and Business Management, Maharishi Markandeshwar (Deemed to be University), Mullana-Ambala, Haryana, India
e-mail: dalipkamboj65@gmail.com

Deepika
UIE-CSE, Chandigarh University, Chandigarh, India

A. Biswas et al. (eds.), *Artificial Intelligence for Societal Issues*, Intelligent Systems Reference Library 231, https://doi.org/10.1007/978-3-031-12419-8_17

17.1 Introduction

In the current scenario, Covid-19 affects a huge population not only in India but across the globe because of its communicable behaviour due to which the rate of spreading this pandemic increases day by day. Some precautionary measures have to be taken to prevent from this pandemic as one should be aware of their surroundings. Till May 2020, there is no medicine or vaccine available to protect from Covid-19 but the existing techniques can be used to save lives. Artificial Intelligent (AI) is one of the upcoming techniques to controls the spread of this virus. Many researchers are doing their research towards this field. Today, the main challenge for the countries is to vaccinate their working population rapidly to prevent the further spread of COVID-19. The countries are planning to unlock their regions from the lockdown and trying to put their efforts to vaccinate their maximum population. The major challenge for largely populated countries is the proper distribution of their vaccine for the classified population. Due to the lack of AI-enabled models for efficient distribution of vaccines, therefore an FDMF-EUV framework is developed by using AI technology which will helpful to manage and prevent the further spread of this virus. The pro-posed framework automatically calculates the experienced-based fuzzy weights for better decision-making. For the experimental analysis and results, the data is to be taken from CoWIN Dashboard [12].

The total population in India is 136 Crores plus. Figure 17.1 shows Vaccination Trends by Dose [12] on different days from the period of January 16, 2021, to June 18, 2021. The vaccination days are shown on the x-axis and the y-axis shows the number of doses in Crore. The total vaccination Doses are 26,07,91,348, Dose One are 21,26,39,943 and Dose Two are 4,81,51,405 were given to the population till June 16, 2021. The above figure shows very less population is vaccinated by Dose

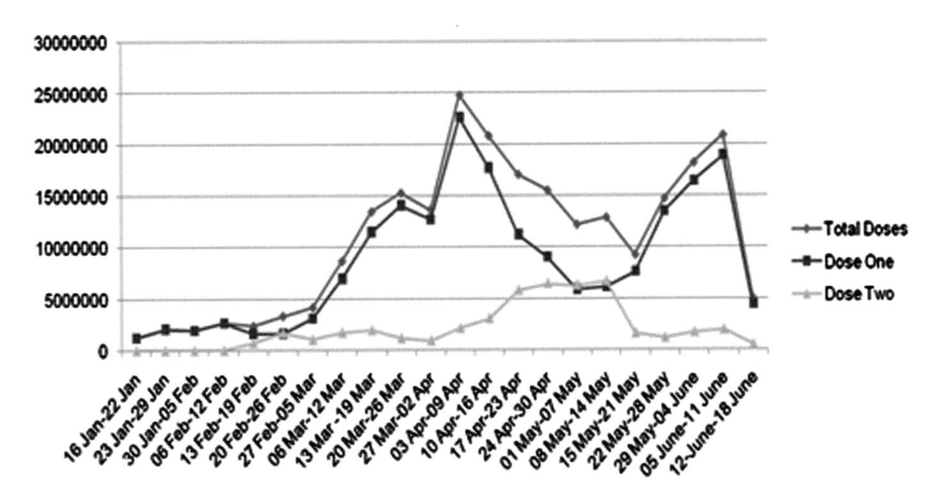

Fig. 17.1 Vaccination trends by dose

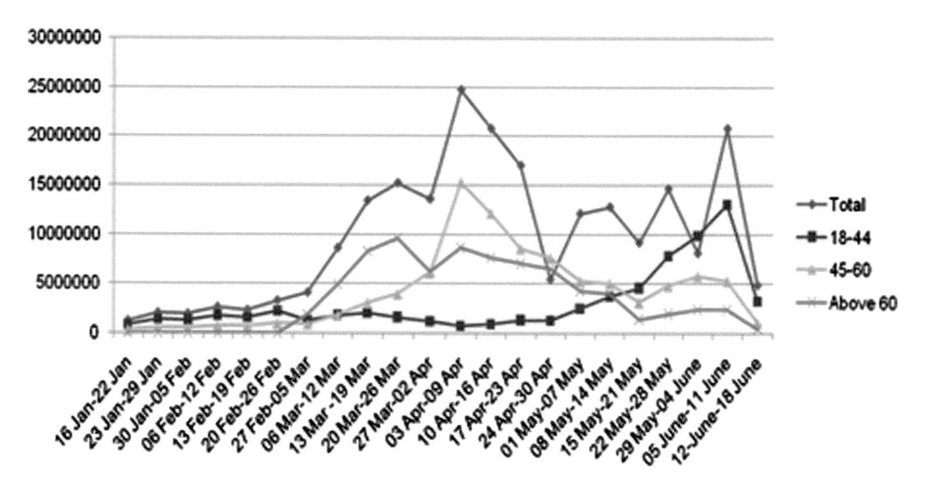

Fig. 17.2 Vaccination trends by age

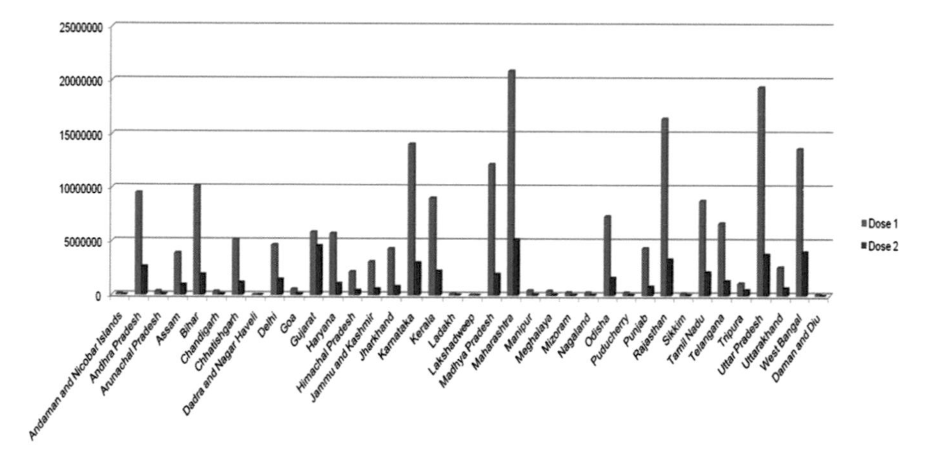

Fig. 17.3 State wise coverage of vaccination

Two. To prevent the further spreading of COVID-19 virus it is very necessary to vaccinate the population with both Doses as soon as possible. Figure 17.2 shows below the Vaccination Trends for different age groups of (18–44) years, (45–60) years and above 60 years from the period of January 16, 2021, to June 18, 2021 whereas Fig. 17.3 depicts the state-wise coverage of vaccination to their population. According to the population of India, the state-wise vaccination coverage should be large for both doses without the wastage of vaccines. It has been quite clear that from Figs. 17.1, 17.2, and 17.3, a novel approach is required for the efficient utilization of COVID-19 vaccine to the largely populated countries from preventing further spread.

The rest of the chapter is organized as follows. Section 17.2 discussed the related work and describe the contribution of various researchers over COVID-19 effective and proper utilization of vaccination for the population. Section 17.3 shows the motivation and contribution of authors. In Sect. 17.4 the novelty of the proposed work is shown. The discussion of results is explained in Sect. 17.5. Finally, the conclusion is presented in Sect. 17.6.

17.2 Background

The literature elaborates the role of AI to prevent the spread of COVID-19 and its proper distribution of coronavirus vaccine. Cloud/fog computing is one of the upcoming techniques to control the spread of this virus [1]. This paper describes a quality of service framework based on fog-assisted IoT which controls its further spread in a cost-effective manner. This framework is used to predict the health of a person and in case of any problem, it generates and circulates an alert message for the infected person to their near ones, doctors, and government health agencies also. The purpose of dispatching the alert message is to outbreak this chronic illness and to take the action on time. The impact of AI against the epidemic covid-19 in the initial stage is described and it has also been concluded that a large amount of data for infected persons was recorded to save lives, trained AI, and reduced financial losses [2]. The COVID-19 pandemic explains the terrible deaths rate and shows the need for more work has been done on data collection and trained AI [3]. After considering the various aspects of AI, the main focus is on the Global Heath of the people in this pandemic situation. There are various types of feature selection algorithms that are used for these datasets [4]. This coronavirus becomes an area of research for the medical field as there is no medicine or vaccination for this disease [5]. To save human lives it is important to detect this virus in the early stages and prevent sufferers. Testing kits are utilized to find the infected persons which control the spread of this terrible disease. A fuzzy expert system diagnostic framework is designed to detect the virus in the initial stages by using major clinical characteristics and symptoms. This model is integrated with MATLAB and tested sample data which brings this system as a tool for early detection of those persons who are the infected ones as well as the health status of the persons. The government, a number of suppliers, and distributors are there who rapidly launch more vaccinations that are under contract [6]. Thus the process of vaccination which facilitates the general public is a difficult process to handle because each person requires two doses of the same vaccine. The vaccination distribution now becomes a wastage and shortage process. As of now, it has been a simple, clear, and timely process to vaccine the general people. The covid-19 vaccination is the latest topic of discussion about the best vaccine to use, its use among people, and the plans for its distribution worldwide, and the price [7]. There are medical practitioners, researchers, scientists, pharmacists, businessmen, and academic institutions who discuss vaccines developed for coronavirus. There are few challenges in this vaccination as it was forward in the third phase for trial and ready for market in terms of efficiency and safety. The availability of this vaccine is the major area of concern as well as the future health catastrophe. In the initial stages of COVID-19, AI and the Internet of Things (IoT) play a vital role to control the spread of this virus. Before this coronavirus, there are few challenges in terms of different age categories of populations and for this, an AI deployment model in healthcare had proposed [8]. A real-time application has been made using AI techniques to control and track the spread of this virus among the huge population as it was spread at a rapid rate. When this vaccination process has been started, the first question is the distribution of this vaccine among the huge population and the scarcity of this vaccine as two doses are provided to complete this vaccination process. In India where the population rate is so much high, it becomes a difficult task to provide covid-19 vaccine for each individual. The government of India sanctioned two vaccines for this virus named Covishield and Covaxin. This vaccination facility is primarily provided to all medical field persons and co-workers in the first phase. In the second phase which was started from 1st March 2021, the persons who are above 60 years of age are facilitated with the vaccines and in the third phase, persons more than 45 years of age are eligible for this dose. In the pathetic situation of COVID-19, the death cases are increased day by day. There are various health organizations, trust, agencies, government aids, and organizations across the world that put their mutual efforts to make this terrible situation in control and save human lives by coordinating with the supply, production, and distribution of vaccines to everyone. It becomes a major challenge in front of these organizations to be aware of these vaccines to the huge population, planning, eligibilities, and coordinating activities of these doses. A complete life cycle to provide this vaccine to common people, keep track of vaccinated people, and kept the entire record at one's place is a complicated task. The updated technologies like Amazon web services (AWS), Cloud, AI, and IoT are used to control this epidemic situation [9]. The launch of coronavirus vaccination becomes one of the historical achievements which are awaited by all over the world and India takes the initiative of introducing this vaccine [10]. Different suppliers developed different vaccines and each has its own efficacy in terms of percentages. The main aim behind these vaccines is to get the trust of a huge population. The government officials set specific guidelines and designed protocols while communicating about this vaccine to the public. This is the reason because of this dense population has faith in the government and gets the vaccination voluntarily [11]. AI is embedded with different latest techniques like machine learning and Global Positioning System (GPS) which prepared the automated systems to track the status of this spreading virus. It has been concluded that after comparing the automated system with the traditional framework, the system provides an accuracy of 98% and 97% in urban and rural areas which helps to control the spread of this worst disease.

During this world wide pandemic [13] this research innovates a new approach by using quantitative and qualitative methods to understand the vaccination intention. Also the proposed approach can help to obtain the perceived effectiveness of COVID-19 measures in the countries fighting against COVID-19 break out. The objective of this analysis is to develop a model SIR [14] for COVID-19 by using fuzzy parameters. The SIR model is developed by taking the following factors into consideration like vaccination, treatment, compliance in executing health protocol and corona virus-load. To access the basis reproduction number and solidity of equilibrium points, generation matrix method is used in this model analysis. The results showed that the differences in transmission of COVID-19 are caused by difference in corona virus load. Correspondingly, the compliance is executing the health protocols and factor s of vaccination have the similar effect in decreasing or stopping the transmission of COVID-19 in Indonesia. Arrangement for mass vaccination [15] against COVID-19 is ongoing in progress in number of countries so that the vaccine will be available for the general public soon. In this pandemic which is in progress needs the usage of conventional and latest temporary vaccination clicks. An effective temporary mass vaccination method among all others is use of drive-through. The result of this model is able to appropriately predict the key product of the simulation tool. So, an online application is developed by using this model which helps mass vaccination planners to obtain the results of different types of drive through mass vaccination facilities very quickly.

A new type of variant of SARS named 2(SARS-COV 2) [16] was discovered in late December 2019. To defeat the continued increase in spread of this new latest pneumonia is the chief task of numerous researchers and scientist. The origin of Artificial Intelligence (AI) has given an enormous contribution to decrease the spread of COVID-19 from three important aspects of identification, prediction and development and propose the ongoing important challenges and possible development direction. The outcome indicates that the proposed model is an effective tool to merge AI technology with the diversity of latest technologies to identify COVID-19 patents. Corona Virus (SARS COV2) 2019 disease [17] is a crucial danger all over the world. The latest achievement in computer application based on Machine Learning, AI and Big data can help in discovery, observing and prediction the severity of COVID-19 pandemic. The aim of this research is to analyse the detection of the COVID-19 by using AI, major challenges and smart health care. Improved and fast perception of this disease is possible by using AI. It helps the healthcare authorities for better decision making by designing the disease venture and periodicity the severity. It is an encouraging mechanization for completed automatic and transparent analyzing system to trace and nurse the patients casually without spreading the virus to others. For handling the COVID 19 patients the title role of AI is discussed in this research paper. It also reduces the workload of medical authorities by providing solution for finding contacts drug evolution.

17.3 Motivation and Contributions

The existing review shows the different research work on COVID-19 such as comparative studies, COVID-19 vaccination frameworks, models, and AI-based decision-making systems etc. Now, many vaccines have been developed by different countries and each country wants to vaccinate their population very quickly but they are facing problems in vaccination according to the priority of the population. For example, assume due to inefficient methodology of vaccination unemployed persons are not the earning source of their families and take the vaccination but employed persons are not able to take the vaccine due to its improper distribution and unavailability. The challenge of designing an efficient framework for efficient utilization of vaccination among the population is to prevent COVID-19 from further spreading which inspires us to design FDMF-EUV framework to be classified population for COVID-19 vaccination.

17.4 A Novel Methodology

This chapter proposed a novel framework which is designed to assign the experienced based fuzzy weight to the classified population on the basis of their importance among individuals for efficient vaccination. The FDMF-EUV framework is divided into four regions as Eligibility and Registration Process, Service Interface, Data Management and Reporting of Vaccine and Health Care Sector and Agencies. The novel framework first checks the eligibility of the users for their vaccination. If the users are eligible then they can register themselves.

The users are categorized into two parts Mobile users and Desktop users. After completion of the eligibility and registration process, the request goes to the service interface, which provides the two types of services i.e. User Interface and Web Interface. The Data Management and Reporting of Vaccine department analyzed the data and assigned the experienced-based weights to the classified population through FDMF-EUV framework as shown in Fig. 17.4. The experienced based weight are given to five criteria from C 1 to C 5 on the basis of the priority of criteria. All the information of vaccine is stored into database and by applying data analytic extract the relevant information and ignore the irrelevant information, which helps in further decision making. The high fuzzy weight indicates high rank and the low fuzzy weight indicate low rank. The designed models use the AI capabilities such as triangular membership function and deep-learning models, which helps in decision making for the health department and the agencies for making better decisions to vaccinate their population and the entire population of any country vaccinates according to their assigned rank. If the rank of the classified population is high it means their priority is high for vaccination and if rank is low it means their priority is less for vaccination. The process of calculate and assigned the rank is automated. Finally, the health care sector will take the decision on the basis of calculated ranks thought proposed system, which helps in vaccination drive to vaccinate the classified populations.

17.5 Discussion of Results

The subjective responses for each criterion are assumed to be from the experts as shown in Table 17.1. The proposed designed system is tested over the different samples of the classified population of India. As shown in Table 17.2, the experienced-based weights from 0.0 to 1.0 and rank from 1 to 9 is assigned to the classified

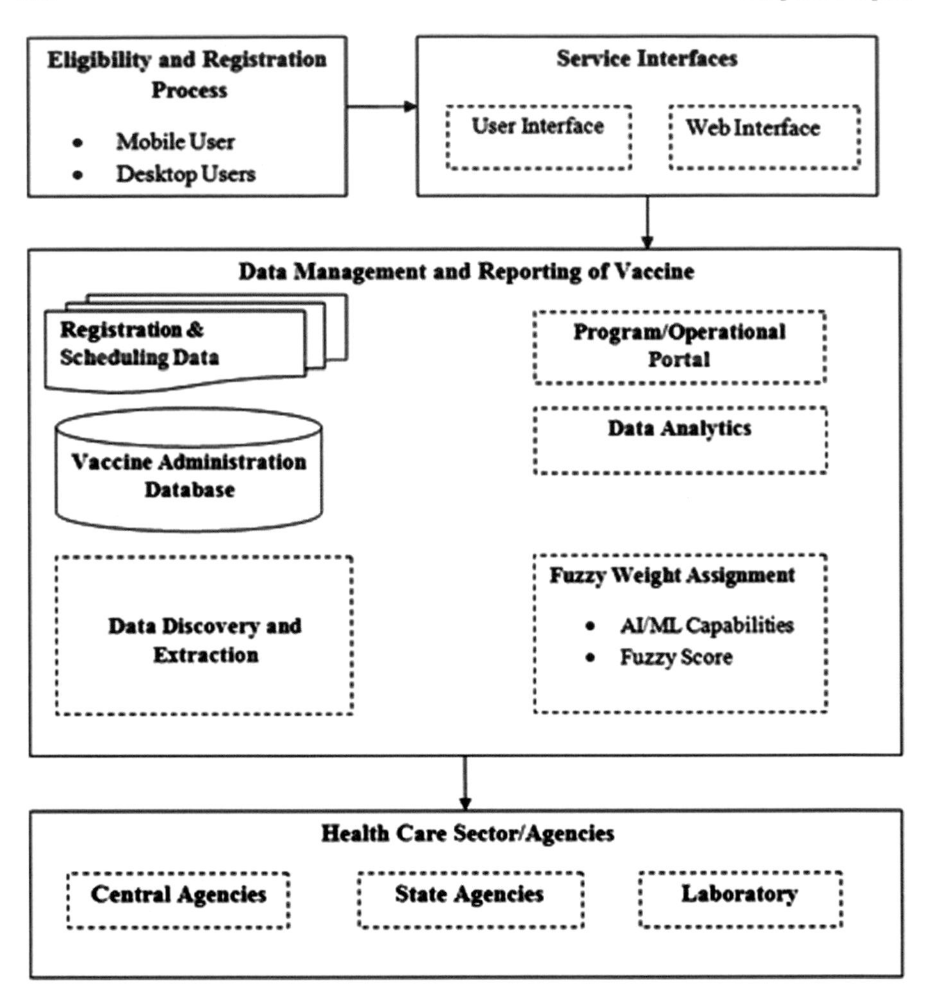

Fig. 17.4 Fuzzy decision making framework for efficient utilization of COVID-19 vaccination (FDMF-EUV)

population and calculated accordingly. The weights are calculated automatically on the basis of expert opinion. The highest rank indicates greater importance for the vaccination and the lowest rank indicates the least importance for the vaccination. The highly ranked population gets the vaccination at the first and the lowest rank population at last. This framework helps to prevent the further spread of COVID-19 due to the efficient utilization of vaccine.

In Figs. 17.5 and 17.6, the x-axis shows the classified population and the y-axis represents the fuzzy weight and percentage of the classified population. The percentage is high in the case of health care workers, a population greater than 50 years and a population less than 50 years with associated comorbidities such as Hyper-

Table 17.1 Experts response

Criteria	Linguistic Term	Numerical Scale	Weight
CR1	Very Important	5	(0.8–1.0)
CR2	Important	4	(0.6–0.8)
CR3	Moderately Important	3	(0.4–0.6)
CR4	Slight important	2	(0.2–0.4)
CR5	Not Important	1	(0.0–0.2)

Table 17.2 Fuzzy weights and ranking for classified population

Population categories	Population classification	Criteria	Fuzzy weight	Rank
Health care workers	Public Heath Carer	CR1	1.0	1
	Private health care	CR1	1.0	1
	ICDS workers	CR1	1.0	1
Frontline workers	Police government	CR1	0.9	2
	Military	CR1	0.9	2
	Home Guards	CR2	0.8	3
	Community Volunteers,	CR2	0.8	3
	Civil Defence Organisation	CR2	0.8	3
	Municipal Workers	CR1	0.9	2
	Revenue Officials Engaged in Surveillance	CR2	0.8	3
	Revenue Officials Engaged in Containment Activities	CR2	0.8	3
Population greater 50 years	All Population	CR1	0.9	2
Population less 50 years	Only those Associated Comorbidities such as Hypertension/Diabetes/HIV/Cancer, etc.	CR1	1.0	1
Essential services	Bank Employees	CR1	0.9	2
	Confectionary Shopkeepers	CR2	0.8	3
	Milk Mans	CR2	0.8	3
Agriculture sector	Farmers	CR1	0.9	2
	Labourers	CR1	0.9	2
Government and private employees	All government and private employees	CR3	0.6	5
Business	Non IT companies	CR3	0.5	6
	IT companies	CR5	0.1	9
Education sector	Administrative staff	CR2	0.7	4
	Teaching staff	CR4	0.3	7
	Non teaching staff	CR4	0.2	8
Unemployed person	All unemployed persons	CR5	0.1	9

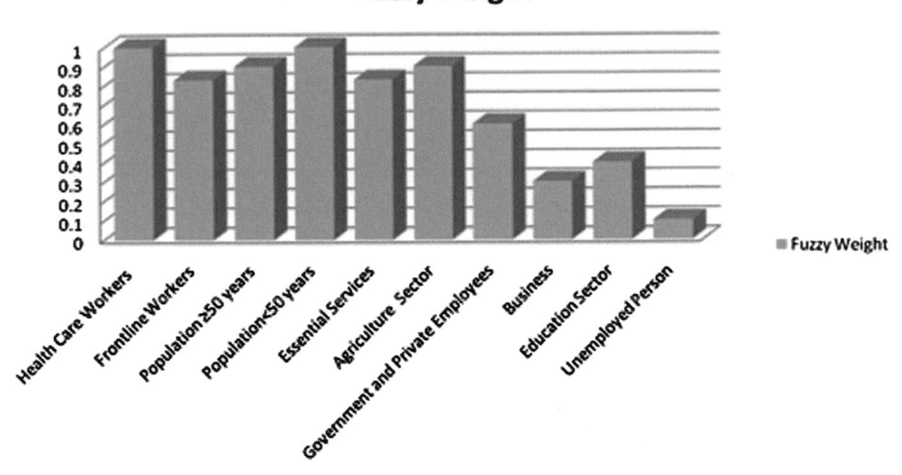

Fig. 17.5 Assigned weight by FDMF-EUV framework

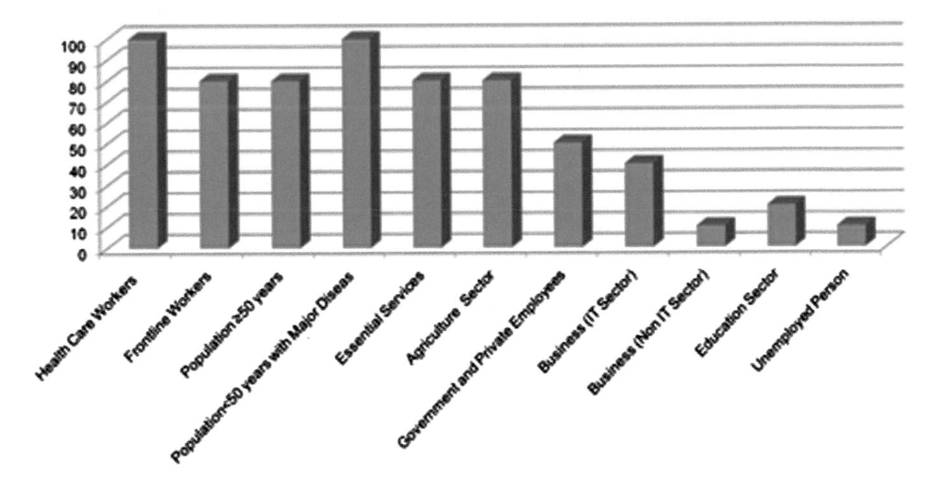

Fig. 17.6 Assigned weight by FDMF-EUV framework

tension/Diabetes/HIV/Cancer etc. which indicates the high priority of vaccination given to this population but the percentage of IT sector and unemployed persons are very less which indicates the low priority of vaccination is provided to as shown in Fig. 17.6. In unlock phases, the proposed method is extremely helpful by vaccinating to those populations who prevent the spread of COVID-19 at maximum level. It has been proved from the above findings that FDMF-EUV framework helps in better decision making for efficient utilization of vaccination to classified population rapidly and prevents the COVID-19 from further spread.

17.6 Conclusion

The findings in this chapter show the importance and use of fuzzy logic for better utilization, distribution, and management of COVID-19 vaccination among the population by using triangular membership function of AI. After the review and comparative analysis of useful and significant findings by using the fuzzy concept of assigning the experienced based fuzzy weight to classified population to prevent the spread of COVID-19 Hence, it has been concluded that the FDMF-EUV framework is helpful for the largely populated country to vaccinate their population efficiently on the basis of the availability of the COVID-19 vaccine. This model contributes its role to prevent the COVID-19 from further spreading with an efficiency of 25–30% by efficient utilization of COVID-19 vaccine. In future the efficient utilization of COVID-19 vaccination can be improved by using latest deep-learning models.

References

1. Singh, P., Kaur, R.: An integrated fog and Artificial Intelligence smart health framework to predict and prevent COVID-19. Glob Transit. **2**, 283–292 (2020). https://doi.org/10.1016/j.glt.2020.11.002
2. Naude, W.: Artificial intelligence vs COVID-19: limitations, constraints and pitfalls. AI Soc. **35**, 761–765 (2020). https://doi.org/10.1007/s00146-020-00978-0
3. Chang, A.C.: Artificial intelligence and COVID-19: present state and future vision. Intell.-Based Med. **3**(4) (2020). https://doi.org/10.1016/j.ibmed.2020.100012
4. Alzubaidi, M.A., Otoom, M., Otoum, N., Etoom, Y., Banihani, R.: A novel computational method for assigning weights of importance to symptoms of COVID-19 patients. Artif. Intell. Med. **112** (2021). https://doi.org/10.1016/j.artmed.2021.102018
5. Ejodamen, P.U., Ekong, V.E.: A fuzzy expert system model for the determination of coronavirus disease risk. Int. J. Mechatronics Electr. Comput. Technol. (IJMEC) **11**(39), 4825–4831 (2021)
6. Mills, M.C., Salisbury, D.: COVID-19 vaccine development to vaccination. J. Nepal Health Res. Counc. **18**(49), 807–809 (2020)
7. Boudjelal, M., Almajed, F., Salman, A.M., Alharbi, N.K., Colangelo, M., Michelotti, J.M., Olinger, G., Baker, M., Hill, A.V., Alaskar, A.: COVID-19 vaccines: global challenges and prospects forum recommendations. Int. J. Infect. Dis. **105**, 448–451 (2021)
8. Kumar, R., Veer, K.: Diabetes & metabolic syn-drome: clinical research & reviews. **15**, 1049–1050 (2021)
9. Amazon Web Services: How cloud, artificial intelligence (AI), and machine learning (ML) technologies can improve the vaccine experience for citizens in COVID-19 and beyond. Digital Strategies for Vaccine Distribution and Ad-ministration (2021)
10. Krishnamurthy, A., Gopinath, K.S.: The big billion Indian COVID 19 vaccine challenge. Indian J. Surg. Oncol. **12**(1) 3–4 (2021). https://doi.org/10.1007/s13193-021-01280-1
11. Dalip, K., Deepika, L.: AI-enabled framework to prevent COVID-19 from further spreading. In: Intelligent Systems and Methods to Combat Covid-19. Springer Briefs in Applied Sciences and Technology. Springer, Singapore (2020). https://doi.org/10.1007/978-981-15-6572-4-4
12. Cowin Homepage: https://dashboard.cowin.gov.in/
13. Nguyen, P.H., Tsai, J.F., Lin, M.H., Hu, Y.C.: A hybrid model with spherical fuzzy-AHP, PLS-SEM and ANN to predict vaccination intention against COVID-19. Mathematics MDPI **9** (2021)

14. Abdy, M., Side, S., Annas, S., Nur, W., Sanusi, W.: An SIR epidemic model for COVID-19 spread with fuzzy parameter: the case of Indonesia. Adv. Differ. Equs. **105** (2021)
15. Asgary, A., Valtchev, S.Z., Chen, M., Najafabadi, M.M., Wu, J.: Artificial intelligence model of drive-through vaccination simulation. Int. J. Environ. Res. Public Health **18**(1) (2021)
16. Peng, Y., Liu, E., Peng, S.: Using artificial intelligence technology to fight COVID-19: a review. Artif. Intell. Rev. (2022)
17. Swayamsiddha, S., Prashant, K., Shaw, D.: The prospective of artificial intelligence in COVID-19 pandemic. Health Technol. **11**, 1311–1320 (2021)

Chapter 18
Covid-19 Diagnosis Based on Fuzzy-Deep Learning Algorithm

Anurag Sinha, Shubham Singh, Md. Ramish, Shubham Kumar, Hassan Raza Mahmood, and Nawaz Khan Choudhury

Abstract Artificial intelligence is rapidly occupying the healthcare sector so as to solve and enhance the accuracy of medical reports, assistance, and many more. Deep learning is a subset of AI that is predominant for building basic models for image classification and segmentation. In this paper, we have used a modified CNN classifier for detecting COVID-19 followed by an x-ray image data set. In this paper, we have also implemented independent and principle component analysis on CNN for optimization of the algorithm by analysing the principles labels of data for model efficiency. Consequently, PCA (principle component analysis) is widely used for dimensionality reduction in machine learning; hence, identifying more progressive labels will increase model accuracy. The Fuzzy interference system is integrated with the deep learning model to make the confidence level of the model more robust and explained in terms of performance metrics such as precision, recall, and cross validation process by 91%.

Keywords Fuzzy logic · Deep learning · Simulation · Medical imaging · Classification

---

A. Sinha (✉)
Department of Information Technology, Amity University Jharkhand, Ranchi, India
e-mail: anuragsinha257@gmail.com

S. Singh
Department of Computer Science, Birla Institute of Technology Mesra, Ranchi, Jharkhand, India

Md. Ramish
Department of Electronics and Communication Engineering, Amity University Jharkhand, Ranchi, India

S. Kumar
Department of Computer Science and Engineering, Amity University Jharkhand, Ranchi, India

H. R. Mahmood
Fast Nukes Cfd Campus Fast-Nu, Fast Square, Faisalabad, Punjab, Pakistan

N. K. Choudhury
Girijananda Chowdhury Institute of Management and Technology, Guwahati, Assam, India

A. Biswas et al. (eds.), *Artificial Intelligence for Societal Issues*, Intelligent Systems Reference Library 231, https://doi.org/10.1007/978-3-031-12419-8_18

18.1 Introduction

The impact of diseases, the effects of immediate environmental change, ecological change, lifestyle and many other factors is growing rapidly. This timeless risk of disarray is extended. Almost 3.4 million individuals kicked the bucket in 2016 from cutting-edge light-deprivation obstructive infection (COPD), which is believed to be impacted by contamination and smoking, and another 400,000 individuals choked from the impacts of asthma. The present clinical foci are a tremendous wellspring of data streaming out of patients [1]. Biomarkers, segment information, and image modalities help and assist clinical experts in finding unmatched diseases, and analysis becomes usable data for trained professionals. Another type (COVID-19) has emerged, depending on the mechanisms by which a man-made conscience can successfully complete a work cycle, leading to widespread redundancies worldwide. The end of 2019 was marked as an example of an infectious event. Corona viruses mainly show up in patients' lungs. Between these lines, if the infection is not expected to resolve at the beginning of the disease, it can seriously damage the lungs. Although the mortality rate of infection is low, it should not be forgotten because of the depth of infectious diseases. The risk of infection becomes more real because good clinical foci cannot support the large number of people who become infected every day. Probably the most common use of neural circuits in computer vision is face recognition. The most well-known artefact is the unit's imperceptible recognition of the faces that are observed in the Deep Faces. We have some network access that needs to be secured and tested, such as the LFW dataset and the MS-Celeb-1 M. This dataset contains a huge number of individual images and some related information such as names, ages, and so on. It is clear that the use of Convolution Nerve Networks (CN) has gained an important and excellent approach to learning about the problems of a positive clinical picture. Various tests have shown that new systems can be integrated so that they can prepare longer, more accurately, and faster. With the proposed expansion of the physical range (DenseNet, VGGNet, and ResNet), a wide range of projects provide a reasonable target for image ranking objects after one has been used, and this ensures that they can be imagined in the deepest selected For wallets, this paper proposes a design layout so that the body can be arranged from box to frame with pictures. The initial element of the learning scene was created using a set of models called ResNet-50, with image development as preparation and abandonment as post-treatment. Recognizing the lines within a given image has always been an engaging aspect of the high learning field in the process of factorization and multiple methods of feature segmentation optimization have been introduced in the past through several investigations. Interface computing and the evolution of computing are two of the emerging areas of optimization that are used in every field of engineering for optimization and problem solving in the best solution. The applications of soft computing and evolutionary algorithms have notably been used in the fields of machine learning and artificial intelligence [3]. Using a soft-computed feature optimization algorithm with a multi-layered platform, the new approach is for optimising the integrated layers within a network architecture that provides a high number of candidates and a population that has the maximum number of epochs collected by replicates. Implementing the number of candidates using genetic algorithms. The accuracy of this model ensures that the snaps network selects the data layer to train and test an ensemble to build a deep learning model [4, 5]. To implement this genetic algorithm based on the multilayer percepts algorithm, we used COVID-19 infected chest, x-ray and lung data to model this based on algorithms, where it gives the highest accuracy. The major findings are as follows:

- Foreseeing the probability of resistance in patients with Covid-19 based on clinical characteristics utilizing verifiable information units and clinical attributes.
- Analyze and distinguish the potential for Covid disease based on parameters such as age and side effects in a number of more extensive age groups.
- The classification of MLP transformers for picture classification is analyzed Data set collected from nearby nursing homes.
- Combine an auto-encoder and CNN to amplify the precision of your expectations.
- Presentation of look optimization and effectiveness calculate examination.

This chapter is organized in this chronological fashion:

I. Sections 18.1 and 18.2 contains overview and objective, scope of the chapter,
II. Section 18.3 contains related literature study done.
III. Fourth section contributes about the literature analysis and comparison of research gap and finding.
IV. Fifth section is all about what techniques are already available for covid-19 recognition.
V. In the 6th section we have proposed our own method for analysis.
VI. last section it discusses about result and conclusion.

Author contribution in this paper is as follows: author 1 and 2 contributed to the dataset preparation and background study and optimization of the algorithm with paper drafting. The third and fourth author of the paper has done all the introductory and theoretical literature investigates. Fourth Author has implemented and simulated the result in programming environment. This work is novel in terms of its algorithmic approach a layer of encoding value is added through model precision.

18.2 Background

The data set has been acquired from the following mentioned sources. The following data set contains the 256 by 256 and 1024 by 1024 pixels of images. It contains over 3000 images having pneumonia and covid-19 and Normal chest X ray image is the format of the data set which is in the format of PNG. The indexes of the data set having several features, attributes Such as Defection rate other kind of biological parameters and metrics, which is measured to detect the Covid-19 infection in the people. The data set is preprocessed and Pre trained using data discretization method. The following data set. Split it into two parts the first is training and the second

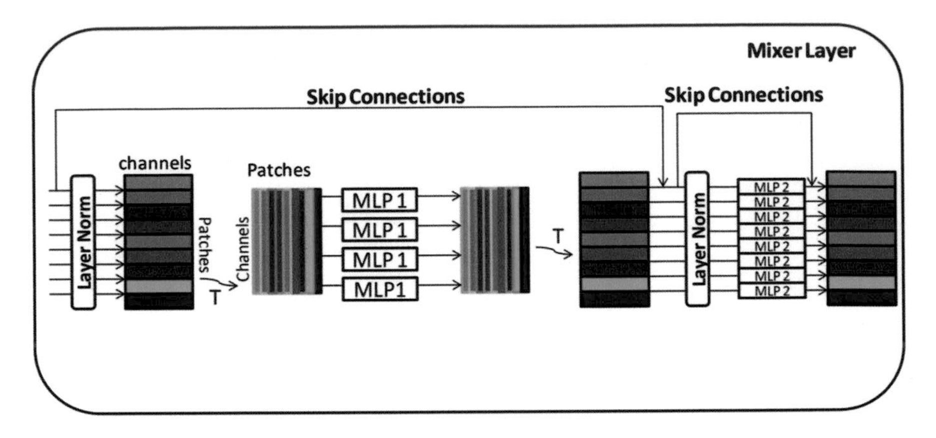

Fig. 18.1 MLP

one is testing for the Ensemble learning process, in which 1800 images is used for training purposes and rest for the testing purposes being ensemble architecture of this particular. Proposed algorithm looks like fiesta cipher structure. But the data set is splitted into two parts and has a round of convolution functions provided by the layers of the fuzzy set and multilayer perceptron to train the model for diagnosis of the COVID-19 [5].

MLP-Mixer: In Fig. 18.1, MLP-Mixer contains two kinds of layers: one with MLPs applied autonomously to picture patches (for example "blending" the per-area highlights), and one with MLPs applied across patches (for example "blending" spatial data). At the point when prepared on enormous datasets, or with current regularization plans, MLP-Mixer achieves cutthroat scores on picture characterization benchmarks, with pre-preparing and derivation cost practically identical to best in class models. We trust that these outcomes sparkle further examination past the domains of grounded CNNs and Transformers. As the historical backdrop of PC vision illustrates, the accessibility of bigger datasets combined with in-wrinkled computational limit regularly prompts a change in perspective. It acknowledges a grouping of directly projected picture patches (additionally alluded to as tokens) formed as a "patches channels" table as an info, and keeps up with this dimensionality. Blender utilizes two sorts of MLP layers: channel-blending MLPs and token-blending MLPs [6].

18.3 Literature Review

In paper [6], Author have proposed the strategy in view of PALM Fuzzy mental guides and fluffy surmising framework a delicate registering methods for customized learning stage in which the psychological modeler is utilized for drafting the fluffy rationale based information portrayal. In paper [2], Author have used transfer learning based VGG16 and mobile net model for covid-19 detection based on x-ray image which gives the accuracy of F1 score, precision, specificity, and sensitivity of 90–97%, respectively. In paper [3], Author have shown six category of data being chosen from big medical x-ray data to train model for classifying Covid inflected image based on Generative Adversarial Networks (GAN). In paper [4], the ResNet152V2 and MobileNetV2, CNN, SVM, LSTM list of Retrained model for classifying deep pneumonia infected image with high RUC curve of 98.8% however this much accuracy of ML sounds very repetitive and unrealistic as deep learning have still Emerge with a problem of selecting an idle layer of convolution. In paper [5], Author asserts that profound learning is adequate for infection forecast highlight extraction strategies for profound learning techniques can in any case assume a significant part in future clinical exploration. However deep learning can be a subset of Conventional medical procedure but can never replace conventional medical process what about the medical pre-test and -medication, an extension of limitation should have been introduced before claiming replacement of conventional medical procedure. In paper [1], Author have shown an framework based on Auto Ml for interpretation of medical report and text with highest precision which can revolutionize patent outcome which up to some extend is considerable but interpretation of text through Imaging can't have the high precision, Natural language based transformer are widely used for understanding word Embedding and optical text recognition. Thus, employing NLP method in medical text recognition can have more accuracy. In paper [7], Author have shown the worth literature draft on AI-based systems in performing medical work in specializations including radiology and its impact on medical professionals which however very applicable to some extent but can't void the relationship between a doctor and patient in which it given in a very literary contrast "As far as farsighted investigation and picture affirmation, AI may in a little while had the chance to be more convincing than specialists, who can't deal with a large number of pictures in any reasonable time span. This has driven to a few concerns that AI-based frameworks will supplant doctors, particularly radiologists". In paper [8], Author have used Fuzzy interference system for primary diagnosis and self assisted diagnosis based on own symptoms for covid-19. In paper [9], Author has proposed HDS thought about against ongoing procedures. Test results have shown that the proposed HDS beats different rivals in wording of the normal worth of exactness, accuracy, review, and F-measure in which it gives about of 97.658%, 96.756%, 96.55%, and 96.615% separately.

We have analysed different algorithm used by the several researcher and findings what exactly the research gap based on relevance level scoring (see Table 18.1).

Table 18.1 Comparative study of related literature

Author	Data source	Types of images	Best model	Accuracy (%)	AUC score and relevancy level
[11]	Iran University of Medical Sciences (IUMS)	CT images	Ensemble	91.94	0.965/8
[12]	"Toy dataset;" "Italian Society of Radiology;" "Shenzhen Hospital X-Ray dataset;" "ChestX-Ray8;" "COVID-CT-Dataset"	X-ray images	Using the GEV activation function for unbalanced data	100	0.88/7
[13]	"COVID-19 CT segmentation dataset;" "Chest X-rays (Radiopaedia)"	grayscale lung CTI images	KNN	87.75	0.987/6
[6]	"COVID-19 hospitals in Shandong Province"	CT images	AD3D-MIL	97.9	0.99/7
[7]	"Radiopaedia and the cancer imaging archive websites"	CT images	LSTM	99.68	0.98/9
[15]	"COVID-chest X-ray;" "SARS-COV-2 CT-scan;" "Chest X-Ray Images (Pneumonia);	X-ray and CT images	VGG-19	95.61	0.97/8
[17]	From the Renmin Hospital of Wuhan University	CT images	Details Relation Extraction neural network	86	0.87/5
[18]	Private	CT images	DenseNet	92	0.98/9
[21]	Private dataset	CT images	Fully connected network and combination of Decision tree and Adaboost	82.9	0.90/8

18.4 Proposed Method

18.4.1 Data Pre-Processing

Data cleaning plans work to "clean" the data by filling in missing characteristics, smoothing rowdy data, perceiving or wiping out abnormalities, and settling anomalies. Expecting that clients acknowledge the data are dirty, they are most likely not going to believe the results of any data mining that has been applied to it. Data decline procures a decreased depiction of the instructive assortment that is significantly more unassuming in volume, yet conveys something almost identical (or essentially the same) consistent results. Information decrease systems incorporate dimensionality decrease and numerosity decrease [9]. In dimensionality decrease, information encoding plans are applied in order to get a decreased or "packed" portrayal of the first information. Models incorporate information pressure procedures, (for example, wavelet changes and head parts examination) too.

As trait subset choice (e.g., eliminating unessential characteristics), and quality development (e.g., where a little arrangement of more helpful properties is gotten from the first set) [4]. Returning to your information, you have chosen, say, that you might want to utilize a distance-based digging calculation for your investigation, like brain organizations, closest neighbor classifiers, or clustering. 1 Such strategies give better results if the information to be dissected have been standardized, or at least, scaled to a more modest reach, for example, [0.0, 1.0] [5]. The client information, for instance, contain the credits age and yearly compensation. The yearly compensation trait generally takes a lot bigger qualities than age. Hence, assuming that the qualities are left normalized, the distance estimations taken on yearly compensation will for the most part offset Distance estimations taken on age. Discretization and idea order age can likewise be helpful, where crude information values for credits are supplanted by ranges or higher applied levels [6]. For instance, crude qualities for age might be Supplanted by more elevated level ideas, like youth, grown-up, or senior [7]. Discretization furthermore, idea ordered progression age are integral assets for information mining in that they permit the mining of information at various degrees of deliberation. Figure 18.2 sums up the information preprocessing steps depicted here. For instance, the expulsion of excess information might be viewed as a type of information cleaning, as well as information decrease [10, 11].

18.4.2 Performance Metrics

This section portrays the appraisal execution of various profound learning models to arrange CXR pictures. Prepared models were legitimate recognized by ten times cross-approval and execution measurements got from the disarray grid were utilized exploratory examination. The disarray network gives a manual for the four bogus. The presence of both FN and FP could influence clinical choices. A PF result happens when a person is erroneously relegated to a class, for instance when it is sound the individual is erroneously delegated a patient with COVID-19. A FN happens when an individual is supposed to fall a specific class is barred from this gathering. The understanding the activity of the various organizations in the test set has been assessed accuracy full scale estimation, F1 score, accuracy, explicitness, responsiveness and coefficient [12, 13].

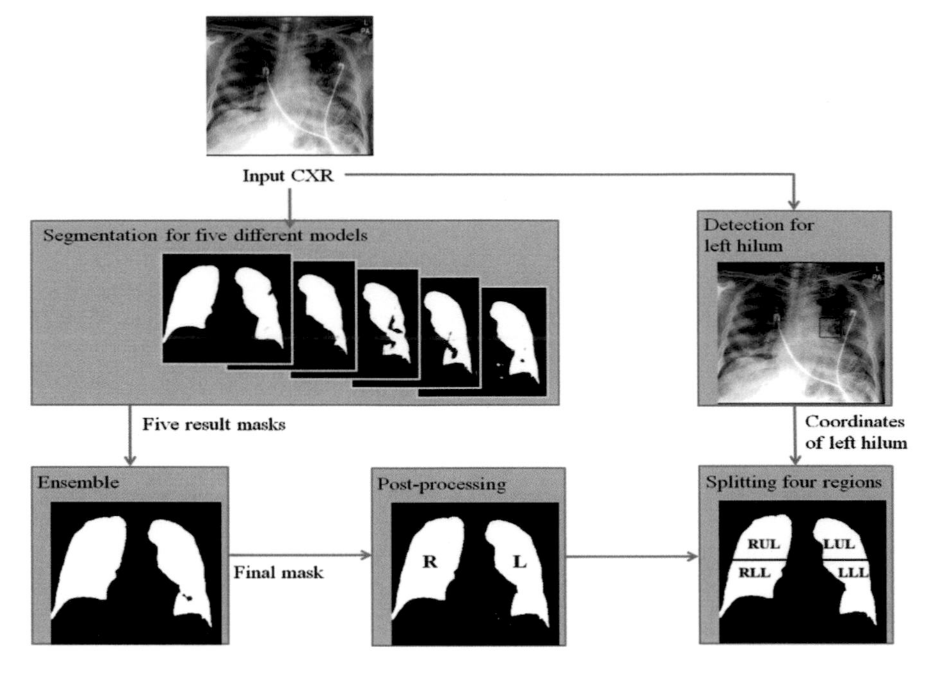

Fig. 18.2 Steps of data pre-processing and region of interest (ROI) extraction

18.4.3 Visualization of Datasets

Initially, an example of the information in this investigation was dissected. Finally, the complete data is dissected. In the attachment, the various tables will provide an overview of the information on lung disease. The actual number shows infections such as pneumonia, covid-19. The cycle is really contorted. In this dataset, the absolute number of guys was more than the complete number of females and the quantity of affirmed cases was essentially bigger than the quantity of guys dissected by system for recognizable proof at various clinical destinations utilizing the information. A significant obstacle in creating huge X-ray beam datasets is the lack of labeling properties for different images. The dataset contains the total of 3000 instances of categorical data to resemble the normalized and reduced data with filled missing values in order to contain model consistency [14]. Ethics approval has been obtained for this information. To submit cases, a subset of the patient's experience is collected through specialists, patient surveys, lab tests, vital symptom estimates, and imaging tests. Considering that our datasets have not been distributed before, let's examine how much each dataset includes patient mortality. Such experiments provide scientists with a fair idea about the properties of the information they collect. Different item selection techniques are accessible to determine the weight of each item in determining the test properties of the dataset. We have chosen search as one of the most commonly used item selection strategies; the meaning of each item (e.g. search)

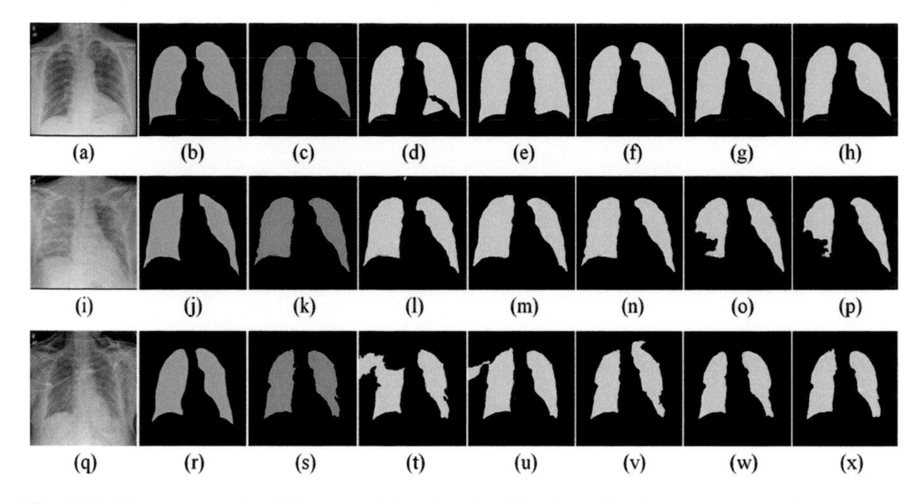

Fig. 18.3 Pre-processed and fine tuned data for classification using image segmentation

is displayed as a bar. The age of the data obtained in contrast to the various highlights is much more important (0.149). Therefore, age has been omitted from Fig. 18.3 to facilitate testing of different highlights. The bar graph shows that malignancy (after age), heart and kidney infection are the second, third and second important factors identified with patient survival. Consequently, patients with constant infection are more powerless against COVID-19. It is essential to take note of that Fig. 18.2 doesn't bar features with zero gained information.

The proposed strategy consists of two approaches, general and media development. The data were collected using CNN. Expand known information using the 10 codes that apply. In this section, we will look at CNN and its code of conduct in a few minutes [12]. CNN is widely used in educational-based films. Due to CNN's entry-level equipment, they receive important information when preparing for the exam. CNN is usually made with minimal editing, aggregation, and full correlation. As shown in Fig. 18.3, the withdrawal ends when the bond and the closing coin are completed [13]. The need for collection reduces the size of the organization without much progress on the image. At CNN, for the most part, the number of episodes is declining as the number of episodes increases. Two of the prices collected are the same as the general collection and the general collection. The codes they use are involved in independent learning because they do not require formal planning information [14]. For the most part, the automated coder packs the input into the shortcut and stores the information by downloading the specified location shortly. As with the basic analysis (PCA), the encoder automatically reduces the voltage step by step. However, PCA relies on direct codecs that will never change the line through large corporations. The standard diagram of the code is shown in Fig. 18.4 [15]. Multilayer perceptron is one of the easiest architecture of the neural networking system. To learn, feed forward and propagate. The data set contains a number of X-ray images. In addition, some additional data can be obtained from the dataset, such

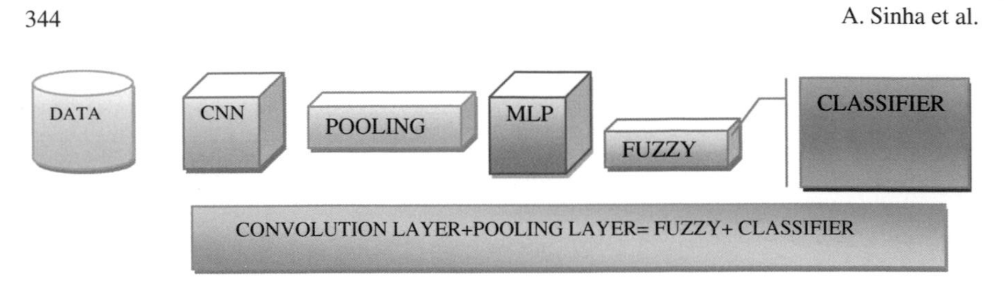

Fig. 18.4 Model architecture fuzzy based neural network

as an age or gender specification. The pre-treatment steps used in this operation are described in the accompanying document [16]. In Fig. 18.2, we have shown a diagram that captures the steps of data preprocessing where the data is distributed among various zones of a category based on its requirements and segmentary bounds, which adheres to the testing and training phasing dimension reduction where images are annotated manually to highlight the masked edges for model training and The Region of Interest (ROI), which is an image processing technique to validate and highlight the pixel region, is used to R and L, which means two divergent masked areas for image segmentation.

Figure 18.3 depicts the outcome of an ensemble-based data preprocessing method as well as the outcome of segmented parts of an image data set that has been split and tested using an image segmentation technique [19, 20]. The efficient net and VGG net classifier are used to train the mask and overlap the errors. Where classes from a to h are segmented using VGG NET, class I to p is segmented via efficient net, and class q to x is segmented using U-net.In terms of learning and predicting, For instance, a positron is considered. The wide range of nodes interconnects neural networks like the tree formed of data and those neural networks. The relationship established between the data is interchanged and transmitted. The threshold in the perceptron layer acts as A mole bit signal passes through from 1 segment to another. The perceptron model can be trained in order to behave like mirror networks. Learning comes under the category of supervised machine learning algorithms. The label mask and labelled data of the COVID-19 multiclass are shown where it depicts the two-fold multiclass of data with a masked portion with pneumonia and without pneumonia. Similarly, the metadata portion is visualised, which shows the metadata category of data collected from various clinics and other sources, like in a pie chart, which shows the clinically approved data of different cohorts and types of x-ray image categories, which fall under different severity ranges and categories like regular, suspected, critical, and so on. In histogram figures, it depicts the quality bound of data and a different pivoted range of folds for mapping different categories [16, 17].

18.4.4 Pre-Processing of Information

The data set contains a number of X-ray images. In addition, some additional data can be obtained from the dataset, such as an age or gender specification. The pre-treatment steps used in this operation are described in the accompanying document [16]. In Fig. 18.2, we have shown a diagram which captures the steps of data preprocessing where the data is distributed among various zones of category based on its requirements and segmentary bounds, which adheres to the testing and training phasing dimension reduction where images are annotated manually to highlight the masked edges for model training and Region of Interest (ROI), which is an image processing technique to validate and highlight the pixel region, is used to R and L, which means two diversed masked areas for image segmentation.

Figure 18.3, depicts the outcome of an ensemble-based data preprocessing method as well as the outcome of segmented parts of an image data set that has been split and tested using an image segmentation technique [19, 20]. The efficient net and VGG net classifier are used to train the mask and overlap the errors. Where classes from a to h are segmented using VGG NET, class I to p is segmented via efficient net, and class q to x is segmented using U-net.

18.4.5 Fuzzy Based Deep Learning Proposed Method for Covid-9 Detection

Multilayer perceptron is one of the easiest architecture of the neural networking system to learn, feed forward and propagate. In terms of learning and predicting image instances a positron is considered, wide range of node interconnected neural networks like the tree formed of data and through those neural networks. The relationship established between the data are interchanged and transmitted by backpropagation function. The threshold in the perceptron layer acts as a mole bit signal which passes through from one segment to another. The perceptron model can be trained in order to behave like mirror networks. Using the Boolean function. Learning comes under the category of supervised machine learning algorithm. Where it produces [18]. The estimated output and learns to produce the output functions, where it urges itself. After repeating the cycle of learning layer by layer convolution. And this learning method is followed by the Delta rule because of its accuracy provided by several jacquards and performance metrics. The current input of the first convolution in the hidden layer is proportional to the desired output and the current input, which is summarised as the rules between the conditioning whether the perceptron is too largc or too small. The different axes of the perceptron shows the equation that separates from pulses and minuses. The fuzzy logic based architecture by having the interconnected dimension of fuzzy neuro-system as shown in Fig. 18.4. The different pulses of the perceptron layer, the equation that separates the pulses of the fuzzy system. The initial segment of the organization: The convolutional network takes an information picture for significant level deliberation by subbing the grouping of convolutional and pooling layers. The second piece of the organization: The fluffy layer performs fundamental appropriation of the information into a foreordained number of groups. The result neurons of the fluffy layer address the enrollment capacity of the fluffy information groups, where The third piece of the organization: In this stage, the qualities are placed as contributions of the classifier to produce the class score as a result of the organization [21, 22].

In Fig. 18.5, the fuzzy layer based neural network are categorized in the structure where the hidden layer and output layer each layer of the nodes represented by a this smaller dots. These lines separate the hidden layer and the fuzzy interference layer to learn and propagate between the fuzzy based prepositional logic to separate the convolution function. In order to maximize the threshold and in-depth feature extraction mode, the back propagation and convergence layer of the fuzzy based neural network choose the desired output for many significant inputs as on the training set [21–23]. Forward propagation for the training phase acts like a pattern which gives the signal to the neural first layer in order to generate the propagation of fuzzy functions. Back propagation and feed forward outputs activations uses the multiplicity of the Delta method and input activation functions. To use the weight in the opposite directions by subtracting the ratio of the fuzzy input [24, 25]. Considering this hypothesis of mapping convolution function into this fuzzy function which uses the optimization layer, where the mapping E depicts the properties of the quality, where the map function X is mapped to the fuzzy equality on X. And the real number XY shows the equality between X&Y. X and y are mapped the definition of the first convolution and the characteristics function, which is shown in the Eq. (18.1).

$$\sum_{i-1}^{i} x \in y \text{ where, } x \text{ and } y = i \tag{18.1}$$

where x is part of y mean the total input concatenated into the value of y as described in the architecture, where layer two shows batch normalization later in fuzzy module an dene layer phase fo model validation.

The purpose of using the fuzzy function inside the neural network is to use the optimization application for the evolving global minimum of the energy functions where the global optimum is optimized using the straightforward fuzzy functions of the weight and edges input in the image dimensions. The miser complexity associated with these fuzzy based optimization neural network algorithm is the high range of possibility of converging the local input to optimized input instead of the global optimum. The fuzzy based neural network is using the precision of mathematics, different functions alluded with the physics system functions. Which uses the different containers of the assumed method like complement union containers, intersections where the optimization of a neural network can be stated as for this system in earlier. System is projected by the different membership functions in the hidden layer where they each of the input function maps each point N to in real discrete interval of the conversations. The set of pair where the number of assets

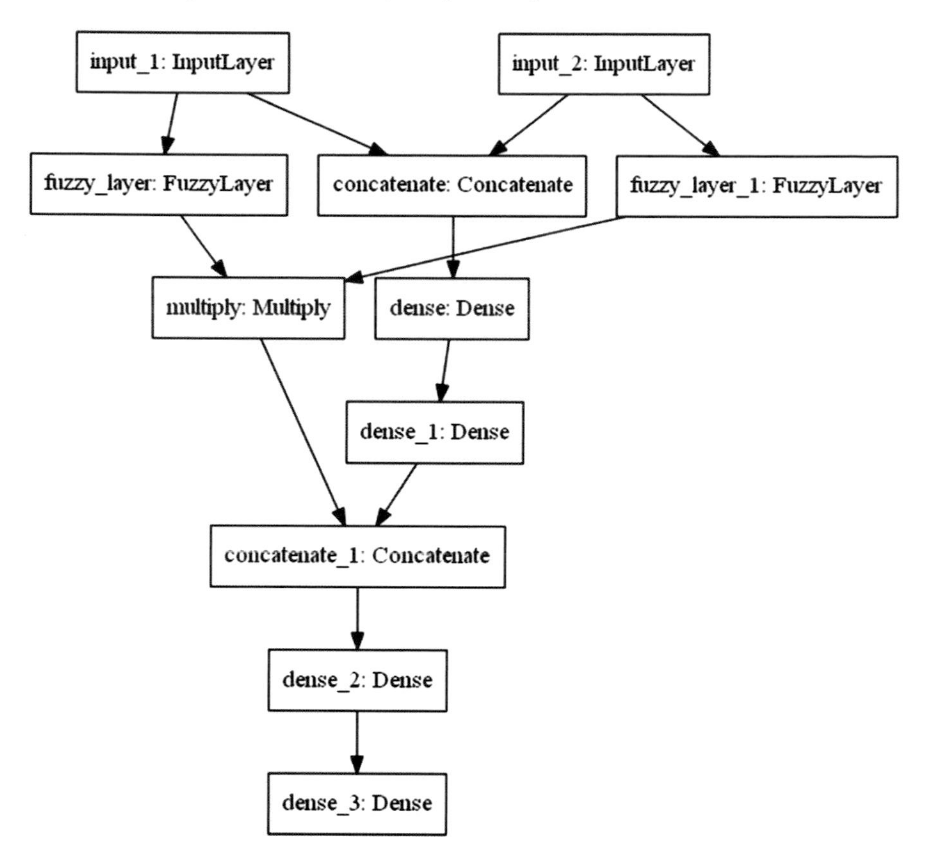

Fig. 18.5 Layer wise model functionality and encoding

uses the different grade of membership from the optimum value. Another function called hedges in the fuzzy value uses the hash functions where the operations. It's quite subjective in terms of the involvement of hedges through the functions which are provided by the efforts to maintain the close pair ship between the fuzzy based neural layers where the initial edges. Depicts the different weight functions the fuzzy based member functions uses the civilian scent of characters. With the set of rules and regulations, define the boundary of the images and tells that the instead of this particular. Combination of the layer is optimized in Convolution layer that can be used for splitting the data into training and testing phase. The crossover, foil and union and associated with associative function, which will this unit layers defines the fuzzy relationship. Which characterizes the classical relation into the partial membership, which describes the degree of two contrary layers acting as a learning phase, where the fuzzy relation between these two convolutions maps the total Cartesian set of the two layer multiplicative product. Where the strength of that particular math layer is expressed by the membership function of the fuzzy relation depicted as into A*b. In Fig. 18.6, we have designed and explained about fuzzy rule based neural system

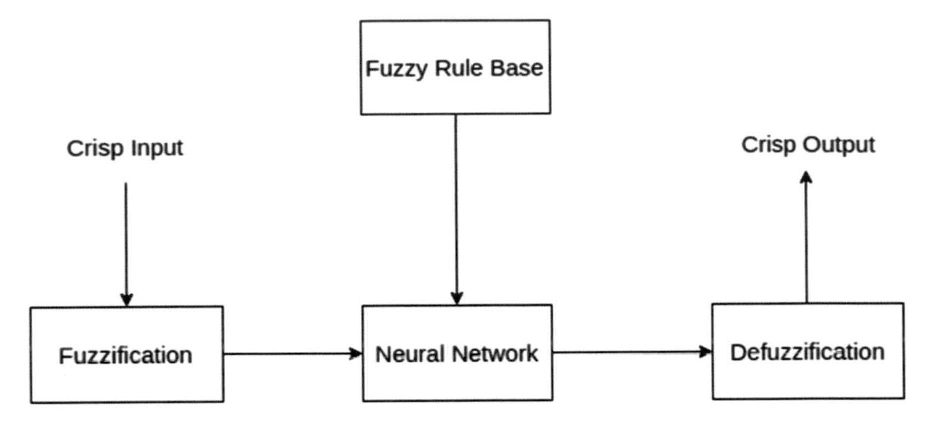

Fig. 18.6 Process of combining fuzzy layer to neural layer

for classification and way of combining fuzzy layer to CNN. Which says A Neuro-Fuzzy Hybrid System depends on the fluffy framework, prepared based on working of Neural Network hypothesis. Here, educational experience works just on the neighborhood data and causes changes just in the basic fluffy frameworks. This crossover framework should be visible as a 3-layer feed forward network.

Layer 1: It contains the info factors, which are utilized with the end goal of fuzzification;

Layer 2: This layer contains the fluffy standards;

Layer 3: It contains the result factors, which are utilized with the end goal of defuzzification. The fluffy sets are encoded as association weight inside the brain network layers for, Giving usefulness in handling. First layer: Here, every neuron sends fresh information signal straightforwardly to the following layer. The neuron addresses the information capacity of predecessor of fluffy rule. Every neuron gets a fresh information and decides how much it has a place with the fluffy set. In the layers, the fluffy sets are addressed utilizing loads though the fluffy information base is addressed utilizing neurons, which conquer the misrepresentations of brain and cross breed frameworks when thought about independently [26, 27].

18.5 Results and Discussion

Figure 18.8, shows that we test with twenty erratic occasions, the informed authority, either a patient or a subject matter expert, just finished records about age, X-transmits, see position, and bearing. We have assessed and perceived the infection of patients going before pushing ahead with the evaluation on additional colossal preliminaries. With a definitive target of the supposition for ailments, we have concluded the Fβ score where β is 0.5. It proposes that actually hanging out there the state of a patient like the tough spot and shock going before the appropriate examination. The majority

Table 18.2 Classification accuracy for data segmentation

LABEL	VGG 16	Efficient net	Inception net	Xcep	U-net	MLP
Overall	7,828,391	0.2096014	0.012559233	0.49621865	0.5678478	7,828,392
COVID-19	0.9967531	0.001842642	0.001403835	0.9737174	0.01207160	0.9967532
Pneumonia	0.9702547	0.008166936	0.021578299	0.86954033	0.0429593	0.9702547
Normal	0.80542165	0.19251792	0.00206049	0.7048106	0.29260245	0.80542165

of the outcomes are by and large something very similar yet there are in this way several cases that aren't right. The particular score for fibrosis tracking down the case. Considering the puzzling demonstration of a plan with fewer cutoff points to set up, this paper pondered this planning among the other CNNs existing models. Table 18.2 shows the approach of the arranging example of the model with the going with limits. Separate the presentation of the proposed technique, we chose the test accuracy, unequivocally, affectability, precision, review, F-measure, G-mean, and MCC. As alluded, a familiar strategy needs with managing an imbalanced class arrangement. Hence, accuracy can't give a genuine show metric and different assessments alluded to above survey our introduced model [28].

Each information variable has two Gaussian enrollments capacities as displayed in Fig. 18.7. The Gaussian enrollment work gaussmf [σ μ] is characterized by its mean μ and standard deviation σ. The fever variable is addressed by the body temperature which ranges somewhere in the range of as introduced in Fig. 18.8 while every one of the excess information factors has a level in the reach from 0 to 5 as demonstrated in Fig. 18.8 [29]. The yield variable reaches from 0 to 100 and it has four Gaussian enrollment capacities; generally safe, medium gamble, high gamble and exceptionally high gamble as introduced [30].

Rule 1: If 3 side effects of Category-1 AND no less than 4 symptoms of Category-2 are available = Very high gamble of COVID-19 contamination.

Rule 2: If 3 side effects of Category-1 AND under 4 side effects of Category-2 are available = High gamble of COVID-19 disease... ... ... This method of information boundary of rules is characterized [31].

In view of the info patient side effects, the induction framework starts a bunch of fluffy guidelines where each standard creates an yield. Fluffy administrator "min" was utilized for creating the yield fluffy set by taking each standard that fulfilled the AND functional rationale for a given arrangement of information values. Then, at that point, the yield fluffy arrangement of each standard was consolidated into a solitary fluffy set by the total cycle. The single fluffy set was defuzzified into a solitary numeric result esteem utilizing the Centroid strategy to decide the rate risk level of being COVID-19 tainted.

A fuzzy relationship measure is proposed in this paper which conquers a portion of the limits of a prior proposed measure. In addition, a likeness rank relationship measure is additionally depicted. The proposed measure can be utilized where the

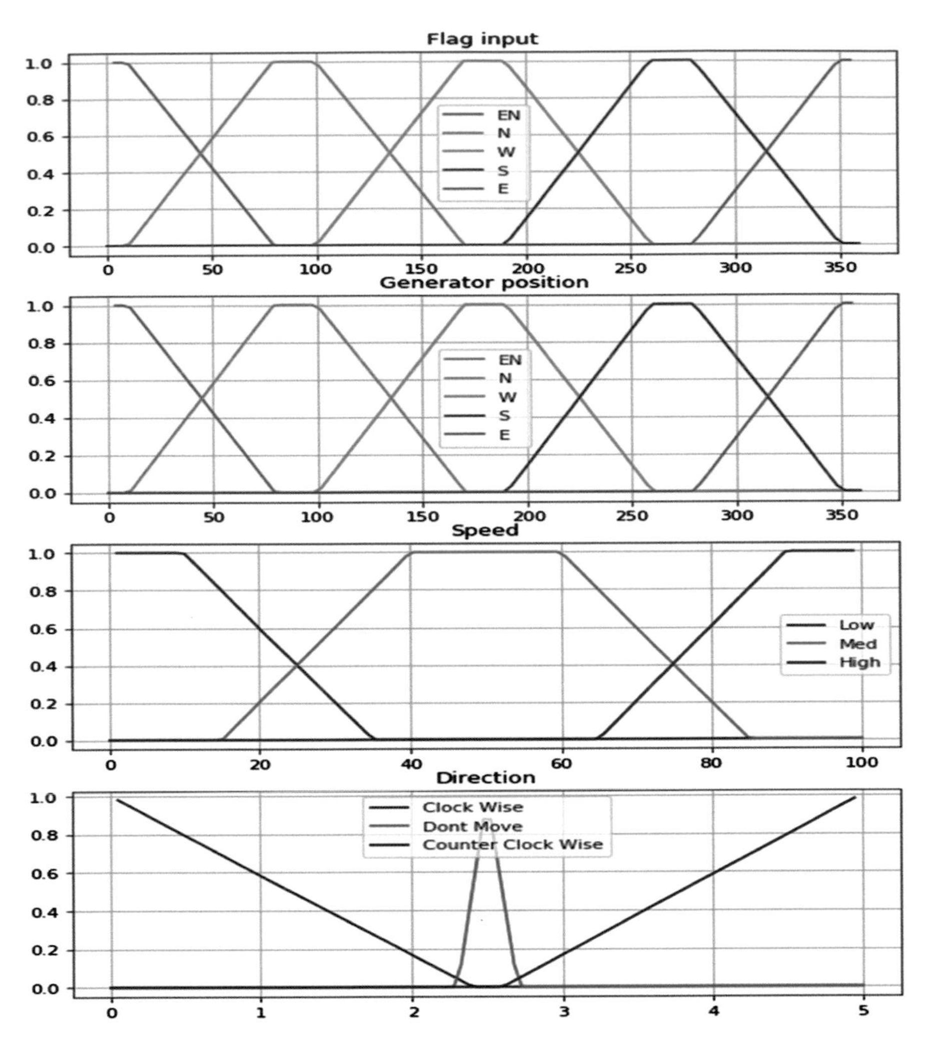

Fig. 18.7 Flag and member function of fuzzy stem

idea of relationship is important yet the presence of uncertainty and vagueness forestalls old style investigation. The fuzzy relationship measure can be utilized in design coordinating on the off chance that appropriate fluffy sets can be developed.

From the picture and example formats. This pertinence of the proposed measure in design matching was tried tentatively by taking a double picture set of manually written as extant in 0–4 scaling, as displayed in Fig. 18.7. Fuzzy set relating to the example formats was built by allotting an enrollment worth to every pixel. The pixels having a place with the foundation of the examples were allotted enrollment esteem 0 and the pixels which had a place with the example were allotted enrollment esteem

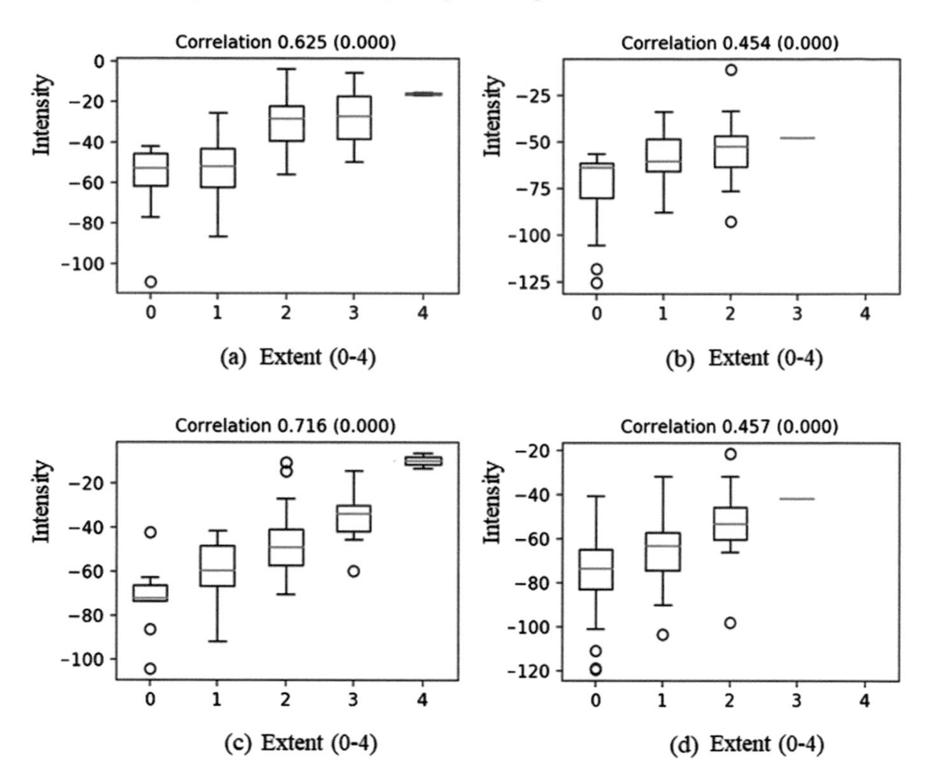

Fig. 18.8 Correlation of fuzzy sets in classification

relative to its dim level. Since the example layouts were double it were effectively built to compare fluffy sets as shown in Fig. 18.8.

To work on the precision of the order of the communication levels, we chose to produce a FIS utilising a Neuro-Fuzzy strategy. Neuro-fuzzy frameworks include a bunch of procedures that offer vigour in the treatment of loose and questionable data that exist in issues connected with this present reality, e.g., acknowledgment of structures, characterization, independent direction, and so on. The fundamental benefit of fuzzy-based neural network frameworks is that they join the learning limit of brain networks with the force of phonetic translation of FIS, permitting the extraction of information for a base of fluffy standards from a large amount of information. Figure 18.9 shows the consequences of this order. The base right cell shows the general exactness, while the section on the most distant right of the plot shows the precision for each anticipated class. The line at the lower part of the plot shows the exactness of each genuine class.

The accuracy plots of setting up the CNN on data are presented in Table 18.2. Furthermore, In Table 18.2, the show estimations are presented as 95% assurance extends which have been enrolled more than 10-overlay cross endorsement. Looking at Fig. 18.8 uncovers that readiness on CT pictures yields extraordinary results. Appropriately, CT pictures can be considered a strong focal point for perseverance

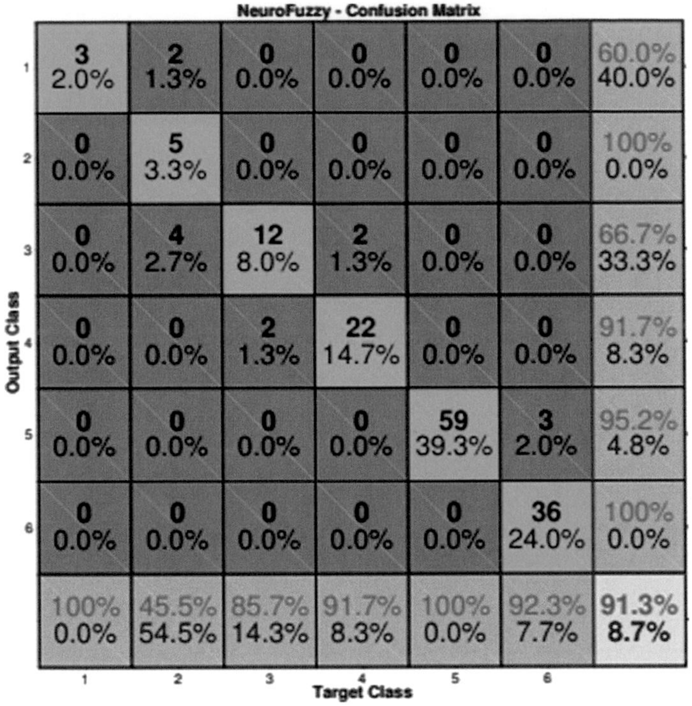

NeuroFuzzy - Confusion Matrix

Output Class \ Target Class	1	2	3	4	5	6	
1	3 2.0%	2 1.3%	0 0.0%	0 0.0%	0 0.0%	0 0.0%	60.0% 40.0%
2	0 0.0%	5 3.3%	0 0.0%	0 0.0%	0 0.0%	0 0.0%	100% 0.0%
3	0 0.0%	4 2.7%	12 8.0%	2 1.3%	0 0.0%	0 0.0%	66.7% 33.3%
4	0 0.0%	0 0.0%	2 1.3%	22 14.7%	0 0.0%	0 0.0%	91.7% 8.3%
5	0 0.0%	0 0.0%	0 0.0%	0 0.0%	59 39.3%	3 2.0%	95.2% 4.8%
6	0 0.0%	0 0.0%	0 0.0%	0 0.0%	0 0.0%	36 24.0%	100% 0.0%
	100% 0.0%	45.5% 54.5%	85.7% 14.3%	91.7% 8.3%	100% 0.0%	92.3% 7.7%	**91.3%** **8.7%**

Fig. 18.9 Confusion matrix plot of target and base class multilevel classification using fuzzy based neural network

chance assumption for COVID-19 patients [30], which shows the comparison of different algorithm used for data segmentation.

In Fig. 18.10, we describe the filtered and detected output image of covid 19 where segments a describe the rule based ROI and b shows detected mask.

In Fig. 18.11 a and b, the predicted normal and covid infected sample of data is shown.

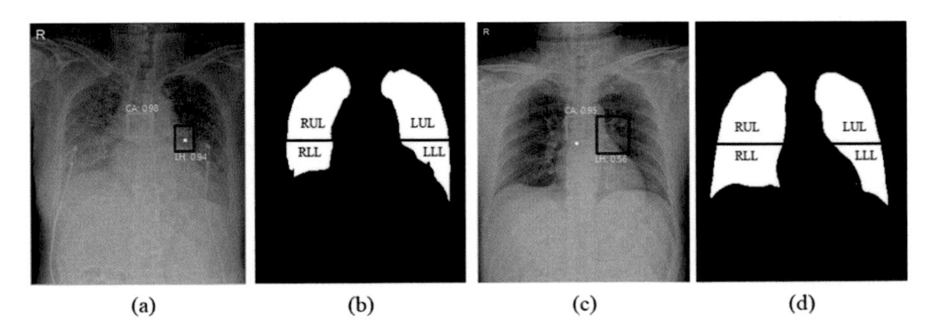

Fig. 18.10 Detected ensemble samples

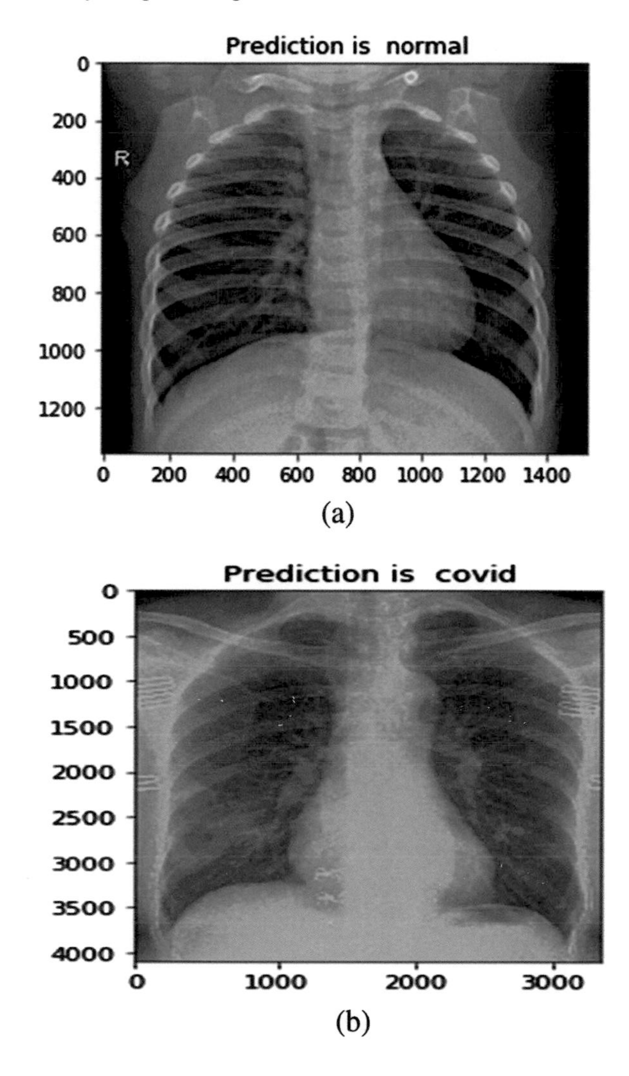

Fig. 18.11 Sample of predicted data

In Fig. 18.12, the overall history of model training curve is depicted, Which projects the total learning rate of spitted data in mean variance time meaning how confidence level of data having error during the initialization fuzzy outlier set analysis before training the data and also testing the second half data using different classifier.

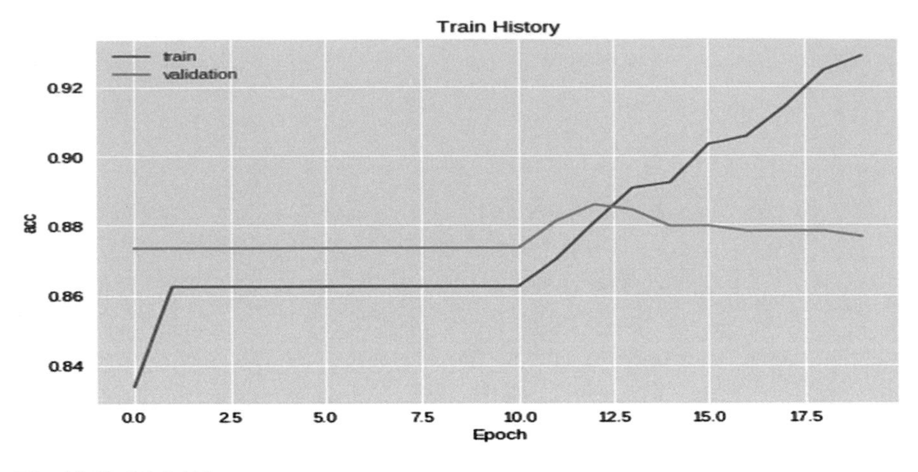

Fig. 18.12 Model history

18.6 Conclusions and Future Scope

We have implemented CNN integrated with auto encoder and Fuzzy system which gives multilayer fully connected layers of convolution for in depth feature extraction and validation to the test cases based on 30 epochs. The overall implemented model gives 97% overall accuracy as compared to the other proposed model exist in the same domain. In future more advance algorithm can be produced using optimization technique of meta-heuristic and bio-inspired algorithm; the algorithm can be optimized by implementation of heuristic algorithm to give more precision and accuracy.

References

1. Mehta, T., Mehendale, N.: Classification of X-ray images into COVID-19, pneumonia, and TB using cGAN and fine-tuned deep transfer learning models. Res. Biomed. Eng. **37**(4), 803–813 (2021). https://doi.org/10.1007/s42600-021-00174-z
2. Elshennawy, N.M., Ibrahim, D.M.: Deep-pneumonia framework using deep learning models based on chest X-Ray images. Diagnostics **10**(9), 649 (2020). https://doi.org/10.3390/diagnostics10090649
3. Shaban, W.M., Rabie, A.H., Saleh, A.I., Abo-Elsoud, M.A.: Detecting COVID-19 patients based on fuzzy inference engine and Deep Neural Network. Appl. Soft Comput. **99**, 106906 (2021). https://doi.org/10.1016/j.asoc.2020.106906
4. Kulkarni, A.J., Siarry, P.: Handbook of AI-based Metaheuristics, 1st edn. Boca Raton, CRC Press, 2021. https://doi.org/10.1201/9781003162841
5. Xie, S., Yu, Z., Lv, Z.: Multi-disease prediction based on deep learning: a survey. Comput. Model. Eng. & Sci. **128**(2), 489–522 (2021). https://doi.org/10.32604/cmes.2021.016728
6. Sweta, S., Lal, K.: Personalized adaptive learner model in E-learning system using FCM and Fuzzy inference system. Int. J. Fuzzy Syst. **19**(4), 1249–1260 (2017). https://doi.org/10.1007/s40815-017-0309-y

7. Shatnawi, M., Shatnawi, A., AlShara, Z., Husari, G.: Symptoms-Based Fuzzy-Logic Approach for COVID-19 Diagnosis. IJACSA **12**, 4 (2021). https://doi.org/10.14569/IJACSA.2021.0120457
8. Ahuja, A.S.: The impact of artificial intelligence in medicine on the future role of the physician. PeerJ **7**, e7702 (2019). https://doi.org/10.7717/peerj.7702
9. Taresh, M.M., Zhu, N., Ali, T.A.A., Hameed, A.S., Mutar, M.L.: Transfer learning to detect COVID-19 automatically from X-Ray images using convolutional neural networks. Int. J. Biomed. Imaging **1–9**, 2021 (2021). https://doi.org/10.1155/2021/8828404
10. Ghaderzadeh, M., Asadi, F.: Deep learning in the detection and diagnosis of COVID-19 using radiology modalities: a systematic review. J. Healthc. Eng.**2021** (2021). https://doi.org/10.1155/2021/6677314
11. Haque, K., Abdelgawad, A.: A deep learning approach to detect COVID-19 patients from chest X-ray Images. Ai **1**(3), 418–435 (2020). https://doi.org/10.3390/ai1030027
12. Horry, M.J., Chakraborty, S., Paul, M., Ulhaq, A., Pradhan, B., Saha, M., Shukla, N.: COVID-19 detection through transfer learning using multimodal imaging data. IEEE Access **8**, 149808–149824 (2020). https://doi.org/10.1109/ACCESS.2020.3016780
13. Irmak, E.: COVID-19 disease severity assessment using CNN model. IET Image Proc. **15**(8), 1814–1824 (2021). https://doi.org/10.1049/ipr2.12153
14. Kassania, S.H., Kassanib, P.H., Wesolowskic, M.J., Schneidera, K.A., Detersa, R.: Automatic detection of coronavirus disease (COVID-19) in X-ray and CT images: a machine learning based approach. Biocybern. Biomed. Eng. **41**(3), 867–879 (2021). https://doi.org/10.1016/j.bbe.2021.05.013
15. Khan, S.H., Sohail, A., Khan, A.: COVID-19 Detection in Chest X-Ray Images using a New Channel Boosted CNN, pp. 1–26 (2020). http://arxiv.org/abs/2012.05073
16. Khozeimeh, F., Sharifrazi, D., Izadi, N.H., Joloudari, J.H., Shoeibi, A., Alizadehsani, R., Gorriz, J.M., Hussain, S., Sani, Z.A., Moosaei, H., Khosravi, A., Nahavandi, S., Islam, S.M.S. CNN AE: Convolution Neural Network combined with Autoencoder approach to detect survival chance of COVID 19 patients (2021). http://arxiv.org/abs/2104.08954
17. Maghdid, H., Asaad, A.T., Ghafoor, K.Z.G., Sadiq, A.S., Mirjalili, S., Khan, M.K.K. Diagnosing COVID-19 pneumonia from x-ray and CT images using deep learning and transfer learning algorithms, p. 26 (2021). https://doi.org/10.1117/12.2588672
18. Mondal, M.R.H., Bharati, S., Podder, P., Podder, P.: Data analytics for novel coronavirus disease. Inform. Med. Unlocked **20**, 100374 (2020). https://doi.org/10.1016/j.imu.2020.100374
19. Narin, A., Kaya, C., Pamuk, Z.: Automatic detection of coronavirus disease (COVID-19) using X-ray images and deep convolutional neural networks. Pattern Anal. Appl. **24**(3), 1207–1220 (2021). https://doi.org/10.1007/s10044-021-00984-y
20. Nur-a-alam, Ahsan, M., Based, M.A., Haider, J., Kowalski, M.: COVID-19 detection from chest X-ray images using feature fusion and deep learning. Sensors**21**(4), 1–30 (2021). https://doi.org/10.3390/s21041480
21. Oyelade, O.N., Ezugwu, A.E., Chiroma, H.: CovFrameNet: an enhanced deep learning framework for COVID-19 detection. IEEE Access (2021). https://doi.org/10.1109/ACCESS.2021.3083516
22. Ozcan, T.: a deep learning framework for coronavirus disease (COVID-19) detection in X-Ray images. 90(352) (2020). https://doi.org/10.21203/rs.3.rs-26500/v1
23. R., D.D.: Deep net model for detection of covid-19 using radiographs based on ROC analysis. J. Innov. Image Process. **2**(3), 135–140 (2020). https://doi.org/10.36548/jiip.2020.3.003
24. Salman, F.M., Abu-Naser, S.S., Alajrami, E., Abu-Nasser, B.S., Ashqar, B.A.M.: COVID-19 Detection using Artificial Intelligence. Int. J. Acad. Eng. Res. 4(3), 18–25 (2020). www.ijeais.org/ijaer
25. Sekeroglu, B., Ozsahin, I.: Detection of COVID-19 from chest X-Ray images using convolutional neural networks. SLAS Technology **25**(6), 553–565 (2020). https://doi.org/10.1177/2472630320958376
26. Sethy, P.K., Behera, S.K., Ratha, P.K., Biswas, P.: Detection of coronavirus disease (COVID-19) based on deep features and support vector machine. Int. J. Math., Eng. Manag. Sci. 5(4), 643–651 (2020). https://doi.org/10.33889/IJMEMS.2020.5.4.052

27. Sharifrazi, D., Alizadehsani, R., Roshanzamir, M., Joloudari, J. H., Shoeibi, A., Jafari, M., Hussain, S., Sani, Z. A., Hasanzadeh, F., Khozeimeh, F., Khosravi, A., Nahavandi, S., Panahi-azar, M., Zare, A., Islam, S.M.S., Acharya, U.R.: Fusion of convolution neural network, support vector machine and Sobel filter for accurate detection of COVID-19 patients using X-ray images. Biomed. Signal Process. Control.68 (2021). https://doi.org/10.1016/j.bspc.2021.102622
28. Tabrizchi, H., Mosavi, A., Vamossy, Z., Varkonyi-Koczy, A.R.: Densely Connected Convolutional Networks (DenseNet) for Diagnosing Coronavirus Disease (COVID-19) from Chest X-ray Imaging, pp. 1–5 (2021). https://doi.org/10.1109/memea52024.2021.9478715
29. Taleb-ahmed, A.: Chest X-ray Databases, pp. 1–20 (2021)
30. Waheed, A., Goyal, M., Gupta, D., Khanna, A., Al-Turjman, F., Pinheiro, P.R.: CovidGAN: data augmentation using auxiliary classifier GAN for improved covid-19 detection. IEEE Access **8**, 91916–91923 (2020). https://doi.org/10.1109/ACCESS.2020.2994762
31. Zhang, Y., Zhang, Y.: Deep Learning for COVID-19 Recognition (2021). https://doi.org/10.20944/preprints202105.0711.v1